Principles of Gene Manipulati

AN INTRODUCTION TO GENETIC ENGINEERING

STUDIES IN MICROBIOLOGY

EDITORS

N. G. CARR

Department of Biochemistry
University of Liverpool

J. L. INGRAHAM

Department of Bacteriology
University of California at Davis

S. C. RITTENBERG

Department of Bacteriology
University of California at Los Angeles

STUDIES IN MICROBIOLOGY VOLUME 2

Principles of Gene Manipulation

AN INTRODUCTION TO GENETIC ENGINEERING

R. W. OLD MA, PhD

Department of Biological Sciences,
University of Warwick,
Coventry

S. B. PRIMROSE BSc, PhD

Searle Research and Development,
High Wycombe

THIRD EDITION

BLACKWELL SCIENTIFIC PUBLICATIONS
OXFORD LONDON EDINBURGH
BOSTON PALO ALTO MELBOURNE

Editorial offices:
Osney Mead, Oxford, OX2 0EL
8 John Street, London, WC1N 2ES
23 Ainslie Place, Edinburgh,
EH3 6AJ
3 Cambridge Center, Suite 208,
Cambridge, Massachusetts 02142,
USA
667 Lytton Avenue, Palo Alto
California 94301, USA
107 Barry Street, Carlton
Victoria 3053, Australia

First published 1980
Second edition 1981
Reprinted 1982, 1983 (twice)
Third edition 1985
Reprinted 1986 (twice), 1987, 1988

Spanish edition 1986
Japanese edition 1987

Typeset by Acorn Bookwork,
Salisbury, Wiltshire
Printed and bound in Great Britain
at the University Press, Cambridge

DISTRIBUTORS

USA
Year Book Medical Publishers
200 North LaSalle Street
Chicago, Illinois 60601

Canada
The C.V. Mosby Company
5240 Finch Avenue East,
Scarborough, Ontario

Australia
Blackwell Scientific Publications
(Australia) Pty Ltd
107 Barry Street
Carlton, Victoria 3053

British Library
Cataloguing in Publication Data

Old, R. W.
Principles of gene
manipulation.—3rd ed.—
(Studies in microbiology; v.2)
1. Genetic engineering
I. Title II. Primrose, S. B.
III. Series
575.1 QH442

ISBN 0-632-01318-4

Contents

Preface to Third Edition

This, the third edition, went to press approximately three years after we submitted the manuscript of the second edition. In the intervening period of time there have been many changes in emphasis in the field of recombinant DNA technology; for example, there have been considerable developments in cloning systems for use in microbial systems other than *E. coli*. In particular, exciting progress is being made in the manipulation of genes in yeast, the construction of synthetic chromosomes being particularly noteworthy. In the study of higher eukaryotes startling discoveries are reported almost weekly. In the case of plant systems, vectors based on the Ti plasmid have been used to obtain expression of foreign genes and with animal systems several new series of vectors have been constructed. New applications of recombinant DNA technology continue unabated, particularly in the diagnosis of genetic disorders and elucidation of potential causes of cancer. Finally, since the last edition first appeared at least one prescription drug (insulin) produced using recombinant DNA technology is now on sale.

The changes in this edition reflect the developments outlined above and many others. In addition we have tried to respond to the helpful criticism provided by colleagues, familiar and unfamiliar, around the world. Readers acquainted with the first two editions will recognize that the arrangement of this book is somewhat different and that it is considerably longer. Nevertheless, we realize that this book is used widely as a teaching aid and consequently we have not changed the level at which the book is written, nor the general style. Teachers should quickly realize that the layout of this edition is such that entire sections or chapters can be excluded if available lecture time does not permit the entire contents of the book to be covered.

Once again we would like to acknowledge the detailed criticism of individual chapters which has been provided by our colleagues. Special thanks are due to Janice Light, Bill

Tacon, Les Bell, Bruce Carter and Roger Hull. Many of their suggestions were incorporated in the final version of the text but on occasion we chose to spurn their proposed modifications; if this has resulted in errors, then the fault is ours. Finally, acknowledgement is due to Maureen Bevan and Dianne Simpson for many long hours spent typing and retyping and to Jill Primrose for preparing the reference section.

R. W. Old
S. B. Primrose

Preface to Second Edition

In the period since we wrote the first edition there have been a number of advances in the field of gene manipulation and these have been consolidated in this edition. By and large the first edition was well received and for this reason we saw no reason to change our basic philosophy, viz. to present the basic principles in sufficient detail to enable the non-specialist reader to understand the current literature. We hope that even specialists will find this book a useful reference source and to make it more appealing to them we have included a number of appendices giving details of restriction enzyme specificities and restriction maps of some of the more common cloning vectors. As before, we assume that the reader has a reasonable working knowledge of molecular biology but because this field grows at an extraordinary rate we have included a greatly enlarged glossary.

In the preface to the first edition we stated '. . . we hope that the content of this book does not date too quickly but that the principles explained herein will provide an introduction to gene manipulation for some time to come.' In this regard we feel we were successful, for with few exceptions the material from the first edition is still factually correct. Indeed, parts of the first edition have been used in this edition. The major changes have been the complete rewriting of the chapter on possible vectors for cloning in plants, the introduction of an entire chapter devoted to cloning in *Bacillus subtilis* and yeast, and an extended final chapter on the newer applications of gene manipulation. Since we have no desire to rewrite this book too frequently we hope that the subject is about to reach the steady-state and that its contents do not date!

Finally, we would like to acknowledge the helpful criticisms of various chapters provided by Alan Kingsman, Jan-Ingmar Flock, Bill Tacon, Roger Hull and Angus Hepburn. Many of the improvements in various chapters are a direct result of their efforts but the errors remain ours. Last but not least we must thank Carolyn Alderson, Dianne Simpson, Malcolm Davies and

Len Bulmer for their skilled assistance in preparing the manuscript and Marilyn Nugent for assistance with proof-reading.

Preface to First Edition

Advances in biology continue to be made at a striking and ever increasing rate. One of the pace-setters is the subject matter of this book, *gene manipulation*, often popularly referred to as *genetic engineering*. A consequence of the phenomenal rate of progress in this subject has been that many biologists have found it impossible to keep pace with current developments, a situation exacerbated by the free use of jargon, and as with all rapidly growing fields it will be some time before comprehensive texts catch up. We have written this book to fill the resultant vacuum.

In the present state of the art, basic techniques are at the point of becoming well established and the trend is towards applying them to solve particular problems. Thus we have endeavoured to give readers enough details of these basic techniques to enable them to follow the current literature and future developments. In using this approach we hope that the content of the book does not date too quickly, but that the principles explained herein will provide an introduction to gene manipulation for some time to come.

The book is based on a series of twenty lectures on gene manipulation given at the University of Warwick to students on biology, microbiology and biochemistry degree courses. It is intended as an introduction to the subject for advanced undergraduates or people already in biological research, and consequently we have assumed that the reader has some prior knowledge of basic molecular biology. The literature has been surveyed up to the end of June 1979. The references cited are intended to point the reader towards the mainstream of the subject and to attribute original results to researchers. However, in a book of this size it is impossible to detail every paper. We have chosen examples from the literature which we feel best illustrate particular topics and hope that we have not offended colleagues whose experiments have not been mentioned.

Finally, it is a pleasure to acknowledge the skilled assistance of Mrs Debbie Bowns and Miss Dianne Simpson who had to interpret our sometimes impenetrable handwriting in producing the typescript; and Malcolm Davies for compiling and checking all the references.

Abbreviations and Conversion Scale

amber (mutation)	*am*
covalently closed circles	ccc
dihydrofolate reductase	DHFR
gene for DNA ligase	*lig*
kilobases	kb
megadaltons	MDal.
molecular weight	mol. wt
plaque-forming unit	pfu
purine	Pu
pyrimidine	Py
resistance (to an antibiotic)	R
sensitivity (to an antibiotic)	S
temperature-sensitive (mutation)	*ts*
thymidine kinase	TK

Abbreviations of antibiotic names

ampicillin	Ap
chloramphenicol	Cm
kanamycin	Km
neomycin	Nm
streptomycin	Sm
sulphonamide	Su
tetracycline	Tc
trimethoprim	Tp

Scale for conversion between kilobase pairs of duplex DNA and molecular weight.

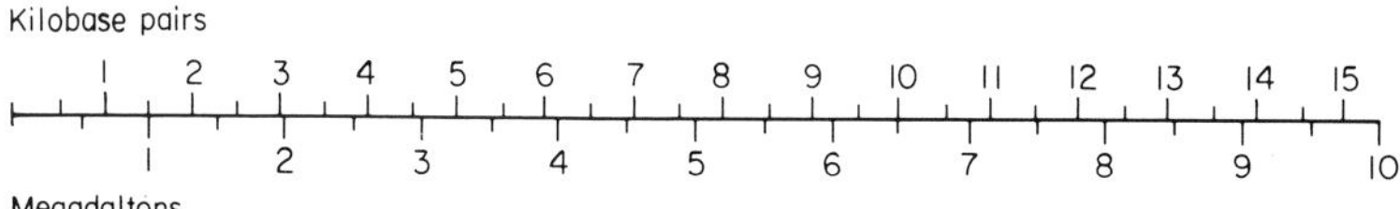

SECTION 1

Introduction to Gene Manipulation

Chapter 1 Basic Techniques

Introduction

Occasionally technical developments in science occur that enable leaps forward in our knowledge and increase the potential for innovation. Molecular biology and biomedical research have recently experienced just such a revolutionary change with the development of gene manipulation. The term gene manipulation can be applied to a variety of sophisticated in-vivo genetics as well as to in-vitro techniques. In fact, in most Western countries there is a precise *legal* definition of gene manipulation as a result of Government legislation to control it. In the United Kingdom gene manipulation is defined as 'the formation of new combinations of heritable material by the insertion of nucleic acid molecules, produced by whatever means outside the cell, into any virus, bacterial plasmid or other vector system so as to allow their incorporation into a host organism in which they do not naturally occur but in which they are capable of continued propagation.' The definitions adopted by other countries are similar and all adequately describe the subject matter of this book.

This legal definition emphasizes the propagation of foreign nucleic acid molecules (the nucleic acid is nearly always DNA) in a different, host organism. The ability to cross natural species barriers and place genes from any organism in an unrelated host organism is one important feature of gene manipulation. A second important feature is the fact that a defined and relatively small piece of DNA is propagated in the host organism. As we shall see, this has far-reaching consequences, for it is then possible to obtain a pure DNA fragment in bulk. This opens the door to a range of molecular biological opportunities including nucleotide sequence determination, site-directed mutagenesis, and manipulation of gene sequences to ensure very high level expression of an encoded polypeptide in a host organism. In addition, the DNA fragment provides a molecular hybridization probe of absolute sequence purity, totally uncontaminated by other sequences from the donor organism.

The initial impetus for gene manipulation *in vitro* came about in the early 1970s with the simultaneous development of techniques for:

1 transformation* of *Escherichia coli*,
2 cutting and joining DNA molecules, and
3 monitoring the cutting and joining reactions.

In order to explain the significance of these developments we must first consider the essential requirements of a successful gene manipulation procedure.

The basic problems

Before the advent of modern gene manipulation methods there had been many early attempts at transforming pro- and eukaryotic cells with foreign DNA. But, in general, little progress could be made. The reasons for this are as follows. Let us assume that the exogenous DNA is taken up by the recipient cells. There are then two basic difficulties. First, where detection of uptake is dependent on gene expression failure could be due to lack of accurate transcription or translation. Second, and more importantly, the exogenous DNA may not be maintained in the transformed cells. If the exogenous DNA is integrated into the host genome, there is no problem. The exact mechanism whereby this integration occurs is not clear and it is usually a rare event. If the exogenous DNA fails to be integrated, it will probably be lost during subsequent multiplication of the host cells. The reason for this is simple. In order to be replicated, DNA molecules must contain an *origin of replication* and in bacteria and viruses there is usually only one per genome. Such molecules are called *replicons*. Fragments of DNA are not replicons and in the absence of replication will be diluted out of their host cells. It should be noted that even if a DNA molecule contains an origin of replication this may not function in a foreign host cell.

There is an additional, subsequent problem. If the early experiments were to proceed, a method was required for

*The sudden change of an animal cell possessing normal growth properties into one with many of the growth properties of the cancer cell is called *growth transformation*. Growth transformation is mentioned in Chapter 11 and should not be confused with bacterial transformation which is described here.

assessing the fate of the donor DNA. In particular, in circumstances where the foreign DNA was maintained because it had become integrated in the host DNA, a method was required for mapping the foreign DNA and the surrounding host sequences.

THE SOLUTIONS: BASIC TECHNIQUES

If fragments of DNA are not replicated, the obvious solution is to attach them to a suitable replicon. Such replicons are known as *vectors* or *cloning vehicles*. Small plasmids and bacteriophages are the most suitable vectors for they are replicons in their own right, their maintenance does not necessarily require integration into the host genome and their DNA can be isolated readily in an intact form. The different plasmids and phages which are used as vectors are described in detail in Chapters 3 and 4. Suffice it to say at this point that initially plasmids and phages suitable as vectors were only found in *Escherichia coli*.

Composite molecules in which foreign DNA has been inserted into a vector molecule are sometimes called DNA *chimaeras* because of their analogy with the Chimaera of mythology—a creature with the head of a lion, body of a goat and the tail of a serpent. The construction of such composite or *artificial recombinant* molecules has also been termed *genetic engineering* or *gene manipulation* because of the potential for creating novel genetic combinations by biochemical means. The process has also been termed *molecular cloning* or *gene cloning* because a line of genetically identical organisms, all of which contain the composite molecule, can be propagated and grown in bulk hence *amplifying* the composite molecule and *any gene product whose synthesis it directs.*

Although conceptually very simple, cloning of a fragment of foreign, or *passenger*, or *target* DNA in a vector demands that the following can be accomplished.

1 The vector DNA must be purified and cut open.

2 The passenger DNA must be inserted into the vector molecule to create the artificial recombinant. DNA joining reactions must therefore be performed. Methods for cutting and joining DNA molecules are now so sophisticated that they warrant a chapter of their own (Chapter 2).

3 The cutting and joining reactions must be readily monitored. This is achieved by the use of gel electrophoresis.

4 Finally, the artificial recombinant must be transformed into *E. coli*, or other host cell. Further details on the use of gel electrophoresis and transformation of *E. coli* are given in the next section. As we have noted, the necessary techniques became available at about the same time and quickly led to many cloning experiments, the first of which were reported in 1972 (Jackson *et al.* 1972, Lobban & Kaiser 1973).

Agarose gel electrophoresis

The progress of the first experiments on cutting and joining of DNA molecules was monitored by velocity sedimentation in sucrose gradients. However, this has been entirely superseded

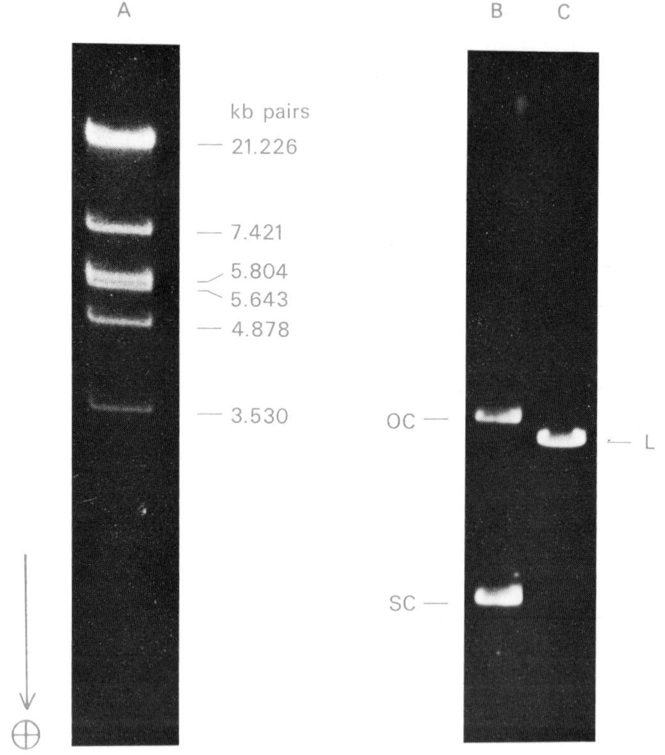

Fig. 1.1 Electrophoresis of DNA in agarose gels. The direction of migration is indicated by the arrow. DNA bands have been visualised by soaking the gel in a solution of ethidium bromide (which complexes with DNA by intercalating between stacked base pairs) and photographing the orange fluorescence which results upon ultraviolet irradiation. (A) Phage λ DNA restricted with *Eco* RI and then electrophoresed in a 1% agarose gel. The λ restriction map is given in Fig. 4.4. (B) Open circular (OC) and super-coiled (SC) forms of a plasmid of 6.4 kb pairs. Note that the compact super-coils migrate considerably faster than open circles. (C) Linear plasmid (L) DNA produced by treatment of the preparation shown in lane B with *Eco* RI for which there is a single target site. Under the conditions of electrophoresis employed here, the linear form migrates just ahead of the open-circular form.

by gel electrophoresis. Gel electrophoresis is not only used as an analytical method, it is routinely used preparatively for the purification of specific DNA fragments. The gel is composed of polyacrylamide or agarose. Agarose is convenient for separating DNA fragments ranging in size from a few hundred to about 20 000 base pairs. Polyacrylamide is preferred for smaller DNA fragments.

A gel is a complex network of polymeric molecules. DNA molecules are negatively charged and under an electric field DNA molecules migrate through the gel at rates dependent upon their sizes: a small DNA molecule can thread its way through the gel easily and hence migrates faster than a larger molecule (Fig. 1.1). Aaij and Borst (1972) showed that the migration rates of the DNA molecules were inversely proportional to the logarithms of the molecular weights. More recently, Southern (1979a,b) has shown that plotting fragment length or molecular weight against the reciprocal of mobility gives a straight line over a wider range than the semi-logarithmic plot. In any event, gel electrophoresis is frequently performed with marker DNA fragments of known size which allow accurate size determination of an unknown DNA molecule by interpolation. A particular advantage of gel electrophoresis is that the DNA bands can be readily detected at high sensitivity. The bands of DNA in the gel are stained with the intercalating dye ethidium bromide (Fig. 1.2), and as little as 0.05 μg of DNA in one band can be detected as visible fluorescence when the gel is illuminated with ultraviolet light.

In addition to resolving DNA fragments of different lengths, gel electrophoresis separates the different molecular configurations of a DNA molecule. The covalently closed circular, the nicked (relaxed) circular and linear forms of a DNA molecule have different mobilities (Fig. 1.1). Readers wishing to

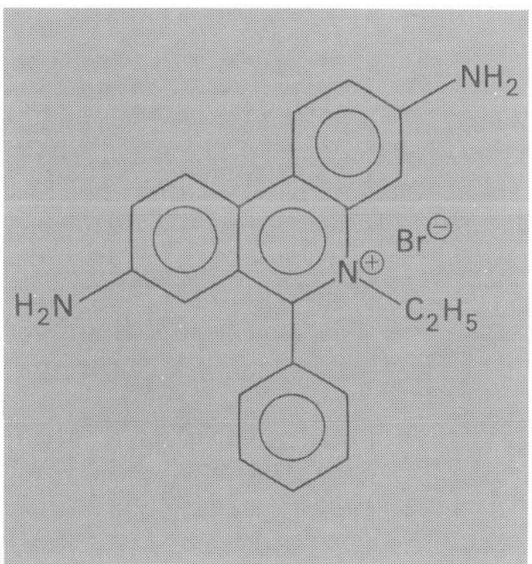

Fig. 1.2 Ethidium bromide.

know more about the factors affecting the electrophoretic mobility in agarose gels of the different conformational isomers of DNA should consult the paper by Johnson and Grossman (1977).

Southern blotting

Frequently it is necessary to know what sequences in a DNA restriction fragment are transcribed into RNA, or to be able to map sequences by hybridization to restriction fragments. Clearly it would be helpful to have a method of detecting fragments in an agarose gel that are complementary to a given RNA or DNA sequence. This can be done by slicing the gel, eluting the DNA and hybridizing to radiolabelled 'probe' DNA or RNA either in solution, or after binding the restriction fragment to filters. This method, which is time consuming and inevitably leads to some loss of resolution, has now been replaced by a neat method described by Southern (1975, 1979b). This method, often referred to as *Southern blotting* is shown in Fig. 1.3.

DNA restriction fragments in the gel are denatured by alkali treatment and the gel is then laid on top of buffer-saturated filter paper. The top surface of the gel is covered with a cellulose nitrate (often called nitrocellulose) filter and overlaid with dry filter paper. Buffer passes through the gel drawn by the dry filter paper and carries the single-stranded DNA with it. When the DNA comes into contact with the cellulose nitrate it binds to it strongly. The DNA fragments can be permanently fixed to the cellulose nitrate by baking at 80 °C. The filter can then be placed in a solution of radioactive RNA or denatured DNA which is complementary in sequence to the blot-transferred DNA. Conditions are chosen so that the radioactive nucleic acid hybridizes with complementary DNA on the cellulose nitrate. After a washing step, the regions of hybridization can be detected autoradiographically by placing the cellulose nitrate in contact with photographic film. The method is extremely sensitive and can even be used to map restriction sites around a single-copy gene sequence

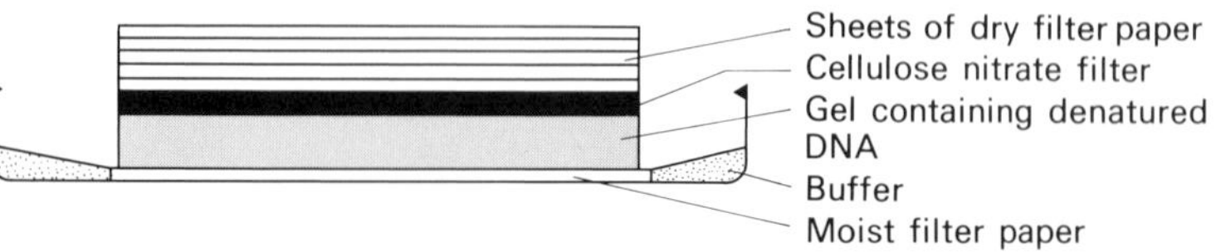

Fig. 1.3 The 'Southern blot' technique. See text for details.

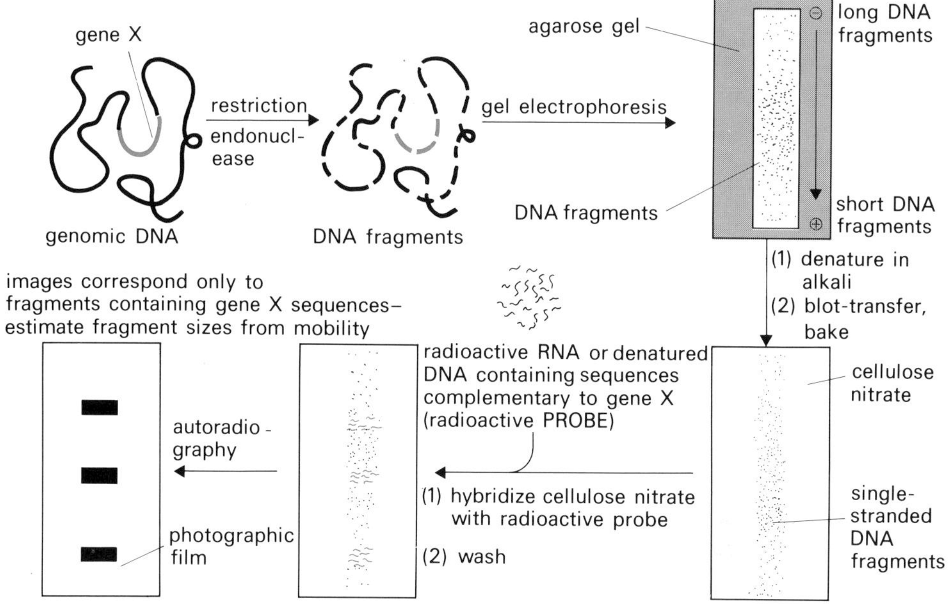

Fig. 1.4 Mapping restriction sites around a hypothetical gene sequence in total genomic DNA by the Southern blot method.

Genomic DNA is cleaved with a restriction endonuclease into hundreds of thousands of fragments of various sizes. The fragments are separated according to size by gel electrophoresis and blot-transferred on to cellulose nitrate paper. Highly radioactive RNA or denatured DNA complementary in sequence to gene X is applied to the cellulose nitrate paper bearing the blotted DNA. The radiolabelled RNA or DNA will hybridize with gene X sequences and can be detected subsequently by autoradiography, so enabling the sizes of restriction fragments containing gene X sequences to be estimated from their electrophoretic mobility. By using several restriction endonucleases singly and in combination, a map of restriction sites in and around gene X can be built up.

in a complex genome (Fig. 1.4). Since the radioactive nucleic acid is used here to search out and detect complementary sequences in the presence of a large amount of non-complementary DNA it is often referred to as the *probe*.

Northern blotting

Southern's technique has been of enormous value, but it was thought that it could not be applied directly to the blot-transfer of RNAs separated by gel electrophoresis since RNA was found not to bind to cellulose nitrate. Alwine *et al.* (1979), therefore, devised a procedure in which RNA bands are blot-transferred from the gel onto chemically reactive paper where they are bound covalently. The reactive paper is prepared by diazotization of aminobenzyloxymethyl-paper which itself can be prepared from Whatman 540 paper by a series of uncomplicated reactions. Once covalently bound, the RNA is available for hybridization with radiolabelled DNA probes. As before, hybridizing bands are located by autoradiography. Alwine's method thus extends that of Southern and for this reason it has acquired the jargon term '*Northern*' blotting!

Because of the firm covalent binding of the RNA to the paper, such blot-transfers are reusable; the probe from previous hybridization reactions having been eluted by washing at a temperature at which hybrids are not stable. Although originally devised for the transfer of RNA bands, the chemically reactive paper is equally effective in binding denatured DNA. In fact, small DNA fragments are more efficiently transferred to the diazotized paper derivative than to nitrocellulose. Extension of the fruitful blot-transfer approach has continued and more recently it has been found that RNA bands *can* indeed be blotted directly onto nitrocellulose under appropriate conditions (Thomas 1980). Because this form of Northern blotting does not require the preparation of reactive paper, it has been widely adopted.

Boxing the compass has continued with the arrival of the term *Western* blot. This was sometimes used to describe the blotting of gel electrophoresed RNA onto nitrocellulose, as just described, but the term is now usually restricted to an application that does not involve nucleic acids, i.e. the detection of specific polypeptides by first electrophoresing them in polyacrylamide gels, blotting them onto nitrocellulose to which they bind, and then detecting them by reaction with a specific, labelled antibody.

Transformation of *E. coli*

Early attempts to achieve transformation of *E. coli* were unsuccessful and it was generally believed that *E. coli* was refractory to transformation. However, Mandel and Higa (1970) found that treatment with $CaCl_2$ allowed *E. coli* cells to take up DNA from bacteriophage λ. A few years later Cohen *et al.* (1972) showed that $CaCl_2$-treated *E. coli* cells are also effective recipients for plasmid DNA. Almost any strain of *E. coli* can be transformed with plasmid DNA, albeit with varying efficiency, whereas until recently it was thought that only *rec*BC$^-$ mutants could be transformed with linear bacterial DNA (Cosloy & Oishi

1973). More recently, Hoekstra *et al.* (1980) have shown that *rec*BC$^+$ cells can be transformed with linear DNA but the efficiency is only 10% of that in otherwise isogenic *rec*BC$^-$ cells. Transformation of *rec*BC$^-$ cells with linear DNA is only possible if the cells are rendered recombination proficient by the addition of a *sbc*A or *sbc*B mutation. The fact that the *rec*BC gene product is an exonuclease explains the difference in transformation efficiency of circular and linear DNA in *rec*BC$^+$ cells.

As will be seen from the next chapter, many bacteria contain restriction systems which can influence the efficiency of transformation. Although the complete function of these restriction systems is not known yet, one role they do play is the recognition and degradation of foreign DNA. For this reason it is usual to use a restrictionless mutant of *E. coli* as a transformable host.

Since transformation of *E. coli* is an essential step in many cloning experiments it is desirable that it be as efficient as possible. Several groups of workers have examined the factors affecting the efficiency of transformation. It has been found that *E. coli* cells and plasmid DNA interact productively in an environment of calcium ions and low temperature (0–5 °C) and that a subsequent heat shock (37–45 °C) is important, but not strictly required. Several other factors, especially the inclusion of metal ions in addition to calcium, have been shown to stimulate the process.

Hanahan (1983) has re-examined factors that affect the efficiency of transformation, and has devised a set of conditions for optimal efficiency (expressed as transformants per μg plasmid DNA) applicable to most *E. coli* K12 strains. Typically, efficiencies of 10^7 or 10^8 transformants/μg can be achieved. Large DNAs transform less efficiently, on a molar basis, than small DNAs. Even with such improved transformation procedures certain potential gene cloning experiments requiring large numbers of clones are not reliable. One approach which can be used to circumvent the problem of low transformation efficiencies is to package recombinant DNA into virus particles *in vitro*. A particular form of this approach, the use of cosmids, is described in detail in Chapter 4.

The biggest problem that prevents analysis of the transformation process is that the maximum level of transformation has not been extended beyond a few percent of the survivors of the $CaCl_2$ treatment. Weston *et al.* (1979) found that when two separate plasmids are employed simultaneously at equally saturating concentrations of DNA the total number of transformants obtained is the same as that when one plasmid alone is used. This confirms that every transformable cell is transformed under these conditions suggesting that only a minor subpopulation of $CaCl_2$-treated cells of *E. coli* are capable of taking up and establishing plasmid molecules. This idea is supported by the work of Jones *et al.* (1981) who found that despite the fact that the transformation frequency increased 100-fold for a 10-fold increase in growth rate, the maximum number of transformable cells never exceeded 1%. Why the frequency of transformation is limited to 1% remains a mystery.

The absolute requirement for $CaCl_2$ for transformation of *E. coli* is probably due to structural alterations in the cell wall which it effects. In almost all Gram-negative bacteria which have been transformed, $CaCl_2$ treatment is a mandatory step. Only a few Gram-negative bacteria, e.g. *Haemophilus* and *Neisseria,* can be transformed without $CaCl_2$ treatment.

Transformation of other organisms

Although *E. coli* often remains the host organism of choice for cloning experiments, many other hosts are now used and with them transformation may still be a critical step. In the case of Gram-positive bacteria the two most important groups of organisms are *Bacillus* sp. and actinomycetes. That *B. subtilis* is naturally competent for transformation has been known for a long time and hence the genetics of this organism are fairly advanced. For this reason *B. subtilis* is a particularly attractive alternative prokaryotic cloning host. The significant features of transformation with this organism are detailed in Chapter 8. Of particular relevance here is that it is possible to transform protoplasts of *B. subtilis,* a technique which leads to improved transformation frequencies. A similar technique is used to transform actinomycetes and recently it has been shown that the frequency can be increased considerably by first entrapping the DNA in liposomes which then fuse with the host cell membrane. This is one aspect of transformation which undoubtedly will receive considerable attention in the near future.

In later chapters we discuss ways in which cloned DNA can be introduced into eukaryotic cells. With animal cells there is no great problem as only the membrane has to be crossed. In the case of yeast, protoplasts are required (Hinnan *et al.* 1978). With higher plants the strategy which has been considered is either to package the DNA in a plant virus or use a bacterial plant pathogen as the donor. It has also been shown that protoplasts prepared from plant cells are competent for transformation.

OTHER TECHNIQUES

Two other basic capabilities must be included in the repertoire of the gene manipulator. The first of these is DNA sequencing. Knowledge of the sequence of a cloned DNA fragment is a prerequisite for planning any substantial manipulation of the DNA; for example, a computer search of the sequence for all known restriction endonuclease target sites will provide a complete and precise restriction map. Importantly, once the sequence has been manipulated *in vitro*, perhaps using in-vitro mutagenesis techniques, it is usually desirable to sequence the product so as to confirm that the required sequence has been obtained. The second capability is gene synthesis. Recent progress in this field has been very rapid. The synthesis of DNA molecules many hundreds of nucleotides in length is now possible. The ability to synthesize polymers up to 20 nucleotides in length quickly and easily has many important applications which are discussed in later chapters.

DNA sequencing by the Maxam and Gilbert method

This method for DNA sequencing makes use of chemical reagents to bring about base-specific cleavage of the DNA. For large-scale sequencing it is now less favoured than the enzymatic, 'dideoxy', method. However, it still finds application and illustrates principles of polyacrylamide gel electrophoresis as applied to sequence determination. Discussion of the dideoxy method is deferred until Chapter 4 because it is nearly always used in combination with phage M13 cloning vectors.

In the Maxam and Gilbert method (Maxam & Gilbert 1977), the starting point is a defined DNA restriction fragment. The DNA strand to be sequenced must be radioactively labelled at one end with a ^{32}P-phosphate group. [A detailed practical account of the entire sequencing procedure, including end-labelling methods, is available (Maxam & Gilbert 1980)]. This DNA can be either in single-stranded or duplex form. The base-specific cleavages depend upon the following points.

1 Chemical reagents have been characterized which alter one or two bases in DNA (Table 1.1). These are base-specific reactions; for example, dimethyl sulphate methylates guanine (at the N7 position).

Table 1.1 Reagents for Maxam and Gilbert DNA sequencing.

Base specificity	Base reaction	Altered base removal	Strand cleavage
1. G	dimethyl sulphate	piperidine	piperidine
2. G + A	acid	acid-catalysed depurination	piperidine
3. T + C	hydrazine	piperidine	piperidine
4. C	hydrazine + NaCl	piperidine	piperidine
5. A > C	NaOH	piperidine	piperidine

2 An altered base can then be removed from the sugar-phosphate backbone of DNA (Table 1.1).

3 The strand is cleaved with piperidine at the sugar residue lacking the base. This cleavage is dependent upon the previous step.

When each of the base-specific reagents is used in a limited reaction with end-labelled DNA, a nested set of end-labelled fragments of different lengths is generated. It is important to emphasize that the base-specific reactions are deliberately limited to give about one, or a few, cleavages per molecule. This is illustrated in Fig. 1.5 where the nested set of fragments produced by the G-specific reaction is given as an example. Sets of fragments are produced by reacting the DNA with each of the reagents separately (Table 1.1). All five reactions (1–5, Table 1.1) may be performed; the fifth reaction gives redundant

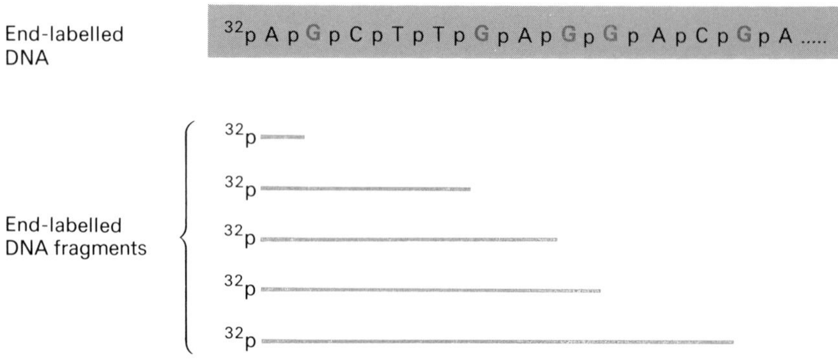

Fig. 1.5 Chemical cleavage of hypothetical DNA at G residues. A nested set of end-labelled DNA fragments is produced by limited reaction of an end-labelled DNA with G-specific reagents. Other fragments are produced but only the terminal fragments bear the label.

information but is confirmatory. These sequencing reactions are analysed by running the 4 or 5 samples side-by-side on a sequencing gel.

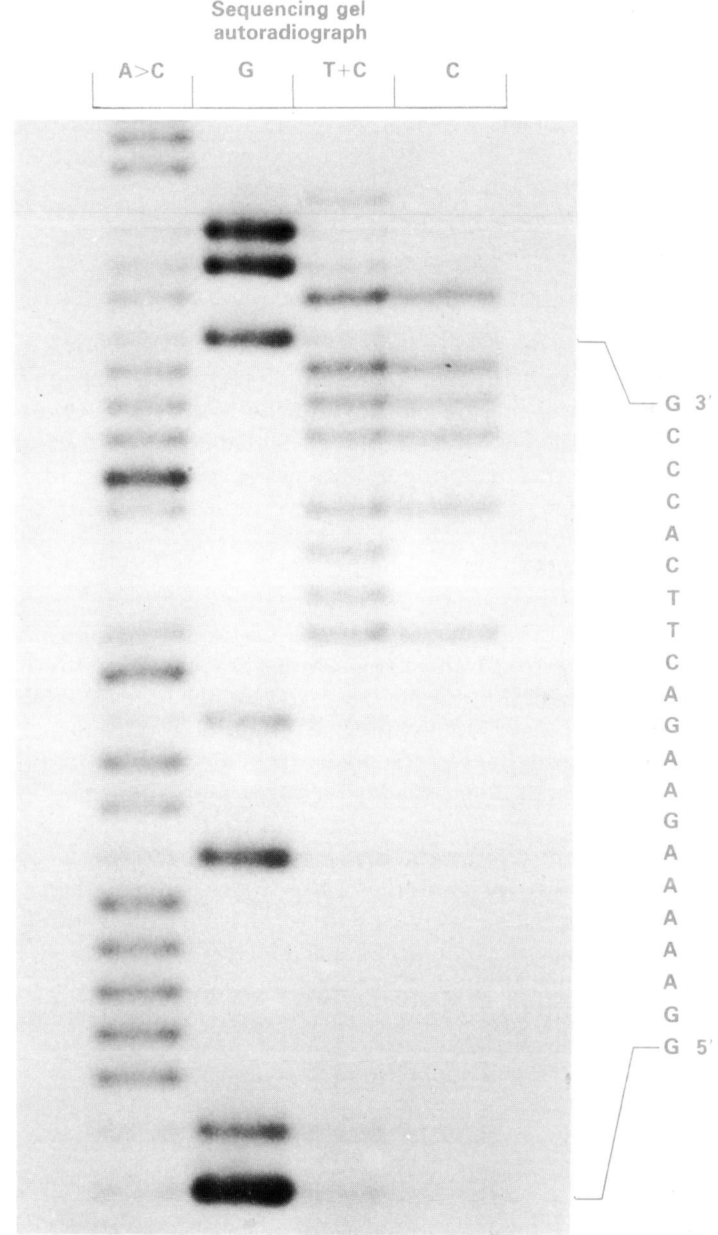

Fig. 1.6. Photograph of the Maxam and Gilbert autoradiograph (courtesy of N. Warburton).

A sequencing gel is a high resolution gel designed to fractionate single-stranded (denatured) DNA fragments on the basis of their length. They routinely contain 6% to 20% polyacrylamide and 7M urea. The urea is a denaturant whose function is to minimize DNA secondary structure effects on electrophoretic mobility. The gel is run at sufficient power to heat up to about 70 °C. This also minimizes DNA secondary structure. The labelled DNA bands obtained after such electrophoresis are revealed by autoradiography on large sheets of X-ray film. The sequence can then be read directly from the sequencing ladders in the adjacent base-specific tracks (Fig. 1.6).

The chemical synthesis of genes

The basic method of gene synthesis is the repetitive formation of an ester linkage between an activated phosphoric acid function of one nucleotide and the hydroxyl group of another nucleoside or nucleotide thus forming the characteristic phosphodiester bridge. The major problem is that deoxyribonucleotides are very reactive molecules having a primary and secondary hydroxyl group, a primary amino group and a phosphate group. Consequently, blocking and deblocking procedures are required and the chemistry involved must not result in scission or alteration of the phosphodiester backbone, the furanose rings, the sugar-purine/pyrimidine bond or the bases themselves—a tall order indeed! The various methods used to synthesize oligonucleotides are given in outline form below. The detailed chemistry is beyond the scope of this book. The reader wishing more information should consult the reviews of Caruthers *et al.* (1982) and Narang (1983).

The original method for synthesizing oligonucleotides is the so-called phosphodiester approach. This method is shown in Fig. 1.7. The 5′-phosphate of one nucleotide, after suitable protection of other functional groups, is condensed with the 3′-hydroxyl of another protected nucleoside or nucleotide using dicyclohexylcarbodiimide or an arylsulphonyl chloride. The chain can then be elongated by further step-wise condensations. The problems with the phosphodiester approach are long reaction times, rapidly decreasing yields as the chain length increases and time-consuming purification procedures. Nevertheless, the method was used successfully to synthesize a biologically active tRNA gene (Khorana 1979).

The phosphotriester approach solves some of the problems of the phosphodiester method by blocking each internucleotide

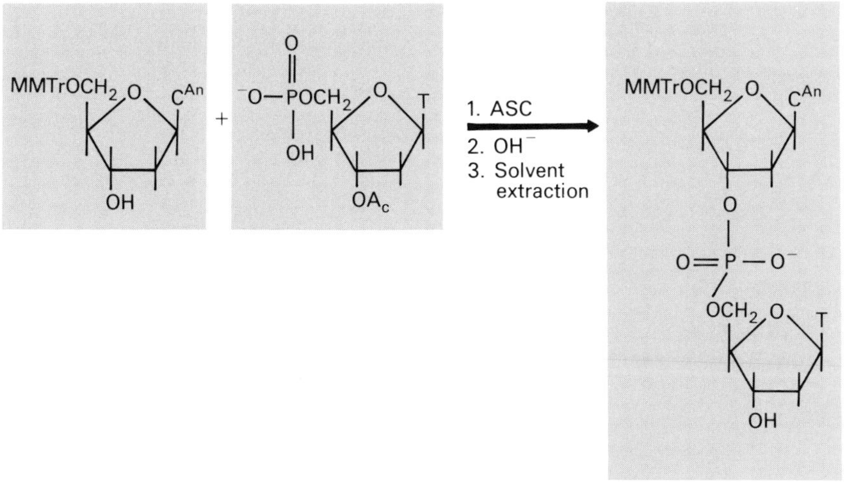

Fig. 1.7 The phosphodiester method of synthesizing oligonucleotides. The various blocking groups shown are: MMTr, monomethoxytrityl; Ac, acetyl; An, anisoyl.

phosphodiester function during the course of building a defined sequence. The basic principle of the method is the use of a totally protected mononucleotide containing a fully masked 3′-phosphotriester group (Fig. 1.8). Using this methodology Edge *et al.* (1981) synthesized 67 oligonucleotides of chain length 10–20 and spliced them together to generate a 517bp interferon-alpha gene.

More recently the phosphite-triester method has been introduced and this makes use of the extreme reactivity of phosphite reagents. In this way the two building blocks are joined in a few minutes compared with the hours required by the phosphotriester approach. The basic principles of the method are shown in Fig. 1.9. Although the phosphite-triester method will work in solution for the condensation of 3 or 4 nucleotides, construction of large oligonucleotides requires the 3′ end of the desired oligonucleotide to be coupled to an insoluble support. Fortunately, immobilization simplifies manipulative procedures and cuts out time-consuming purification steps following each cycle of condensation. Appropriately blocked mononucleotides are added sequentially and reagents, starting materials and by-products removed by filtration. At the conclusion of the synthesis the deoxyoligonucleotide is chemically freed of blocking groups, hydrolysed from the support and purified by electrophoresis or HPLC. By

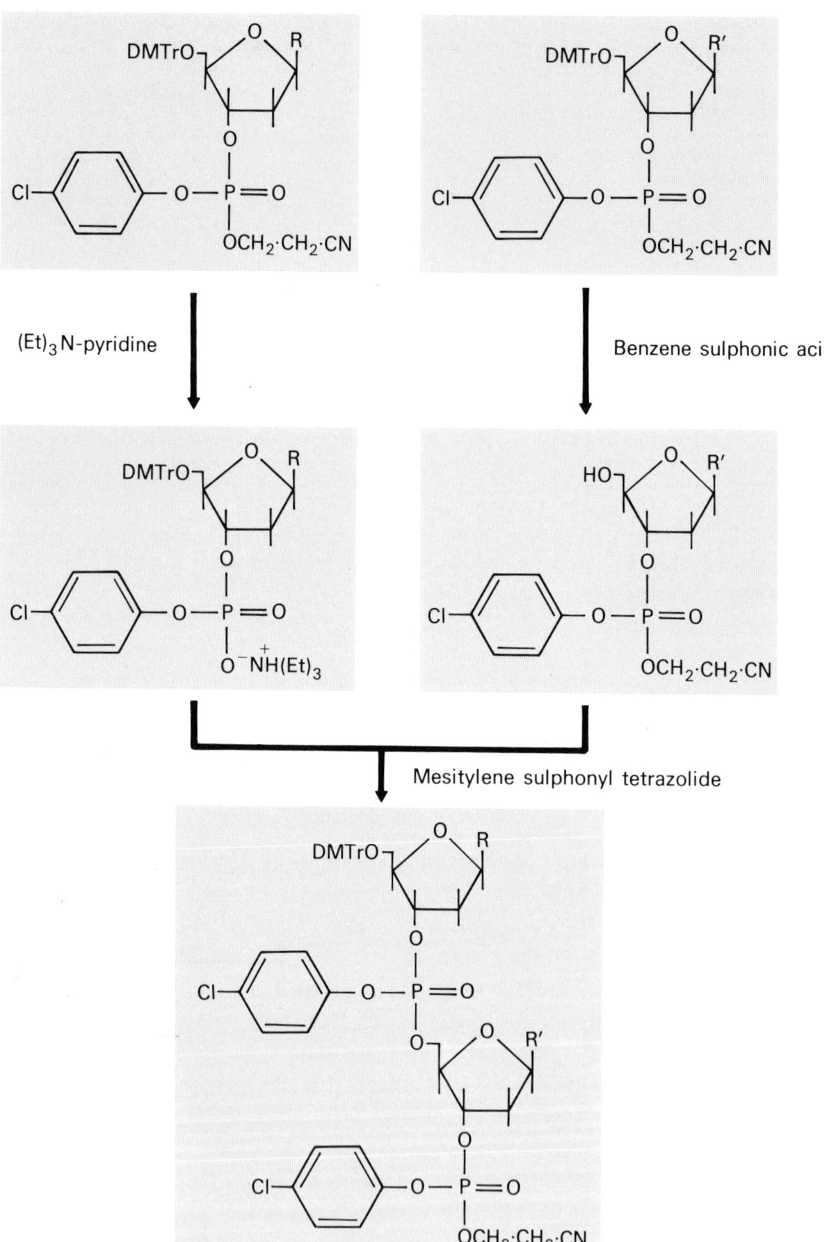

Fig. 1.8 The phosphotriester method of synthesizing oligonucleotides. The letter R or R′ represent any one of the four bases; adenine, guanine, cytosine or thymine. DMTr represents the dimethoxytrityl blocking group.

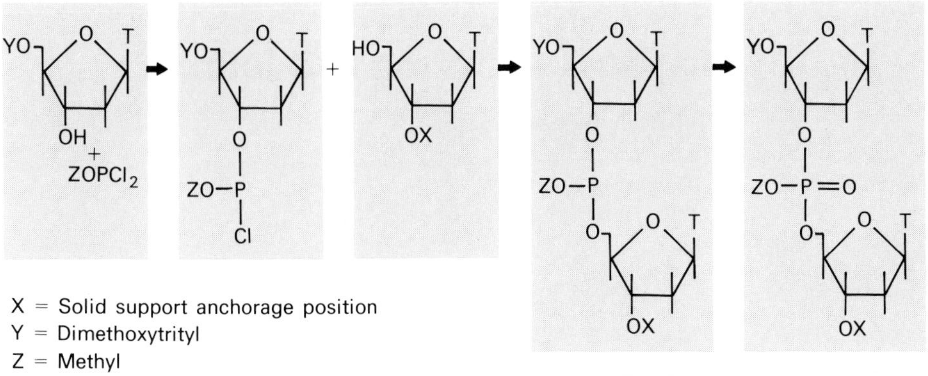

X = Solid support anchorage position
Y = Dimethoxytrityl
Z = Methyl

Fig. 1.9 The phosphite-triester method of synthesizing oligonucleotides.

immobilizing the polymer support carrying the initiating deoxynucleotide in a column, the filtration steps can be replaced by a simple washing procedure and this lends itself to a fully automatic synthesis. Automatic gene synthesizers consist simply of reagent reservoirs whose contents are added to or removed from the immobilized protected oligonucleotide via valves controlled by a microcomputer.

Chapter 2 Cutting and Joining DNA Molecules

CUTTING DNA MOLECULES

It is worth recalling that prior to 1970 there was simply no method available for cutting a duplex DNA molecule into discrete fragments. DNA biochemistry was circumscribed by this impasse. It became apparent that the related phenomena of host-controlled restriction and modification might lead towards a solution to the problem when it was discovered that restriction involves specific endonucleases. The favourite organism of molecular biologists, *E. coli* K12, was the first to be studied in this regard, but turned out to be an unfortunate choice. Its endonuclease is perverse in the complexity of its behaviour. The breakthrough in 1970 came with the discovery in *Haemophilus influenzae* of an enzyme that behaves more simply. Present-day DNA technology is totally dependent upon our ability to cut DNA molecules at specific sites with restriction endonucleases. An account of host-controlled restriction and modification therefore forms the first part of this chapter.

Host-controlled restriction and modification

Host-controlled restriction and modification are most readily observed when bacteriophages are transferred from one bacterial host strain to another. If a stock preparation of phage λ, for example, is made by growth upon *E. coli* strain C and this stock is then titred upon *E. coli* C and *E. coli* K, the titres observed on these two strains will differ by several orders of magnitude, the titre on *E. coli* K being the lower. The phage are said to be *restricted* by the second host strain (*E. coli* K). When those phage that do result from the infection of *E. coli* K are now replated on *E. coli* K they are no longer restricted; but if they are first cycled through *E. coli* C they are once again restricted when plated upon *E. coli* K (Fig. 2.1). Thus the efficiency with which phage λ plates upon a particular host strain depends upon the strain on which it was last propagated. This non-heritable

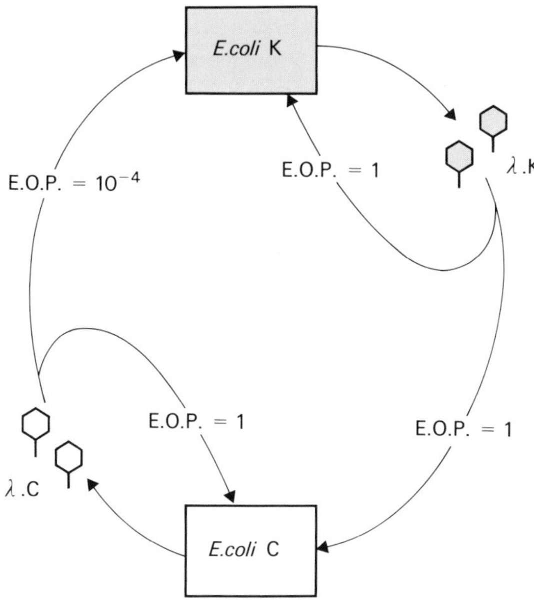

Fig. 2.1 Host-controlled restriction and modification of phage λ in *E. coli* strain K, analysed by efficiency of plating (E.O.P.). Phage propagated by growth on strains K or C (i.e. λ.K or λ.C) have E.O.P.s on the two strains as indicated by arrows. *E. coli* C has no known restriction and modification system.

change conferred upon the phage by the second host strain (*E. coli* K) that allows it to be replated on that strain without further restriction is called modification.

The restricted phages adsorb to restrictive hosts and inject their DNA normally. When the phage are labelled with ^{32}P it is apparent that their DNA is degraded soon after injection (Dussoix & Arber 1962) and the endonuclease that is primarily responsible for this degradation is called a *restriction endonuclease* or restriction enzyme (Lederberg & Meselson 1964). The restrictive host must, of course, protect its own DNA from the potentially lethal effects of the restriction endonuclease and so its DNA must be appropriately modified. Modification involves methylation of certain bases at a very limited number of sequences within DNA which constitute the recognition sequences for the restriction endonuclease. This explains why phage that survive one cycle of growth upon the restrictive host can subsequently reinfect that host efficiently; their DNA has been replicated in the presence of the modifying methylase and so it, like the host DNA, becomes methylated and protected from the restriction system.

Although phage infection has been chosen as our example to illustrate restriction and modification, these processes can occur whenever DNA is transferred from one bacterial strain to another. Conjugation, transduction, transformation and transfection are all subject to the constraint of host-controlled restriction. The genes that specify host-controlled restriction and modification systems may reside upon the bacterial chromosome itself or may be located on a plasmid or prophage such as P1.

The restriction endonuclease of *E. coli* K was the first to be isolated and studied in detail. Meselson and Yuan (1968) achieved this by devising an ingenious assay in which a fractionated cell extract was incubated with a mixture of unmodified and modified phage λ DNAs which were differentially radiolabelled—one with ^{3}H, the other with ^{32}P—so that they could be distinguished. After incubation, the DNA mixture was analysed by sedimentation through a sucrose gradient where the appearance of degraded unmodified DNA in the presence of undegraded modified DNA indicated the activity of restriction endonuclease.

The enzyme from *E. coli* K, and the similar one from *E. coli* B, were found to have unusual properties. In addition to magnesium ions, they require the cofactors ATP and S-adenosyl-methionine, and DNA degradation *in vitro* is accompanied by hydrolysis of the ATP in amounts greatly exceeding the stoichiometry of DNA breakage (Bickle *et al.* 1978). In addition, the enzymes are now known to interact with an unmodified *recognition* sequence in duplex DNA and then surprisingly, to track along the DNA molecule. After travelling for a distance corresponding to between 1000 and 5000 nucleotides the enzyme cleaves one strand only of the DNA at an apparently random site, and makes a gap about 75 nucleotides in length by releasing acid-soluble oligonucleotides. There is no evidence that the enzyme is truly catalytic, and having acted once in this way a second enzyme molecule is required to complete the double-strand break (Rosamond *et al.* 1979). Enzymes with these properties are now known as type I restriction endonucleases. Like all restriction endonucleases they recognize specific nucleotide sequences. However, they are not particularly useful for gene manipulation since their cleavage sites are non-specific. Their biochemistry still presents many puzzles; for instance, the precise role of S-adenosyl-methionine remains unclear.

While these bizarre properties of type I restriction enzymes were being unravelled, a restriction endonuclease from *H. influenzae* Rd was discovered (Kelly & Smith 1970, Smith

& Wilcox 1970) that was to become the prototype of a large number of restriction endonucleases—now known as type II enzymes—that have none of the unusual properties displayed by type I enzymes and which are fundamentally important in the manipulation of DNA. The type II enzymes recognize a particular target sequence in a duplex DNA molecule and break the polynucleotide chains within, or near to, that sequence to give rise to discrete DNA fragments of defined length and sequence. In fact, the activity of these enzymes is often assayed and studied by gel electrophoresis of the DNA fragments which they generate (see Fig. 1.1). As expected, digests of small plasmid or viral DNAs give characteristic simple DNA band patterns.

Very many type II restriction endonucleases have now been isolated from a wide variety of bacteria. In a recent review, Roberts (1984) lists 475 enzymes that have been at least partially characterized, and the number continues to grow as more bacterial genera are surveyed for their presence. It is worth noting that many so-called restriction endonucleases have not formally been shown to correspond with any genetically identified restriction and modification system of the bacteria from which they have been prepared: it is usually assumed that a site-specific endonuclease which is inactive upon host DNA and active upon exogenous DNA is, in fact, a restriction endonuclease.

Nomenclature

The discovery of a large number of restriction enzymes called for a uniform nomenclature. A system based upon the proposals of Smith and Nathans (1973) has been followed for the most part. The proposals were as follows.

1 The species name of the host organism is identified by the first letter of the genus name and the first two letters of the specific epithet to form a three-letter abbreviation in italics: for example, *Escherichia coli* = *Eco* and *Haemophilus influenzae* = *Hin*.

2 Strain or type identification is written as a subscript, e.g. $\textit{Eco}_{\mathrm{K}}$. In cases where the restriction and modification system is genetically specified by a virus or plasmid, the abbreviated species name of the host is given and the extrachromosomal element is identified by a subscript, e.g. $\textit{Eco}_{\mathrm{PI}}$, $\textit{Eco}_{\mathrm{RI}}$.

3 When a particular host strain has several different restriction and modification systems, these are identified by Roman numerals, thus the systems from *H. influenzae* strain Rd would

Table 2.1 Target sites for some restriction endonucleases. A comprehensive list is given in Appendix 2.

Anabaena variabilis	*Ava* I	C↓$\binom{C}{T}$CG$\binom{A}{G}$G	
Bacillus amyloliquefaciens H	*Bam* HI	G↓GATCC	
Bacillus globigii	*Bgl* II	A↓GATCT	
Escherichia coli RY13	*Eco* RI	G↓A$\overset{*}{\mathrm{A}}$TTC	1,4
Escherichia coli R245	*Eco* RII	↓CC$\binom{A}{T}$GG	2
Haemophilus aegyptius	*Hae* III	GG↓$\overset{*}{\mathrm{C}}$C	
Haemophilus gallinarum	*Hga* I	GACGC	3
Haemophilus haemolyticus	*Hha* I	G$\overset{*}{\mathrm{C}}$G↓C	
Haemophilus influenzae Rd	*Hind* II	GT$\binom{C}{T}$↓$\binom{A}{G}$$\overset{*}{\mathrm{A}}$C	
	Hind III	$\overset{*}{\mathrm{A}}$↓AGCTT	
Haemophilus parainfluenzae	*Hpa* I	GTT↓AAC	
	Hpa II	C↓$\overset{*}{\mathrm{C}}$GG	
Klebsiella pneumoniae	*Kpn* I	GGTAC↓C	
Moraxella bovis	*Mbo* I	↓GATC	
Providencia stuartii	*Pst* I	CTGCA↓G	
Serratia marcescens	*Sma* I	CCC↓GGG	
Streptomyces stanford	*Sst* I	GAGCT↓C	
Xanthomonas malvacearum	*Xma* I	C↓CCGGG	

Source: Roberts (1984). Recognition sequences are written from 5′→3′, only one strand being given, and the point of cleavage is indicated by an arrow. Bases written in parentheses signify that either base may occupy that position. Where known, the base modified by the corresponding specific methylase is indicated by an asterisk.
$\overset{*}{\mathrm{A}}$ is N^6-methyladenine, $\overset{*}{\mathrm{C}}$ is 5-methylcytosine.

Notes

1, 2. The names of these two enzymes are anomalous. The genes specifying the enzymes are borne on two Resistance Transfer Factors which have been classified separately. Hence RI and RII.

3. *Hga* I is a Type II restriction endonuclease, cleaving as indicated:

5′ GACGCNNNNN↓
3′ CTGCGNNNNN NNNNN↑

where N is any nucleotide.

4. Under certain conditions (low ionic strength, alkaline pH or 50% glycerol) the *Eco* RI specificity is reduced so that only the internal tetranucleotide sequence of the canonical hexanucleotide is necessary for recognition and cleavage. This is so-called *Eco* RI* (RI-star) activity. It is inhibited by parachloromercuribenzoate, whereas *Eco* RI activity is insensitive (Tikchonenko *et al.* 1978). Many other enzymes exhibit star activity, i.e. reduced specificity, under suboptimal conditions.

be Hin_dI, Hin_dII, Hin_dIII, etc. These Roman numerals should not be confused with those in the classification of restriction enzymes into type I, etc.

4 All restriction enzymes have the general name endonuclease R, but, in addition, carry the system name, e.g. endonuclease R.Hin_dIII. Similarly, modification enzymes are named methylase M followed by the system name. The modification enzyme from *H. influenzae* Rd corresponding to endonuclease R.Hin_dIII is designated methylase M.Hin_dIII.

In practice this system of nomenclature has been simplified further:

(a) subscripts are typographically inconvenient: the whole abbreviation is now usually written on the line;

(b) where the context makes it clear that restriction enzymes only are involved, the designation endonuclease R. is omitted. This is the system used in Table 2.1, which lists some of the more commonly used restriction endonucleases. A more extensive list is given in Appendix 2.

Target sites

The vast majority of, but not all, type II restriction endonucleases recognize and break DNA within particular sequences of tetra-, penta-, hexa- or hepta-nucleotides which have an axis of *rotational symmetry*; for example, *Eco* RI cuts at the positions indicated by arrows in the sequence

```
            axis of symmetry
          ↓ *    |
    5'—G A A     |   T T C—
    3'—C T T     |   A A G—
                 |   *  ↑
```

giving rise to termini bearing 5′-phosphate and 3′-hydroxyl groups. Such sequences are sometimes said to be *palindromic* by analogy with words that read alike backwards and forwards. (However, this term has also been applied to sequences such as

5′—A G C C G A—
3′—T C G G C T—

which are palindromic *within one strand*, yet do not have an axis of rotational symmetry.) If the sequence is modified by methylation so that 6-methyladenine residues are found at *one* or *both* of the positions indicated by asterisks then the sequence is resistant to endonuclease R.*Eco* RI. The resistance of the half-methylated site protects the bacterial host's own duplex DNA from attack immediately after semi-conservative replication of the fully methylated site until the modification methylase can once again restore the daughter duplexes to the fully methylated state.

We can see that *Eco* RI makes single-strand breaks four nucleotide pairs apart in the opposite strands of its target sequence, and so generates fragments with protruding 5′-termini. These DNA fragments can associate by hydrogen bonding between overlapping 5′-termini, or the fragments can circularize by intramolecular reaction, and for this reason the fragments are said to have *sticky* or *cohesive* ends (Fig. 2.2). In principle, DNA fragments from diverse sources can be joined by means of the cohesive ends, and it is possible, as we shall see later, to seal the remaining nicks in the two strands to form an intact *artificially recombinant* duplex DNA molecule.

It is clear from Table 2.1 that not all type II enzymes cleave their target sites like *Eco* RI. Some enzymes (e.g. *Pst* I) produce fragments bearing 3′-cohesive ends. Others (e.g. *Hae* III) make even cuts giving rise to flush- or blunt-ended fragments with no cohesive end at all. Some enzymes recognize tetranucleotide sequences, others recognize longer sequences, and this of course determines the average fragment length produced. We would expect any particular tetranucleotide target to occur about once every 4^4 (i.e. 256) nucleotide pairs in a long random

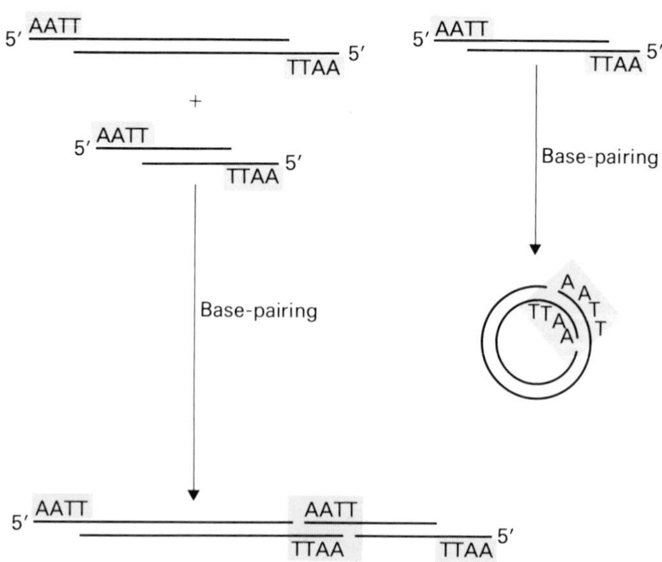

Fig. 2.2 Cohesive ends of DNA fragments produced by digestion with *Eco* RI.

DNA sequence, assuming all bases are equally frequent. Any particular hexanucleotide target would be expected to occur once in every 4^6 (i.e. 4096) nucleotide pairs. Some enzymes (e.g. *Sau* 3AI) recognize a tetranucleotide sequence that is included within the hexanucleotide sequence recognized by a different enzyme (e.g. *Bam* HI). The cohesive termini produced by these enzymes are such that fragments produced by *Sau* 3AI will cohere with those produced by *Bam* HI. If the fragments are then covalently joined, the 'hybrid site' so produced will be once again sensitive to *Sau* 3AI, but may not constitute a target for *Bam* HI; this will depend upon the nucleotides adjacent to the original *Sau* 3AI site (Fig. 2.3). Several other combinations of enzymes have this property.

From Table 2.1 we can also see that *Hin*d II, the first type II enzyme to be discovered, is an example of an enzyme recognizing a sequence with some ambiguity; in this case all three sequences corresponding to the structure given in Table 2.1 are substrates. There are also several known examples of enzymes from different sources which recognize the same target. They are *isoschizomers*. Some pairs of isoschizomers cut their target at different places (e.g. *Sma* I, *Xma* I).

In our discussion of the phenomena of restriction and modification of phage λ by *E. coli* K, we saw that methylation was the basis of modification in that system. In the wide variety of type II restriction enzymes now known there are some curious and useful examples of the influence of methyl groups at restriction sites. The enzymes *Hpa* II and *Msp* I are isoschizomers with the target sequence CCGG. *Hpa* II will not cut the target when it contains 5-methylcytosine as indicated by the asterisk C$\overset{*}{\text{C}}$GG (indeed this is the product of M. *Hpa* II). However, *Msp* I is known to be indifferent to methylation at

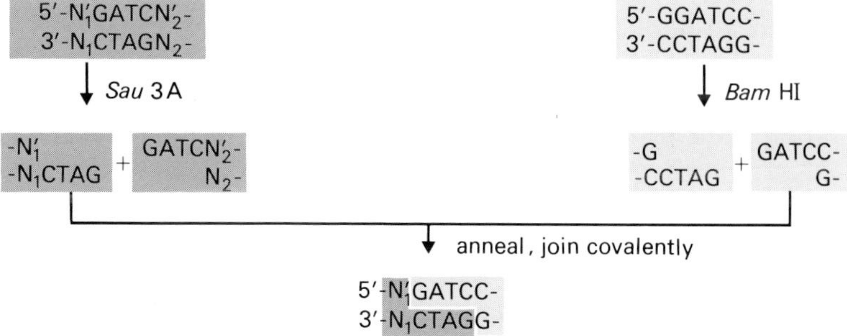

Fig. 2.3 Production of a hybrid site by cohesion of complementary sticky ends generated by *Sau* 3A and *Bam* HI.

this nucleotide: it cleaves whether or not this C residue is methylated. Now it has been found that over 90% of the methyl groups in genomic DNA of many animals, including vertebrates and echinoderms, occur as 5-methylcytosine in the sequence CG. Many of these methyl groups occur at *Msp* I sites, and their presence can be detected by comparing digests of the DNA generated by *Hpa* II and *Msp* I. Indeed methylation at *Msp* I sites around a single gene can be investigated in detail by combining the use of these two enzymes with the Southern-blot technique (Bird & Southern 1978, Razin & Riggs 1980).

Recently a new kind of type II restriction endonuclease has been identified. Enzymes of this type make breaks in the two strands at *measured distances* to one side of their asymmetric target sequence (e.g. *Hga* I; see Table 2.1). Finally, type III restriction enzymes have been classified as those which cleave DNA at well-defined sites, require ATP and Mg^{2+}, but which have only a partial requirement for (i.e. are stimulated by) S-adenosylmethionine. In these respects they have properties intermediate between type I and type II enzymes (Table 2.2).

The wide variety of properties exhibited by restriction endonucleases described in the preceding paragraphs has provided great scope for ingenious and resourceful gene manipulators. This will be apparent from examples in following chapters.

What is the function of restriction endonucleases *in vivo*? Clearly host-controlled restriction acts as a mechanism by which bacteria distinguish self from non-self. It is analogous to an immunity system. Restriction is moderately effective in preventing infection by some bacteriophages. It may be for this reason that the T-even phages (T2, T4 and T6) have evolved with glucosylated hydroxymethylcytosine residues replacing cytosine in their DNA, so rendering it resistant to many restriction endonucleases. The restriction and glucosylation modification of T-even phage DNA is beyond the scope of this book. For a detailed discussion the reader is referred to Kornberg (1980). However, it is worth noting that a mutant strain of T4 is available which does have cytosine residues in its DNA and is therefore amenable to conventional restriction methodology (Velten *et al.* 1976, Murray *et al.* 1979, Krisch & Selzer 1981). As an alternative to the unusual DNA structure of the T-even phages, other mechanisms appear to have evolved in T3 and T7 for overcoming restriction *in vivo* (Spoerel *et al.* 1979). In spite of this evidence we may be mistaken in concluding that immunity to phage infection is the sole or main function of restriction endonucleases in nature.

Table 2.2 Characteristics of restriction endonucleases (Yuan 1981, Iida *et al.* 1982, Hadi *et al.* 1982).

	Type I	Type II	Type III
1. Restriction and modification activities	Single multifunctional enzyme	Separate endonuclease and methylase	Separate enzymes with a subunit in common
2. Protein structure of restriction endonuclease	3 different subunits	Simple	2 different subunits
3. Requirements for restriction	ATP, Mg^{2+} S-adenosyl-methionine	Mg^{2+}	ATP, Mg^{2+} (S-adenosyl-methionine)
4. Sequence of host specificity sites	*Eco* B: $TGAN_8TGCT$ *Eco* K: $AACN_6GTGC$	rotational symmetry	*Eco* P1: AGACC *Eco* P15: CAGCAG
5. Cleavage sites	Possibly random, at least 1000 bp from host specificity site	At or near host specificity site	24–26 bp to 3′ of host specificity site
6. Enzymatic turnover	No	Yes	Yes
7. DNA translocation	Yes	No	No
8. Site of methylation	Host specificity site	Host specificity site	Host specificity site

N = any nucleotide

Mechanical shearing of DNA

In addition to digesting DNA with restriction endonucleases to produce discrete fragments, there are a variety of treatments which result in non-specific breakage. Non-specific endonucleases and chemical degradation can be used but the only method that has been much applied to gene manipulation involves mechanical shearing.

The long, thin threads which constitute duplex DNA molecules are sufficiently rigid to be very easily broken by shear forces in solution. Intense sonication with ultrasound can reduce the length to about 300 nucleotide pairs. More controlled shearing can be achieved by high-speed stirring in a blender. Typically, high mol. wt DNA is sheared to a population of molecules with a mean size of about 8 kb pairs by stirring at 1500 rev/min for 30 min (Wensink *et al.* 1974). Breakage occurs essentially at random with respect to DNA sequence. The termini consist of short single-stranded regions which may have to be taken into account in subsequent joining procedures.

JOINING DNA MOLECULES

Having described the methods available for cutting DNA molecules we must consider the ways in which DNA fragments can be joined to create artificially recombinant molecules. There are currently three methods for joining DNA fragments *in vitro*. The first of these capitalizes on the ability of DNA ligase to join covalently the annealed cohesive ends produced by certain restriction enzymes. The second depends upon the ability of DNA ligase from phage T4-infected *E. coli* to catalyse the formation of phosphodiester bonds between blunt-ended fragments. The third utilizes the enzyme terminal deoxynucleotidyltransferase to synthesize homopolymeric 3′-single-stranded tails at the ends of fragments. We can now look at these three methods a little more deeply.

DNA ligase

E. coli and phage T4 encode an enzyme, DNA ligase, which seals single-stranded nicks between adjacent nucleotides in a duplex DNA chain (Olivera *et al.* 1968, Gumport & Lehman 1971). Although the reactions catalysed by the enzymes of *E. coli* and T4-infected *E. coli* are very similar, they differ in their cofactor requirements. The T4 enzyme requires ATP, whilst the *E. coli* enzyme requires NAD^+. In each case the cofactor is split

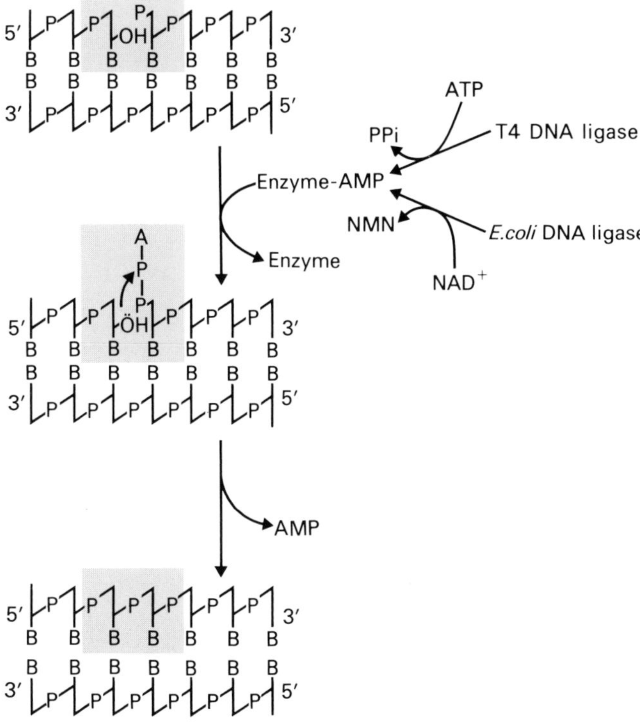

Fig. 2.4. Action of DNA ligase. An enzyme-AMP complex binds to a nick bearing 3′-OH and 5′-P groups. The AMP reacts with the phosphate group. Attack by the 3′-OH group on this moiety generates a new phosphodiester bond which seals the nick.

and forms an enzyme-AMP complex. The complex binds to the nick, which must expose a 5′-phosphate and 3′-OH group, and makes a covalent bond in the phosphodiester chain as shown in Fig. 2.4.

When termini created by a restriction endonuclease which creates cohesive ends associate, the joint has nicks a few base pairs apart in opposite strands. DNA ligase can then repair these nicks to form an intact duplex. This reaction, performed *in vitro* with purified DNA ligase, is fundamental to many gene manipulation procedures, such as that shown in Fig. 2.5.

The optimum temperature for ligation of nicked DNA is 37 °C, but at this temperature the hydrogen-bonded joint between the sticky ends is unstable. *Eco* RI-generated termini associate through only four A.T base pairs and these are not sufficient to resist thermal disruption at such a high temperature. The optimum temperature for ligating the cohesive termini is therefore a compromise between the rate of enzyme

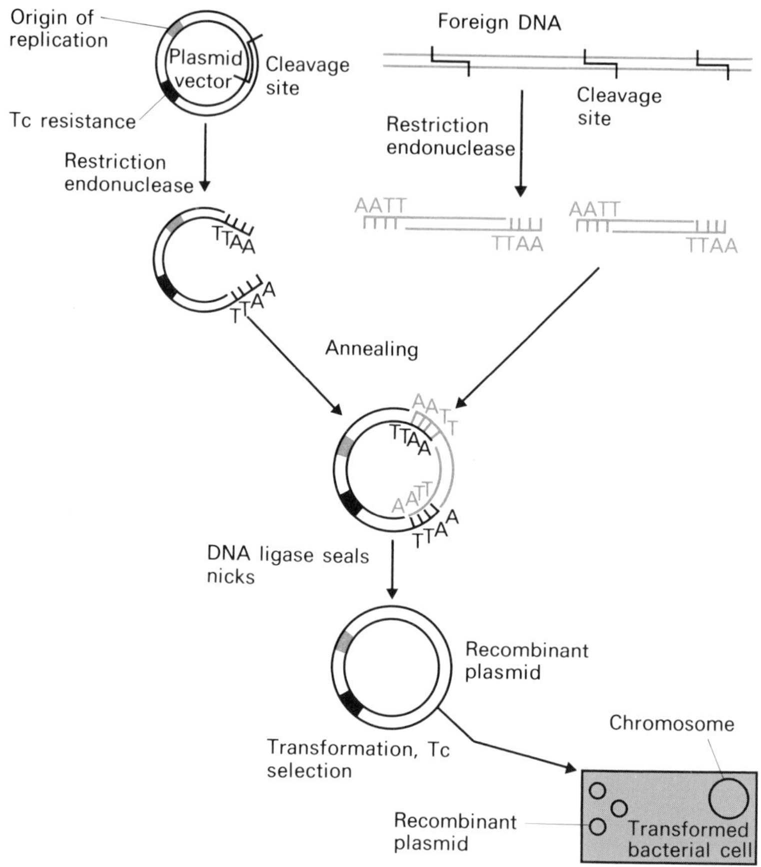

Fig. 2.5 Use of DNA ligase to create a covalent DNA recombinant joined through association of termini generated by *Eco* RI.

action and association of the termini, and has been found by experiment to be in the range 4–15 °C (Dugaicyzk *et al.* 1975, Ferretti & Sgaramella 1981).

The ligation reaction can be performed so as to favour the formation of recombinants. First, the population of recombinants can be increased by performing the reaction at a high DNA concentration; in dilute solutions *circularization* of linear fragments is relatively favoured because of the reduced frequency of intermolecular reactions. Second, by treating linearized plasmid vector DNA with alkaline phosphatase to remove 5′-terminal phosphate groups, both recircularization and plasmid dimer formation are prevented (Fig. 2.6). In this case, circularization of the vector can occur only by insertion of non-phosphatase-treated foreign DNA which provides one 5′-

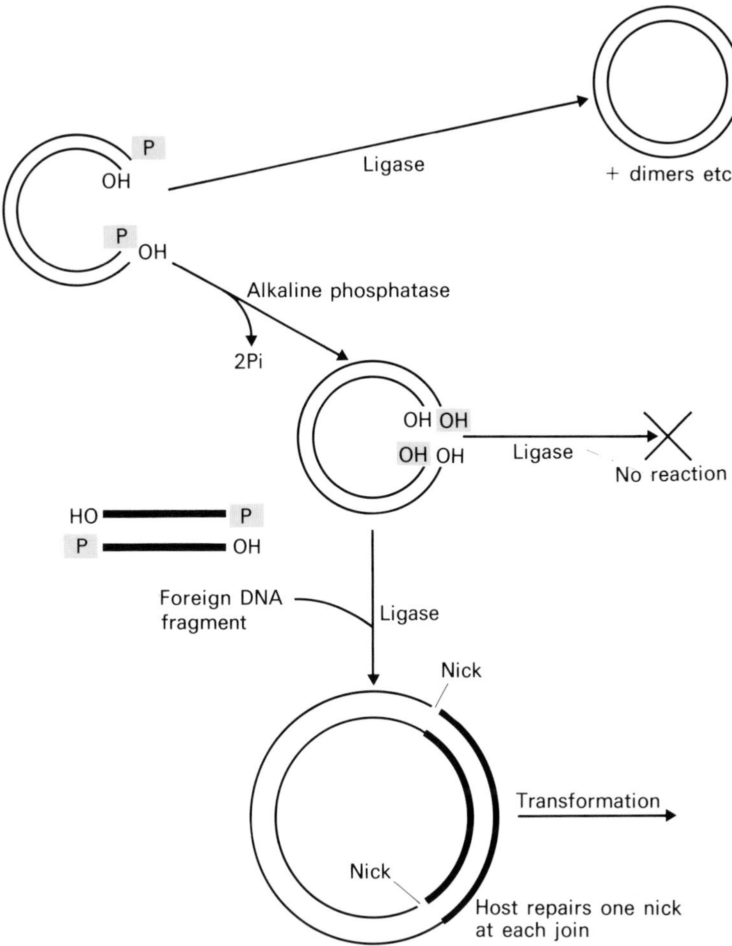

Fig. 2.6 Application of alkaline phosphatase treatment to prevent recircularization of vector plasmid without insertion of foreign DNA.

terminal phosphate at each join. One nick at each join remains unligated, but after transformation of host bacteria, cellular repair mechanisms reconstitute the intact duplex.

Joining DNA fragments with cohesive ends by DNA ligase is a relatively efficient process which has been used extensively to create artificial recombinants. A modification of this procedure depends upon the ability of T4 DNA ligase to join blunt-ended DNA molecules (Sgaramella 1972). The *E. coli* DNA ligase will not catalyse blunt ligation except under special reaction conditions of macromolecular crowding (Zimmerman *et al.* 1983). Blunt ligation is most usefully applied to joining blunt-ended fragments via *linker* molecules; for example,

Scheller *et al.* (1977) have synthesized self-complementary decameric oligonucleotides which contain sites for one or more restriction endonucleases. One such molecule is shown in Fig. 2.7. The molecule can be ligated to both ends of the foreign DNA to be cloned, and then treated with restriction endonuclease to produce a sticky-ended fragment which can be incorporated into a vector molecule that has been cut with the same restriction endonuclease. Insertion by means of the linker creates restriction enzyme target sites at each end of the foreign DNA and so enables the foreign DNA to be excised and recovered after cloning and amplification in the host bacterium.

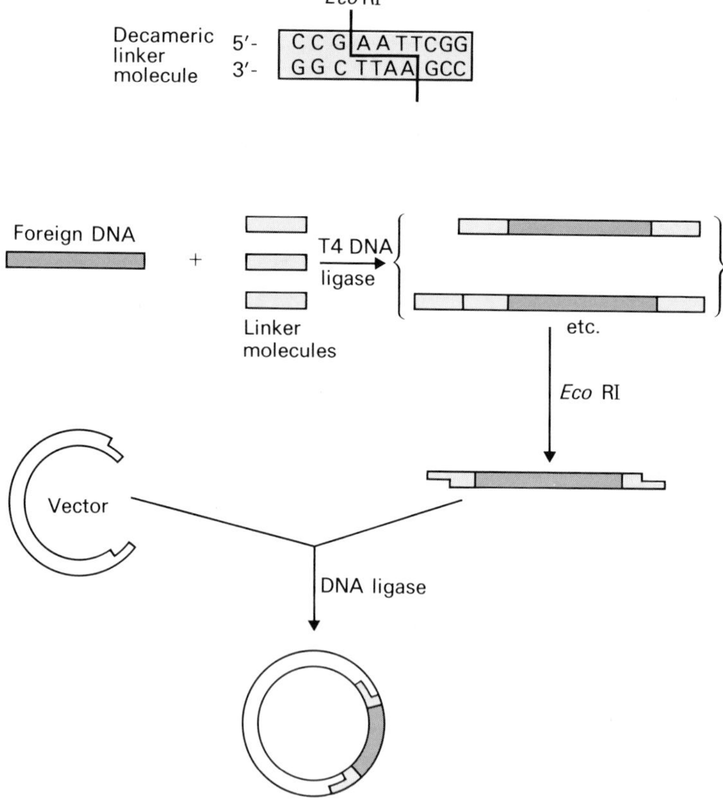

Fig. 2.7 A decameric linker molecule containing an *Eco* RI target site is joined by T4 DNA ligase to both ends of flush-ended foreign DNA. Cohesive ends are then generated by *Eco* RI. This DNA can then be incorporated into a vector that has been treated with the same restriction endonuclease.

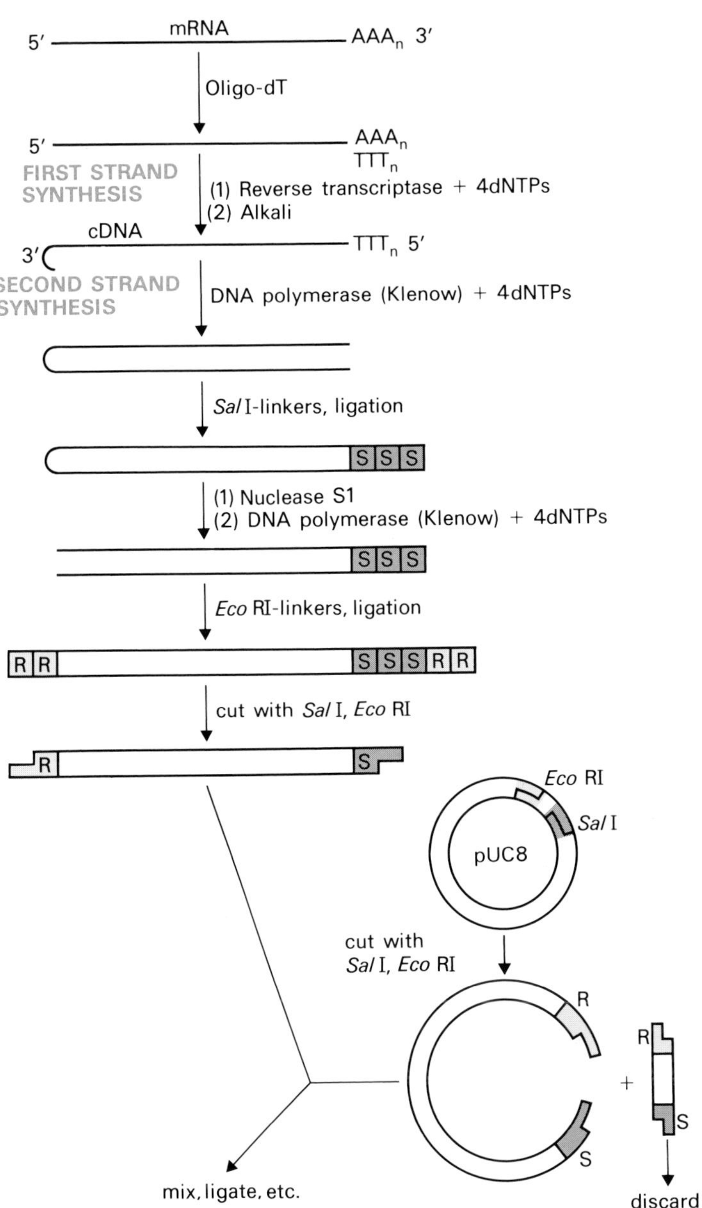

Fig. 2.8 Double-linkers. mRNA is copied into double-stranded cDNA and *Sal* I-linkers (S) are added. The hairpin loop formed by self-priming of second strand synthesis is removed by nuclease S1. In this procedure, any raggedness left by nuclease S1 (i.e. short single-strand projections at the terminus) is removed by polishing with Klenow polymerase. *Eco* RI-linkers are then ligated to the duplex molecule, and cohesive termini revealed by restriction with *Sal* I and *Eco* RI. The cDNA plus linkers is then ligated into a vector cut with the same two enzymes.

Double-linkers

Plasmid vectors have been derived which contain a set of closely clustered cloning sites. An example of such a vector is pUC8, which is described in more detail in Appendix 6. This vector has been used to clone duplex cDNA molecules by the double-linker approach (Kurtz & Nicodemus 1981, Helfman *et al.* 1983), in which *different* linker molecules are added to the opposite ends of the cDNA (Fig. 2.8). This has the following advantages.

1 The problem of vector reclosure without insertion of foreign DNA is overcome. Partly for this reason the method is efficient, i.e. cDNAs have been cloned which were derived from rare mRNA molecules in the starting population.

2 The use of linkers, rather than homopolymers, is desirable when expression from a vector-borne promoter is sought (see Chapter 7).

3 The orientation of the inserted DNA is fixed.

Adaptors

It may be the case that the restriction enzyme used to generate the cohesive ends in the linker will also cut the foreign DNA at internal sites. In this situation the foreign DNA will be cloned as two or more subfragments. One solution to this problem is to choose another restriction enzyme, but there may not be a suitable choice if the foreign DNA is large and has sites for several restriction enzymes. Another solution is to methylate internal restriction sites with the appropriate modification methylase. An example of this is described in Chapter 5. Alternatively, a general solution to the problem is provided by chemically synthesized adaptor molecules which have a *preformed* cohesive end (Wu *et al.* 1978). Consider a blunt-ended foreign DNA containing an internal *Bam* HI site (Fig. 2.9), which is to be cloned in a *Bam* HI-cut vector. The *Bam* adaptor molecule has one blunt end, bearing a 5′-phosphate group, and a Bam cohesive end which is not phosphorylated. The adaptor can be ligated to the foreign DNA ends, without any risk of adaptor self-polymerization. The foreign DNA plus added adaptors is then phosphorylated at the 5′-termini and ligated into the *Bam* HI site of the vector. If the foreign DNA were to be recovered from the recombinant with *Bam* HI, it would be obtained in two fragments. However, the adaptor is designed to contain two other restriction sites (*Sma* I, *Hpa* II) which may enable the foreign DNA to be recovered intact.

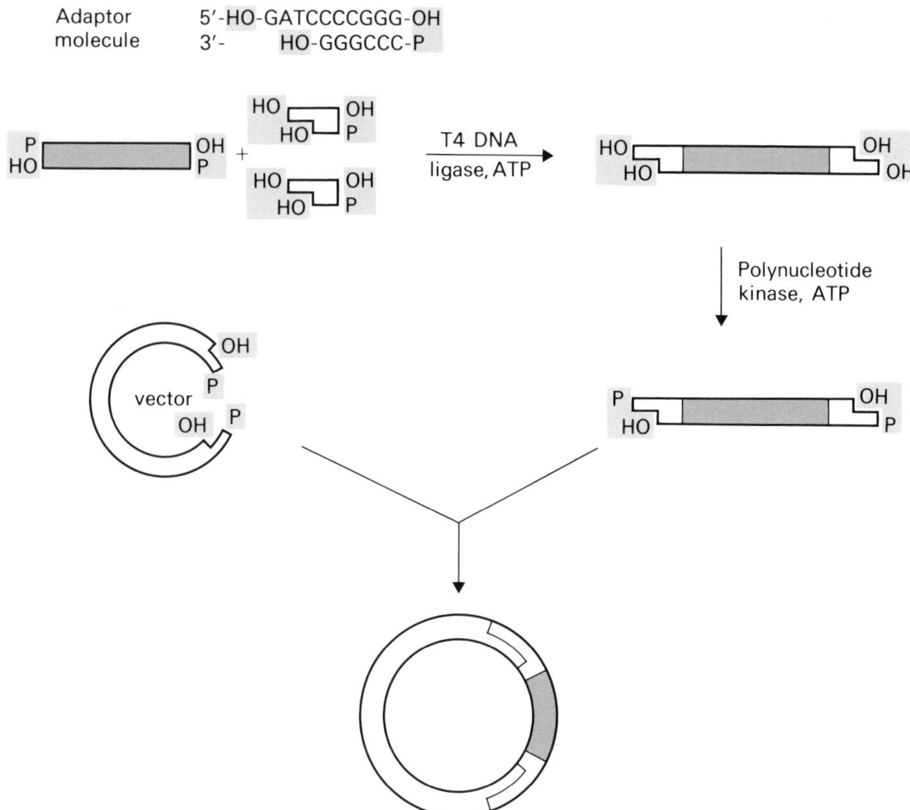

Fig. 2.9 Use of a *Bam* HI adaptor molecule. A synthetic adaptor molecule is ligated to the foreign DNA. The adaptor is used in the 5′-hydroxyl form to prevent self-polymerization. The foreign DNA plus ligated adaptors is phosphorylated at the 5′-termini and ligated into the vector previously cut with *Bam* HI.

Homopolymer tailing

A general method for joining DNA molecules makes use of the annealing of complementary homopolymer sequences. Thus, by adding oligo (dA) sequences to the 3′ ends of one population of DNA molecules, and oligo (dT) blocks to the 3′ ends of another population, the two types of molecule can anneal to form mixed dimeric circles (Fig. 2.10).

An enzyme purified from calf-thymus, terminal deoxynucleotidyl-transferase, provides the means by which the homopolymeric extensions can be synthesized, for if presented with a single deoxynucleotide triphosphate it will repeatedly add nucleotides to the 3′-OH termini of a population of DNA

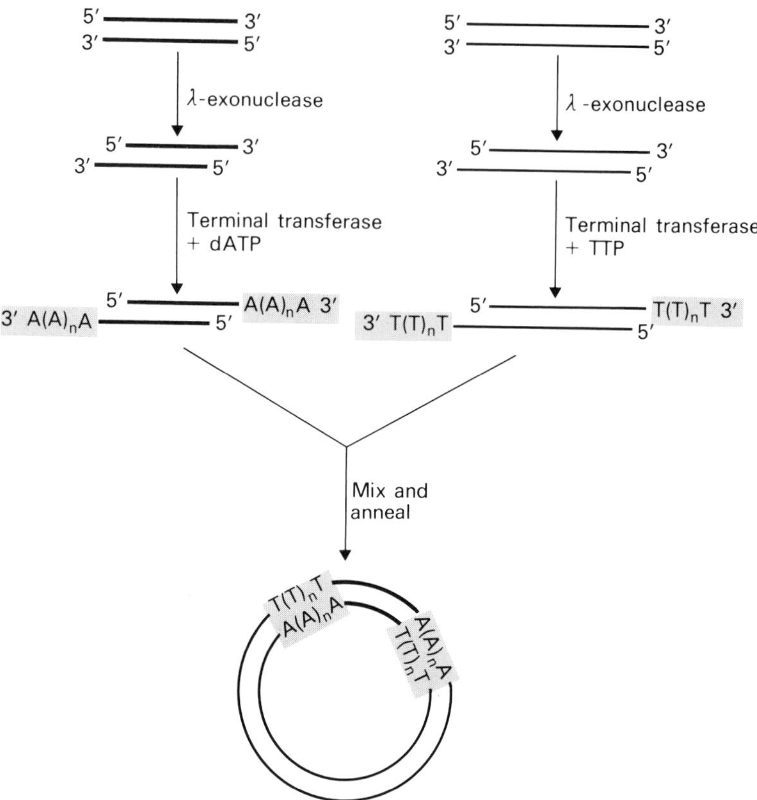

Fig. 2.10 Use of calf-thymus terminal deoxynucleotidyl-transferase to add complementary homopolymer tails to two DNA molecules.

molecules (Chang & Bollum 1971). DNA with exposed 3′-OH groups, such as arise from pretreatment with phage λ exonuclease or restriction with an enzyme such as *Pst* I, is a very good substrate for the transferase. However, conditions have been found in which the enzyme will extend even the shielded 3′-OH of 5′-cohesive termini generated by *Eco* RI (Roychoudhury *et al.* 1976, Humphries *et al.* 1978).

The terminal transferase reactions have been recently characterized in detail with regard to their use in gene manipulation (Deng & Wu 1981, Michelson & Orkin 1982). Typically, 10–40 homopolymeric residues are added to each end.

In 1972, Jackson *et al.* were among the first to apply the homopolymer method when they constructed a recombinant in which a fragment of phage λ DNA was inserted into SV40 DNA. In their experiments, the single-stranded gaps which remained in the two strands at each join were repaired *in vitro* with DNA polymerase and DNA ligase so as to produce covalently closed circular molecules, which were then used to transfect susceptible mammalian cells (see Chapter 11).

Subsequently, the homopolymer method, employing either dA.dT or dG.dC homopolymers, has been applied extensively in constructing recombinant plasmids for cloning in *E. coli.* Commonly, the annealed circles are used directly for transformation with repair of the gaps occurring *in vivo.* In the example which follows we shall see how homopolymer tailing can be applied to cloning DNA copies of eukaryotic messenger RNA and how a careful choice of which homopolymers are used can be important.

Cloning cDNA by homopolymer tailing

If we wish to construct a clone containing sequences derived from eukaryotic mRNA, we must first obtain the sequence in DNA form. We can do this by making a complementary (cDNA) copy of the mRNA, using the enzyme reverse transcriptase, which is a type of DNA polymerase found in retroviruses, and whose function is to synthesize DNA upon an RNA template.

Like other true DNA polymerases, reverse transcriptase can only synthesize a new DNA strand if provided with a growing point in the form of a pre-existing primer which is base-paired with the template and bears a free 3′-OH group. Fortunately, most eukaryotic mRNAs occur naturally in a polyadenylated form with up to 200 adenylate residues at their 3′-termini and hence we can provide a primer simply by hybridizing a short oligo (dT) molecule with this poly (A) sequence. The primer is then suitably located for synthesis of a complete cDNA by reverse transcriptase in the presence of all four deoxynucleoside triphosphates (Fig. 2.11).

The immediate product of the reaction is an RNA–DNA hybrid. The RNA strand can then be destroyed by alkaline hydrolysis, to which DNA is resistant, leaving a single-stranded cDNA which can be converted into the double-stranded form in a second DNA polymerase reaction. This reaction depends upon the observation that cDNAs can form a transient self-priming structure in which a hairpin loop at the 3′-terminus is stabilized by enough base-pairing to allow initiation of second-strand synthesis. Once initiated, subsequent synthesis of the second strand stabilizes the hairpin (Efstratiadis *et al.* 1976, Higuchi *et al.* 1976). The hairpin and any single-stranded DNA at the other end of the cDNA molecule are then trimmed away by treatment with the single-

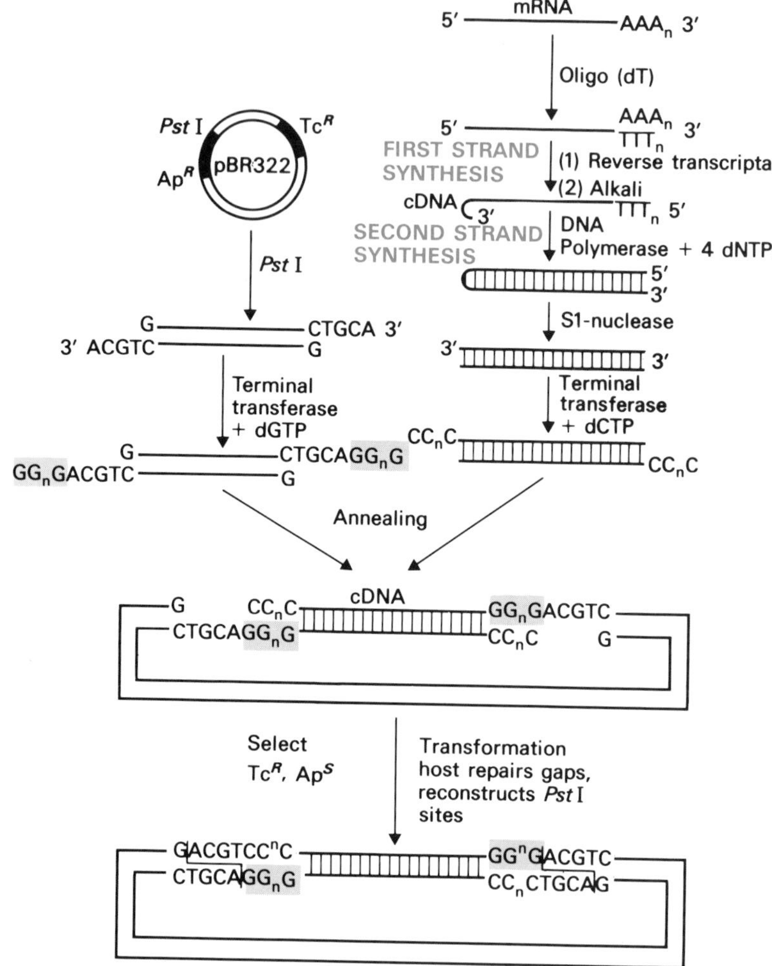

Fig. 2.11 Synthesis of a cDNA copy of a polyadenylated mRNA and insertion into a vector molecule by homopolymer tailing. See text for explanation.

strand-specific nuclease S1, giving rise to a fully duplex molecule.

In our example (Fig. 2.11) the duplex cDNA is tailed with oligo (dC) and annealed with the pBR322 vector which has been cut open with *Pst* I and tailed with oligo (dG). It will be seen that these homopolymers have been chosen so that *Pst* I target sites are reconstructed in the recombinant molecule, thus providing a simple means for excising the inserted sequences after amplification (Smith *et al.* 1979). This can be accomplished in another way, by constructing dA.dT joins. In that case the homopolymeric regions will have a lower melting temperature than the rest of the recombinant molecule, and so under partially denaturing conditions can be cleaved by nuclease S1 to release the inserted sequence (see p. 247 for example).

Full-length cDNA cloning

In the previous cDNA cloning scheme second strand synthesis is self-primed, resulting in the formation of a duplex cDNA with a hairpin loop that is subsequently removed by nuclease S1. This step necessarily leads to the loss of a certain amount of sequence corresponding to the 5′ end of the mRNA, and unless the nuclease S1 is very pure, there can be adventitious damage to the duplex cDNA.

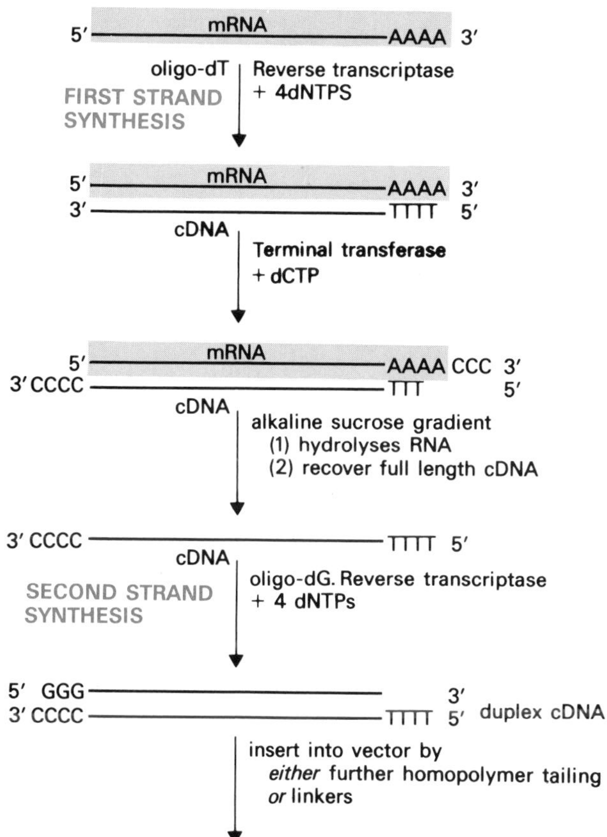

Fig. 2.12 Improved method for full-length duplex cDNA synthesis. The first strand is tailed with oligo(dC) so as to allow priming of the second strand synthesis by oligo (dG).

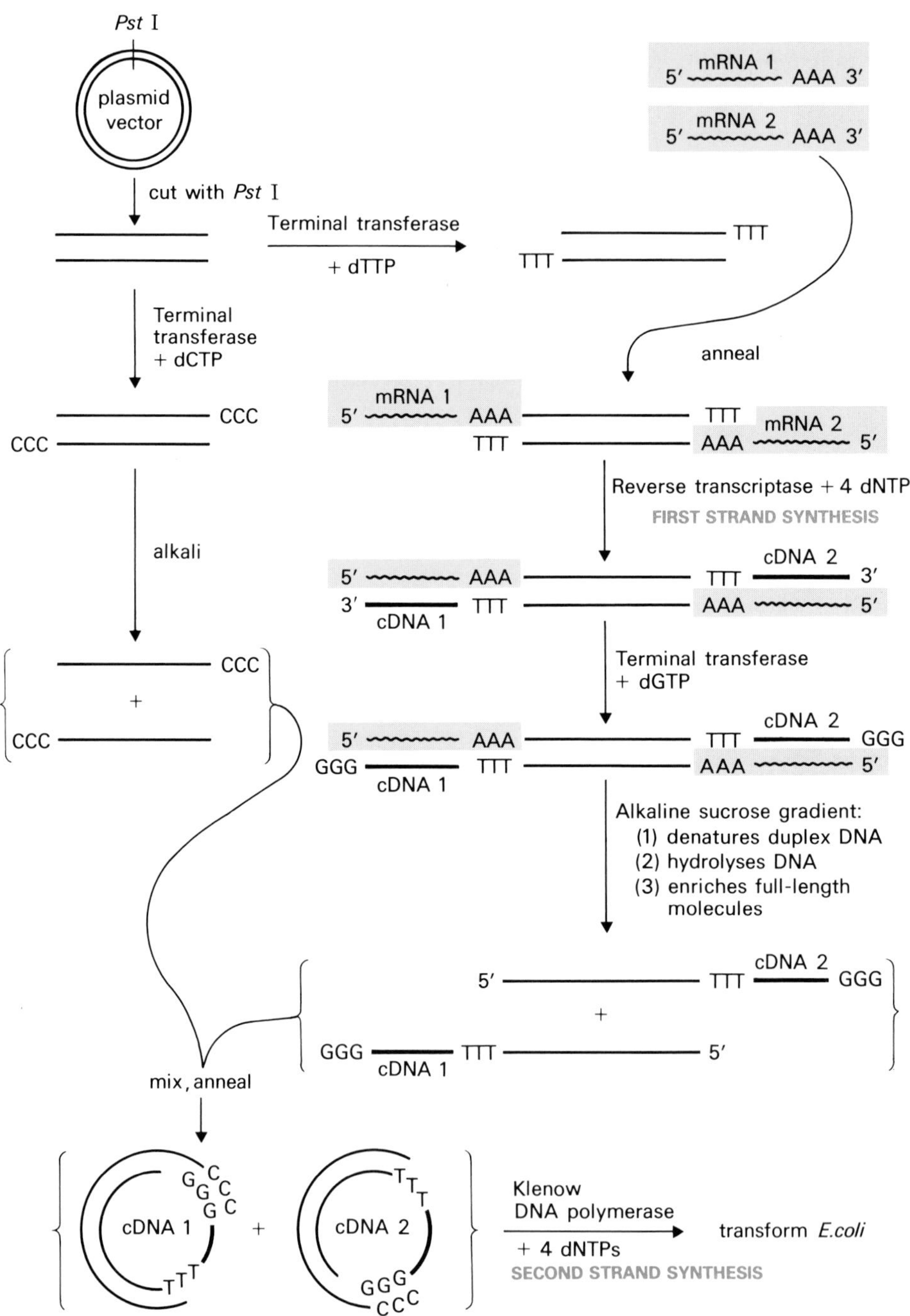

Pst I
plasmid vector
mRNA 1
5′ AAA 3′
mRNA 2
5′ AAA 3′
cut with Pst I
Terminal transferase
+ dTTP
TTT
TTT
Terminal transferase + dCTP
anneal
CCC
CCC
mRNA 1
5′ AAA
TTT
TTT
mRNA 2
AAA 5′
Reverse transcriptase + 4 dNTPs
FIRST STRAND SYNTHESIS
alkali
5′ AAA
3′ TTT
cDNA 1
TTT
cDNA 2
3′
AAA 5′
CCC
+
CCC
Terminal transferase
+ dGTP
5′ AAA
GGG TTT
cDNA 1
TTT
cDNA 2
GGG
AAA 5′
Alkaline sucrose gradient:
(1) denatures duplex DNA
(2) hydrolyses DNA
(3) enriches full-length molecules
5′ TTT cDNA 2 GGG
+
GGG cDNA 1 TTT 5′
mix, anneal
cDNA 1
GGCGCC
TTT
+
cDNA 2
TTT
GGGCCC
Klenow
DNA polymerase
+ 4 dNTPs
SECOND STRAND SYNTHESIS
transform E.coli

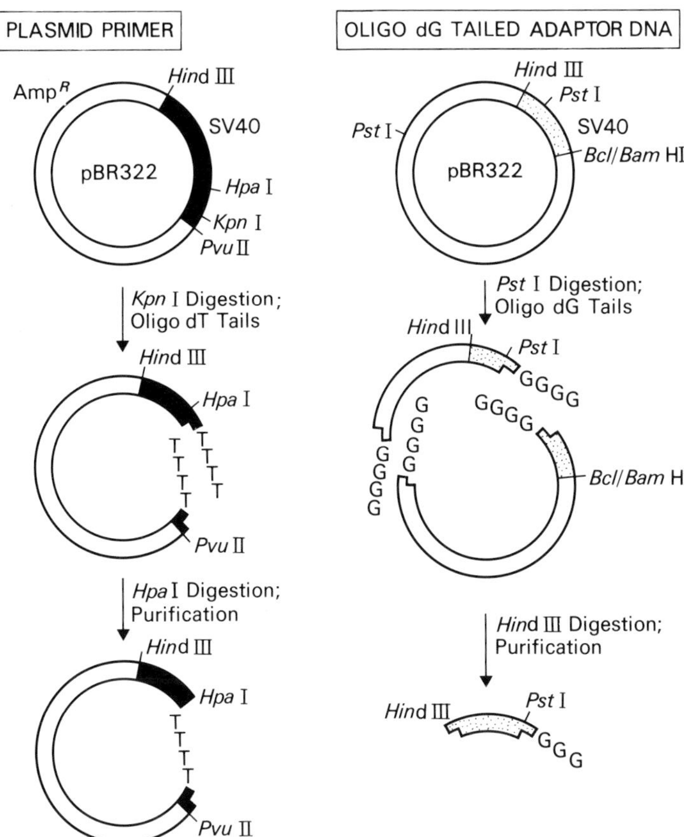

Fig. 2.14 The Okayama and Berg method of cDNA cloning. (a) Preparation of plasmid primer and adaptor DNA. The unshaded portion of each ring is pBR322 DNA, and the shaded or stippled segments are from SV40 DNA.

Fig. 2.13 (*facing page*) Efficient full-length cDNA cloning (Heidecker & Messing 1983). The mRNA is annealed to linearized and oligo-dT tailed plasmid DNA, which then primes synthesis of the first cDNA strand. Oligo-dG tails are added to the cDNA-plasmid molecules, which are then centrifuged through an alkaline sucrose gradient. This step removes small molecules, hydrolyses the mRNA and separates the two cDNAs which were formerly attached to the same duplex plasmid. Denatured, oligo-dC tailed plasmid DNA is added (in excess) and conditions adjusted to favour circularization by the complementary homopolymer tails. The excess oligo-dC tailed plasmid may simply renature, but cannot circularize. The circular molecules have a free 3′-hydroxyl on the oligo-dC tail which primes second strand synthesis of the cDNA to create duplex recombinant plasmids which transform *E. coli*. Clones can be obtained with the cDNA inserted in both orientations.

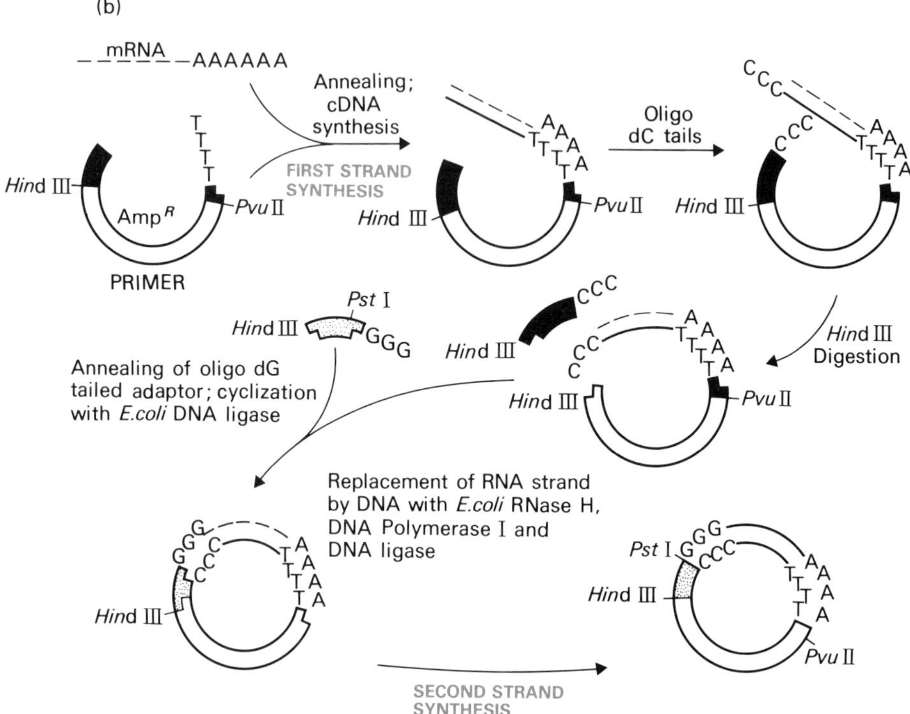

Fig. 2.14 (b) Steps in the construction of plasmid-cDNA recombinants. The designations for the DNA segments are as mentioned in (a).

dC tailing of single-stranded cDNA followed by oligo-dG priming of second strand synthesis (Fig. 2.12) does not lead to hairpin formation, nuclease S1 treatment is not required, and consequently is an effective method for generating full-length cDNA clones (Land *et al.* 1981). Two further methods, shown schematically in Figs 2.13 and 2.14, have been devised to eliminate the use of nuclease S1. Additionally, in both methods the oligo-dT sequence for priming the first strand cDNA synthesis is linked to the vector DNA in a prior reaction. Both methods have been reported to promote full length cDNA cloning with a very high efficiency (Okayama & Berg 1982, Heidecker & Messing 1983). It is thought that full length reverse transcripts are obtained *preferentially* because in each case an RNA–DNA hybrid molecule, which is the result of first strand synthesis, is the substrate for a terminal transferase reaction. A cDNA that does not extend to the end of the mRNA will present a shielded 3′-hydroxy group, which is a poor substrate for tailing.

SECTION 2

Cloning in *E. coli*

Chapter 3 Plasmids as Cloning Vehicles for Use in *E. coli*

Basic properties of plasmids

Plasmids are widely used as cloning vehicles but before discussing their use in this context it is appropriate to review some of their basic properties. Plasmids are replicons which are stably inherited in an extra-chromosomal state. It should be emphasized that extra-chromosomal nucleic acid molecules are not necessarily plasmids, for the definition given above implies genetic homogeneity, constant monomeric unit size and the ability to replicate independently of the chromosome. Thus the heterogeneous circular DNA molecules which are found in *Bacillus megaterium* (Carlton & Helinski 1969) are not necessarily plasmids. The definition given above, however, does include the prophages of those temperate phages, e.g. P1, which are maintained in an extra-chromosomal state, as opposed to those such as λ (see Chapter 4) which are maintained by integration into the host chromosome. Also included are the replicative forms of the filamentous coliphages which specify the continued production and release of phage particles without concomitant cell lysis.

Plasmids are widely distributed throughout the prokaryotes, vary in size from less than 1×10^6 daltons to greater than 200×10^6, and are generally dispensable. Some of the phenotypes which these plasmids confer on their host cells are listed in Table 3.1. Plasmids to which phenotypic traits have not yet been ascribed are called *cryptic* plasmids.

Plasmids can be categorized into one of the two major types—conjugative or non-conjugative—depending on whether or not they carry a set of transfer genes, called the *tra* genes, that promotes bacterial conjugation. Plasmids can also be categorized on the basis of their being maintained as multiple copies per cell (*relaxed* plasmids) or as a limited number of copies per cell (*stringent* plasmids). Generally, conjugative plasmids are of a relatively high mol. wt and are present as 1–3 copies per chromosome whereas non-con-

Table 3.1 Phenotypic traits exhibited by plasmid-carried genes.

Antibiotic resistance	Heavy-metal resistance
Antibiotic production	Bacteriocin production
Degradation of aromatic compounds	Induction of plant tumours
Haemolysin production	Hydrogen sulphide production
Sugar fermentation	Host-controlled restriction and modification
Enterotoxin production	

jugative plasmids are of low mol. wt and present as multiple copies per cell (Table 3.2). An exception is the conjugative plasmid R6K which has a mol. wt of 25×10^6 daltons and is maintained as a relaxed plasmid.

Plasmid *incompatibility* is the inability of two different plasmids to coexist in the same host cell in the absence of selection pressure. The term incompatibility can only be used when it is certain that entry of the second plasmid has taken place and that DNA restriction is not involved. Groups of plasmids which are mutually incompatible are considered to belong to the same incompatibility group. Currently, over 25 incompatibility groups have been defined among plasmids of *E. coli* and 7 for plasmids of *Staphylococcus aureus.* Plasmids belonging to incompatibility classes P, Q and W, are termed *promiscuous* for they are capable of promoting their own transfer to a wide range of Gram-negative bacteria (groups P

Table 3.2 Properties of some conjugative and non-conjugative plasmids of Gram-negative organisms.

Plasmid	Size (Mdal.)	Conjugative	No. of plasmid copies/ chromosome equivalent	Phenotype
Col E1	4.2	no	10–15	Colicin E1 production
RSF 1030	5.6	no	20–40	ampicillin resistance
clo DF13	6	no	10	cloacin production
R6K	25	yes	13–38	ampicillin and streptomycin resistance
F	62	yes	1–2	—
RI	62.5	yes	3–6	multiple drug resistance
Ent P 307	65	yes	1–3	enterotoxin production

and W) and of being stably maintained in these diverse hosts. Such promiscuous plasmids thus offer the potential of readily transferring cloned DNA molecules into a wide range of genetic environments (see p. 156).

An extremely useful article explaining the terminology used in plasmid genetics is that of Novick *et al.* (1976). A much fuller discussion of the topics outlined above is provided by Falkow (1975) and Broda (1979).

The purification of plasmid DNA

An obvious prerequisite for cloning in plasmids is the purification of the plasmid DNA. Although a wide range of plasmid DNAs are now routinely purified the methods used are not without their problems. Undoubtedly the trickiest stage is the lysis of the host cells; both incomplete lysis and total dissolution of the cells result in greatly reduced recoveries of plasmid DNA. The ideal situation occurs when each cell is just sufficiently broken to permit the plasmid DNA to escape without too much contaminating chromosomal DNA. Provided the lysis is done gently most of the chromosomal DNA which is released will be of high mol. wt and can be removed, along with cell debris, by high speed centrifugation to yield a *cleared lysate*. The production of satisfactory cleared lysates from bacteria other than *E. coli*, particularly if large plasmids are to be isolated, is frequently a combination of skill, luck and patience.

Many methods are available for isolating pure plasmid DNA from cleared lysates but only two will be described here. The first of these is the 'classical' method and is due to Vinograd (Radloff *et al.* 1967). This method involves isopycnic centrifugation of cleared lysates in a solution of CsCl containing ethidium bromide (EtBr). EtBr binds by intercalating between the DNA base pairs and in so doing causes the DNA to unwind. A covalently closed circular (ccc) DNA molecule such as a plasmid has no free ends and can only unwind to a limited extent thus limiting the amount of EtBr bound. A linear DNA molecule, such as fragmented chromosomal DNA, has no such topological constraints and can therefore bind more of the EtBr molecules. Because the density of the DNA/EtBr complex decreases as more EtBr is bound, and because more EtBr can be bound to a linear molecule than a covalent circle, the covalent circle has a higher density at saturating concentrations of EtBr. Thus covalent circles (i.e. plasmids) can be separated from linear chromosomal DNA (Fig. 3.1).

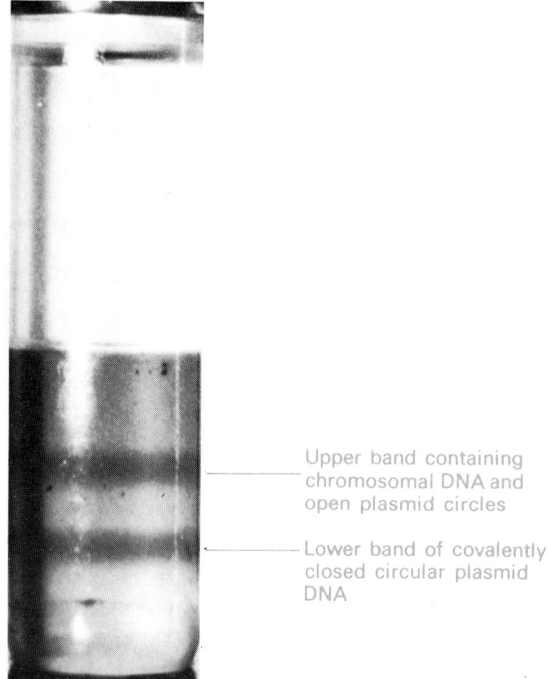

Fig. 3.1 Purification of ColEI kan^R plasmid DNA by isopycnic centrifugation in a CsC1-EtBr gradient. (Photograph by courtesy of Dr G. Birnie.)

Currently the most popular method of extracting and purifying plasmid DNA is that of Birnboim and Doly (1979). This method makes use of the observation that there is a narrow range of pH (12.0–12.5) within which denaturation of linear DNA, but not covalently closed circular DNA, occurs. Plasmid-containing cells are treated with lysozyme to weaken the cell wall and then lysed with sodium hydroxide and sodium dodecyl sulphate. Chromosomal DNA remains in a high mol. wt form but is denatured. Upon neutralization with acidic sodium acetate the chromosomal DNA renatures and aggregates to form an insoluble network. Simultaneously, the high concentration of sodium acetate causes precipitation of protein-SDS complexes and of high mol. wt RNA. Provided the pH of the alkaline denaturation step has been carefully controlled the covalently closed circular plasmid DNA molecules will remain in a native state and in solution while the contaminating macromolecules co-precipitate. The precipitate can be removed by centrifugation and the plasmid concentrated by ethanol precipitation. If necessary, the plasmid DNA can be purified further by gel filtration.

Desirable properties of plasmid cloning vehicles

An ideal cloning vehicle would have the following three properties:

1 low mol. wt;

2 ability to confer readily selectable phenotypic traits on host cells;

3 single sites for a large number of restriction endonucleases, preferably in genes with a readily scorable phenotype.

The advantages of a low mol. wt are several. First, the plasmid is much easier to handle, i.e. it is more resistant to damage by shearing, and is readily isolated from host cells. Second, low mol. wt plasmids are usually present as multiple copies (see Table 3.2) and this not only facilitates their isolation but leads to gene dosage effects for all cloned genes. Finally, with a low mol. wt there is less chance that the vector will have multiple substrate sites for any restriction endonuclease (see below).

After a piece of foreign DNA is inserted into a vector the resulting chimaeric molecules have to be transformed into a suitable recipient. Since the efficiency of transformation is so low it is essential that the chimaeras have some readily scorable phenotype. Usually this results from some gene, e.g. antibiotic resistance, carried on the vector but could also be produced by a gene carried on the inserted DNA.

One of the first steps in cloning is to cut the vector DNA and the DNA to be inserted with either the same endonuclease or ones producing the same ends. If the vector has more than one site for the endonuclease, more than one fragment will be produced. When the two samples of cleaved DNA are subsequently mixed and ligated the resulting chimaeras will, in all probability, lack one of the vector fragments. It is advantageous if insertion of foreign DNA at endonuclease-sensitive sites inactivates a gene whose phenotype is readily scorable, for in this way it is possible to distinguish chimaeras from cleaved plasmid molecules which have self-annealed. Of course, readily detectable insertional inactivation is not essential if the vector and insert are to be joined by the homopolymer tailing method (see p. 37) or if the insert confers a new phenotype on host cells.

Some examples will be presented which illustrate the points

raised above but first we shall consider how some of the common plasmids rate as cloning vehicles.

Usefulness of 'natural' plasmids as cloning vehicles

The term 'natural' is used loosely in this context to describe plasmids which were not constructed *in vitro* for the sole purpose of cloning. Col E1 is a naturally occurring plasmid which specifies the production of a bacteriocin, colicin E1. By necessity this plasmid also carries a gene which confers on host cells immunity to colicin E1. RSF 2124 is a derivative of Col E1 which carries a transposon specifying ampicillin resistance. The exact origin of pSC101 is not clear but for the purposes of this discussion we shall consider it to be a 'natural' plasmid. Details of these plasmids are shown in Table 3.3.

To clone DNA in pSC101, the plasmid DNA and the DNA to be inserted are digested with *Eco* RI, for example, and treated with DNA ligase. The ligated molecules are then used to transform a suitable recipient to tetracycline-resistance. Unfortunately there is no easy genetical method of distinguishing chimaeras from reconstituted vector DNA unless the insert confers a new phenotype on the transformants. Two examples of the use of pSC101 for cloning DNA are presented in the next section. When using *Eco* RI, cloning with Col E1 as the vector is a little simpler. Transformants are selected on the basis of immunity to colicin E1 and chimaeras recognized by their inability to produce colicin E1. Unfortunately, screening for immunity to

Table 3.3 Properties of some 'natural' plasmids used for cloning DNA.

Plasmid	Size (Mdal.)	Single sites for endonucleases	Marker for selecting transformants	Insertional inactivation of
pSC101	5.8	*Xho* I, *Eco* RI *Pvu* II, *Hinc* II *Hpa* I	tetracycline resistance	—
		Hind III, *Bam* HI *Sal* I	—	tetracycline resistance
Col E1	4.2	*Eco* RI	immunity to colicin E1	colicin E1 production
RSF 2124	7.4	*Eco* RI, *Bam* HI	ampicillin resistance	colicin E1 production

colicin E1 is not technically simple and plasmid RSF 2124 is more useful in this respect since transformants are selected by virtue of their ampicillin resistance.

Col E1 and plasmids derived from it (see later) have two distinct advantages over pSC101. They have a higher copy number and they can be enriched with chloramphenicol. When chloramphenicol is added to a late log-phase culture of a Col E1-containing strain of *E. coli,* chromosome replication ceases because of the need for continued protein synthesis. However, the cessation of protein synthesis has no effect on Col E1 replication such that after 10–12 hours over 50% of the DNA in the cells is plasmid DNA (Hershfield *et al.* 1974). Since there may be 1000–3000 copies of the plasmid in each cell it is easy to see why chloramphenicol enrichment is a useful step in plasmid isolation.

Example of the use of pSC101 for cloning.
1. Expression of *Staphylococccus* plasmid genes in *E. coli*

For this experiment Chang and Cohen (1974) considered *S. aureus* plasmid p1258 (mol. wt 20×10^6) as being particularly

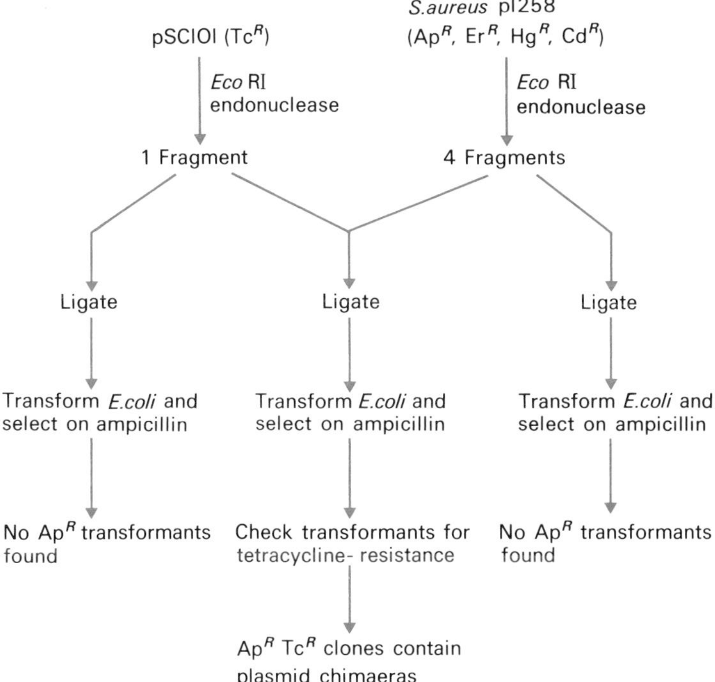

Fig. 3.2 Insertion of an *S. aureus* gene specifying ampicillin-resistance into plasmid pSC101.

appropriate for experiments involving interspecies genome construction since it carries several different genetic determinants that were potentially detectable in *E. coli*. Moreover, agarose gel electrophoresis indicated that this plasmid is cleaved by the *Eco* RI restriction endonuclease into four easily identifiable fragments. Molecular chimaeras containing DNA derived from both *Staphylococcus* and *E. coli* were constructed by ligation of a mixture of *Eco* RI-cleaved pSC101 and p1258 DNA and then were used to transform a restrictionless strain of *E. coli* (Fig. 3.2). *E. coli* transformants that expressed the penicillin resistance determinant carried by the *Staphylococcus* plasmid were selected and checked for tetracycline resistance.

Caesium chloride gradient analysis of one ampicillin-resistant, tetracycline-resistant chimaera showed that its buoyant density was intermediate to the buoyant densities of the parental plasmids. In addition, treatment of this chimaera with *Eco* RI produced two fragments, one the size of *Eco* RI cleaved pSC101 and the other the size of one of the *Eco* RI fragments of p1258.

Example of the use of pSC101 for cloning. 2. Cloning of *Xenopus* DNA in *E. coli*

This experiment by Morrow *et al.* (1974) involved the construction *in vitro* of plasmid chimaeras composed of both prokaryotic and eukaryotic DNA, and the recovery of recombinant DNA molecules from transformed *E. coli* in the absence of selection for genetic determinants carried by the eukaryotic DNA. The amplified ribosomal DNA from *Xenopus laevis* oocytes was used as the source of eukaryotic DNA for these experiments since this DNA can be purified readily and had been well characterized. In addition, the repeat unit of *X. laevis* rDNA is susceptible to cleavage by *Eco* RI resulting in the production of discrete fragments that can be linked to the pSC101 vector.

In this experiment (Fig. 3.3) a mixture of *Eco* RI-cleaved pSC101 DNA and *X. laevis* rDNA was ligated and was used to transform *E. coli* to tetracycline-resistance. Fifty-five separate transformants were selected and their plasmid DNA extracted and analysed. All 55 plasmids gave a fragment of mol. wt 5.8×10^6 on *Eco* RI digestion and 13 of them yielded additional fragments corresponding in size to *Eco* RI-produced fragments of *X. laevis* rDNA. It was indeed fortuitous that such a high percentage (23.6%) of clones contained chimaeric molecules.

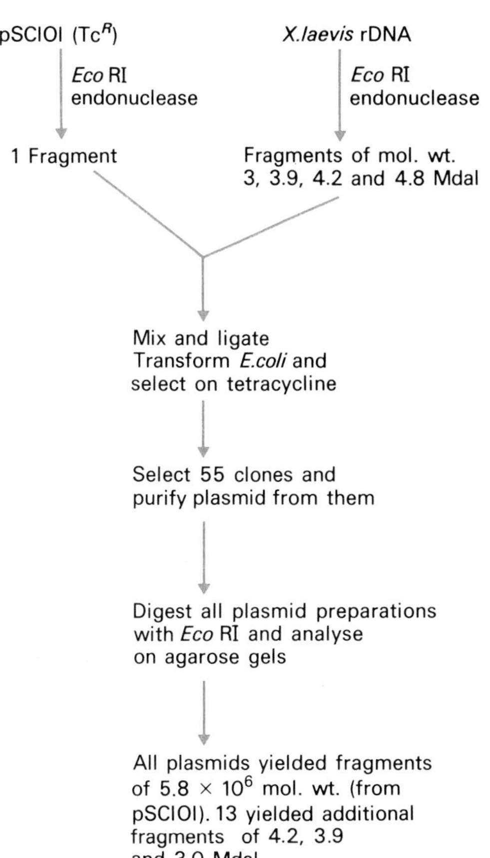
pSCIOI (Tc^R)
Eco RI endonuclease
1 Fragment
X.laevis rDNA
Eco RI endonuclease
Fragments of mol. wt. 3, 3.9, 4.2 and 4.8 Mdal
Mix and ligate
Transform *E.coli* and select on tetracycline
Select 55 clones and purify plasmid from them
Digest all plasmid preparations with *Eco* RI and analyse on agarose gels
All plasmids yielded fragments of 5.8×10^6 mol. wt. (from pSCIOI). 13 yielded additional fragments of 4.2, 3.9 and 3.0 Mdal

Fig. 3.3 Cloning of genes from *Xenopus laevis* in *Escherichia coli* with the aid of pSC101.

Construction and characterization of a new cloning vehicle: pBR322

Although pSC101, Col E1 and RSF 2124 can be used to clone DNA they suffer from a number of disadvantages as outlined above. For this reason considerable effort has been expended on constructing, *in vitro*, superior cloning vehicles. Undoubtedly the most versatile and widely used of these artificial plasmid vectors is pBR322. Plasmid pBR322 contains the Ap^R and Tc^R genes of RSF 2124 and pSC101 respectively, combined with replication elements of pMB1, a Col E1-like plasmid (Fig. 3.4a). The origins of pBR322, and its progenitor pBR313, are shown in Fig. 3.4b and details of its construction can be found in the papers of Bolivar *et al.* (1977a,b).

Plasmid pBR322 has been completely sequenced. The original published sequence (Sutcliffe 1979) was 4362 base pairs long. More recently (Backman & Boyer 1983, Peden 1983) the sequence has been revised by the inclusion of an additional CG base pair at position 526 thus increasing the size of the plasmid to 4363 base pairs. The most useful aspect of the DNA sequence is that it totally characterizes pBR322 in terms of its restriction sites such that the exact length of every fragment can be calculated (see Appendix 3). These fragments can serve as DNA markers for sizing any other DNA fragment in the range of several base pairs up to the entire length of the plasmid.

There are now 20 known enzymes which cleave pBR322 at unique sites (Fig. 3.5). The target sites of six of these enzymes (*Eco* RV, *Bam* HI, *Sph* I, *Sal* I, *Xma* III and *Nru* I) lie within the Tc^R gene and there are sights for a further two (*Cla* I and *Hind* III) within the promoter of that gene. There are unique sites for three enzymes (*Pst* I, *Pvu* I and *Sca* I) within the Ap^R gene. Thus cloning in pBR322 with the aid of any one of these 11 enzymes will result in insertional inactivation of either the Ap^R or the Tc^R marker. However, cloning in the other nine unique sites does not permit the easy selection of recombinants because neither of the antibiotic resistance determinants is inactivated.

Following manipulation *in vitro E. coli* cells transformed

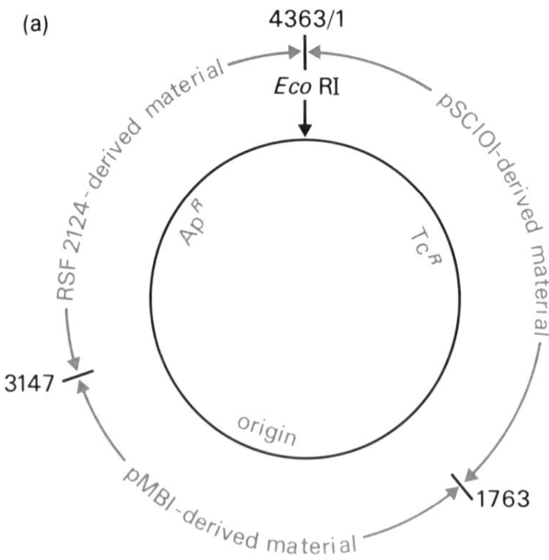

Fig. 3.4 The origins of plasmid pBR322. (a) The boundaries between the pSC101, pMB1 and RSF 2124-derived material. The numbers indicate the positions of the junctions in base-pairs from the unique *Eco* RI site.

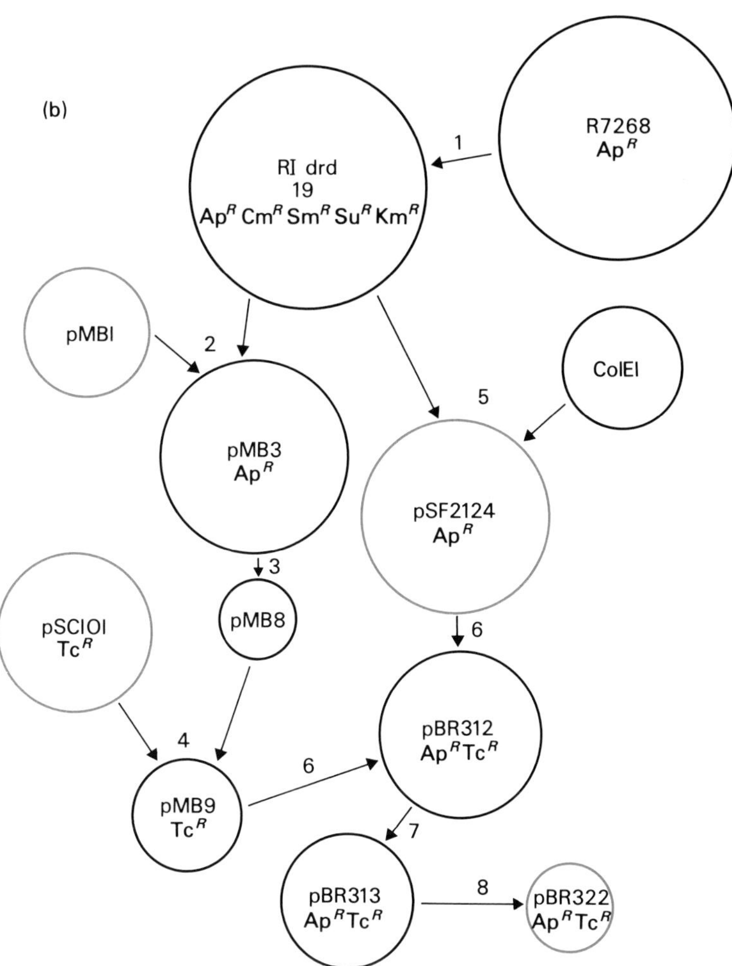

Fig. 3.4 (*cont.*) The origins of plasmid pBR322. (b) R7268 was isolated in London in 1963 and later renamed R1. 1 A variant, R1 *drd*19, which was derepressed for mating transfer, was isolated. 2 The Ap transposon, Tn3, from this plasmid was transposed onto pMB1 to form pMB3. 3 This plasmid was reduced in size by Eco RI* rearrangement to form a tiny plasmid, pMB8, which carries only colicin immunity. 4 Eco RI* fragments from pSC101 were combined with pMB8 opened at its unique Eco RI site and the resulting chimaeric molecule rearranged by Eco RI* activity to generate pMB9. 5 In a separate event, the Tn3 of R1*drd*19 was transposed to Col E1 to form pSF2124. 6 The Tn3 element was then transposed to pMB9 to form pBR312. 7 Eco RI* rearrangement of pBR312 led to the formation of pBR313, from which, 8, two separate fragments were isolated and ligated together to form pBR322. During this series of constructions, R1 and Col E1 served only as carriers for Tn3. (Reproduced by courtesy of Dr G. Sutcliffe and Cold Spring Harbor Laboratory.)

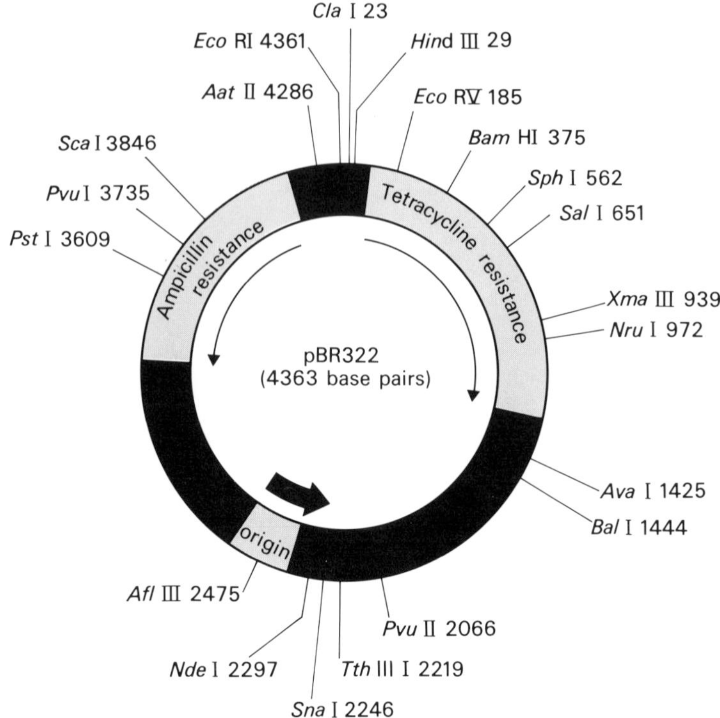

Fig. 3.5 The structure of pBR322 showing the unique cleavage sites. The numbers relate to the coordinate of the base which corresponds to the 5′ nucleotide of each recognition sequence. The thin arrows inside the circle show the direction of transcription of the Ap^R and Tc^R genes. The thick arrow shows the direction of DNA replication. Details of the other endonuclease cleavage sites are given in Appendix 3.

with plasmids with inserts in the Tc^R gene can be distinguished from those cells transformed with recircularized vector. The former are Ap^R and Tc^S whereas the latter are both Ap^R and Tc^R. In practice, transformants are selected on the basis of their Ap resistance and then replica-plated onto Tc-containing media to identify those that are Tc^S. Cells transformed with pBR322 derivatives carrying inserts in the Ap^R gene can be identified more readily (Boyko & Ganschow 1982). Detection is based upon the ability of β-lactamase produced by Ap^R cells to convert penicillin to penicilloic acid which in turn binds iodine. Transformants are selected on rich medium containing soluble starch and Tc. When colonized plates are flooded with an indicator solution of iodine and penicillin β-lactamase-producing (Ap^R) colonies clear the indicator solution whereas Ap^S colonies do not.

The *Pst* I site in the Ap^R gene is particularly useful because the 3′ tetranucleotide extensions formed on digestion are ideal substrates for terminal transferase. Thus this site is excellent for cloning by the homopolymer tailing method described in the previous chapter (see p. 37). If oligo (dG.dC) tailing is used, the *Pst* I site is regenerated (see Fig. 2.11) and the insert may be cut out with that enzyme. In addition, the *Pst* I site is particularly useful for obtaining expression of cloned genes and this aspect is covered in detail in Chapter 7 (p. 127).

Example of the use of plasmid pBR322 as a vector.
1. Isolation of DNA fragments which carry promoters

Cloning into the *Hin*d III site of pBR322 generally results in loss of tetracycline resistance. However, in some recombinants Tc^R is retained or even increased. This is because the *Hin*d III site lies within the promoter rather than the coding sequence. Thus whether or not insertional inactivation occurs depends on whether the cloned DNA carries a promoter-like sequence able to initiate transcription of the Tc^R gene. Widera *et al.* (1978) have used this technique to search for promoter-containing fragments.

Four structural domains can be recognized within *E. coli* promoters (see Chapter 7). These are:

1 position 1, the purine initiation nucleotide from which RNA synthesis begins;
2 position −6 to −12, the Pribnow box;
3 the region around base pair −35;
4 the sequence between base pairs −12 and −35.

Although the *Hin*d III site lies within the Pribnow box (Rodriguez *et al.* 1979) the box is re-created on insertion of a

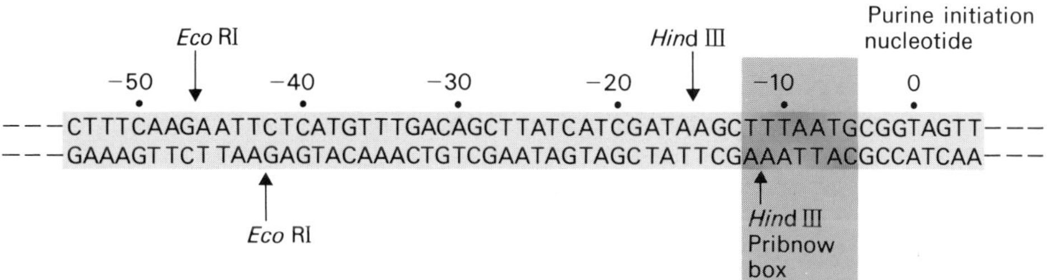

Fig. 3.6 DNA base sequence of the promoter of the tetracycline-resistance gene. The bases are numbered on the basis of the purine initiation nucleotide being in position + 1. (In the conventional pBR322 map the bases are numbered from the *Eco* RI site.) The arrows indicate the positions of cleavage by restriction endonucleases *Eco RI* and *Hind* III.

foreign DNA fragment (Fig. 3.6). Thus, when insertional inactivation occurs it must be the region from −13 to −40 which is modified.

Example of the use of plasmid pBR322 as a vector. 2. Expression in *E. coli* of a chemically synthesized gene for the hormone somatostatin

There are two reasons for selecting this particular early experiment to illustrate the use of pBR322 as a vector. First, the fact that gene manipulation had been successfully used to obtain a functional gene product from a chemically synthesized gene indicated that the potential of genetic engineering was not just a dream but had already been realized. Second, numerous elegant 'tricks' were used to ensure success and they warrant detailed examination for they illustrate the versatility of the basic techniques of gene manipulation.

The rationale of the experiment was in fact to show that recombinant DNA technology can be used to fuse chemically synthesized genes to plasmid elements for expression in *E. coli* or other bacteria. As a model, Itakura *et al.* (1977) designed and synthesized a gene for the small hormone somatostatin. The somatostatin 'gene' was chosen because somatostatin is a small polypeptide of known amino acid composition, there are sensitive radioimmune and biological assays, and, being a hormone it was of intrinsic biological interest.

The somatostatin gene was synthesized chemically (see p. 16) such that there was an *Eco* RI site at one end and a *Bam* HI site at the other (Fig. 3.7). Also, a methionine codon preceded the normal NH_2-terminal amino acid of somatostatin and the COOH-terminal amino acid was followed by two stop codons.

In the first part of the experiment three new plasmids were created from pBR322. The control region of the *lac* operon, comprising the *lac* promoter, catabolite gene activator-protein binding site, the operator, the ribosome binding site, and the first seven triplets of the β-galactosidase structural gene, were

```
Eco RI
             Met Ala Gly Cys Lys Asn Phe Phe Trp Lys Thr Phe Thr Ser Cys Stop Stop
5' AATTC ATG GCT GGT TGT AAG AAC TTC TTT TGG AAG ACT TTC ACT TCG TGT TGA TAG
           G TAC CGA CCA ACA TTC TTG AAG AAA ACC TTC TGA AAG TGA AGC ACA ACT ATCCTAG 5'
                                                                              Bam HI
```

Fig. 3.7 Sequence of the chemically synthesized somatostatin gene.

inserted into the *Eco* RI site of pBR322 to create pBH10. Plasmid pBH10 has two *Eco* RI sites and one of these was removed to generate pBH20. Finally, the synthetic somatostatin gene was inserted next to the *lac* control gene to yield pSOM I.

The formation of pBH10 (Fig. 3.8)

Digestion of pBR322 with *Eco* RI produces a linear duplex with short, single-stranded tails. These tails were converted to

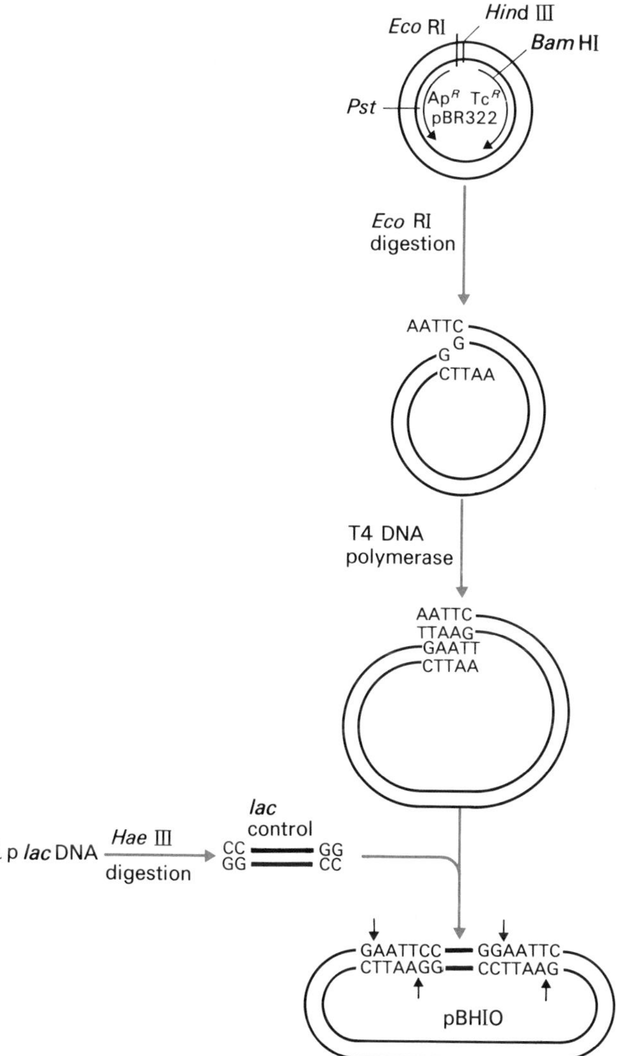

Fig. 3.8 The formation of pBH10 by insertion of the *lac* control region into pBR322. Note the generation of two *Eco* RI sites, one on either side of the *lac* control region.

duplexes by treatment with T4 DNA polymerase to yield a blunt-ended molecule. The endonuclease *Hae* III which produces blunt ends, was used to excise the *lac* control region from bacteriophage λ*plac*. With the aid of T4 DNA ligase, which permits ligation of blunt-ended molecules, the *lac* control region was joined to the blunt-ended pBR322 derivative. It should be noted that this procedure results in the formation of two *Eco* RI sites, one on either side of the *lac* control region.

After ligation, the mixture was transformed into *E. coli* and selection made for blue colonies on medium containing ampicillin, tetracycline and the chromogenic substrate 5-bromo-4-chloro-3-indolyl-β-D-galactoside (Xgal). The rationale for this was as follows. Xgal is not an inducer of β-galactosidase but is cleaved by β-galactosidase releasing a blue indolyl derivative. Since Xgal is not an inducer, only mutants constitutive for β-galactosidase produce blue colonies on medium containing Xgal. Plasmid pBH10 is maintained as a relaxed plasmid, i.e. multiple copies per cell. Thus, cells carrying pBH10 have multiple copies of the *lac* control region and can titrate out all the repressor produced by the single chromosomal *lac* I gene leading to a constitutive phenotype.

Clearly there are two possible orientations for the insertion of the *Hae* III fragment but these can be distinguished by the location of an asymetrically placed *Hha* site relative to the *Hind* III site (Fig. 3.9).

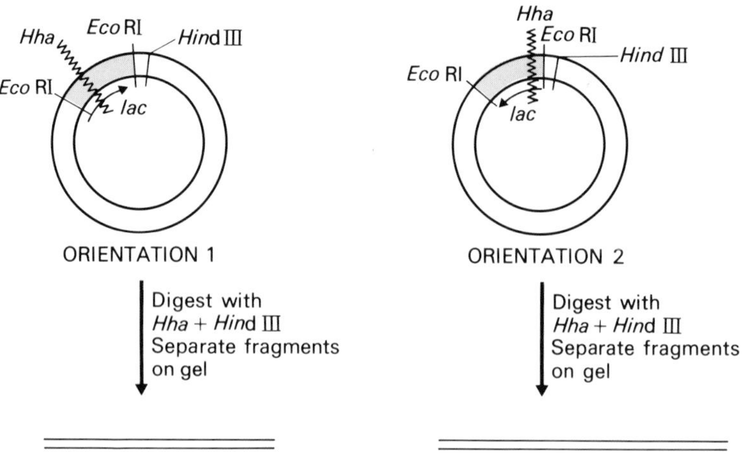

Fig. 3.9 Determination of the orientation of the insert in pBH10 by double digestion with *Hha* I and *Hind* III. Note particularly the size of the smaller DNA fragment produced in each case.

The formation of pBH20 (Fig. 3.10)

Plasmid pBH10 has two *Eco* RI sites and it was desirable to retain only one of them, that located between the *lac* control region and the Tc^R gene. The *lac* and Tc^R promoters in pBH10 are only 40 base pairs apart and addition of RNA polymerase effectively protects this site from *Eco* RI digestion. This is because in the absence of ribonucleoside triphosphates the

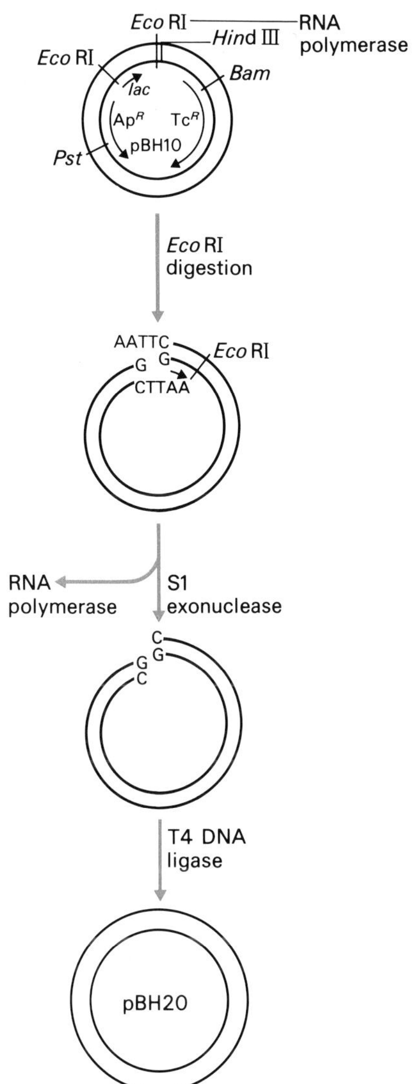

Fig. 3.10 The formation of pBH20 by the removal of one of the *Eco* RI sites of pBH10.

RNA polymerase does not initiate transcription and remains firmly bound to the promoters. The second *Eco* RI site, however, is exposed. After addition of RNA polymerase, pBH10 was cleaved with *Eco* RI and the single-stranded tails removed with S1 exonuclease. The linear duplex so produced was circularized with T4 DNA ligase to generate pBH20.

The formation of pSom I (Fig. 3.11)

Plasmid pBH20 was digested with a mixture of *Eco* RI and *Bam* HI to produce a large and small DNA fragment. The small fragment was discarded and the large fragment treated with alkaline phosphatase to prevent subsequent self-ligation (see p. 33). T4 DNA ligase was then used to join the synthetic somatostatin gene to the large fragment of pBH20 and transformants were selected by virtue of their ampicillin resistance.

The DNA sequence of pSom I indicated that the clone carrying this plasmid should produce a peptide containing somatostatin, but no somatostatin was found. However, in reconstruction experiments it was observed that exogenous somatostatin was degraded rapidly in *E. coli* extracts. Thus the failure to find somatostatin activity could be accounted for by intracellular degradation by endogenous proteolytic enzymes. Such proteolytic degradation might be prevented by attachment of the somatostatin to a large protein, e.g. β-galac-

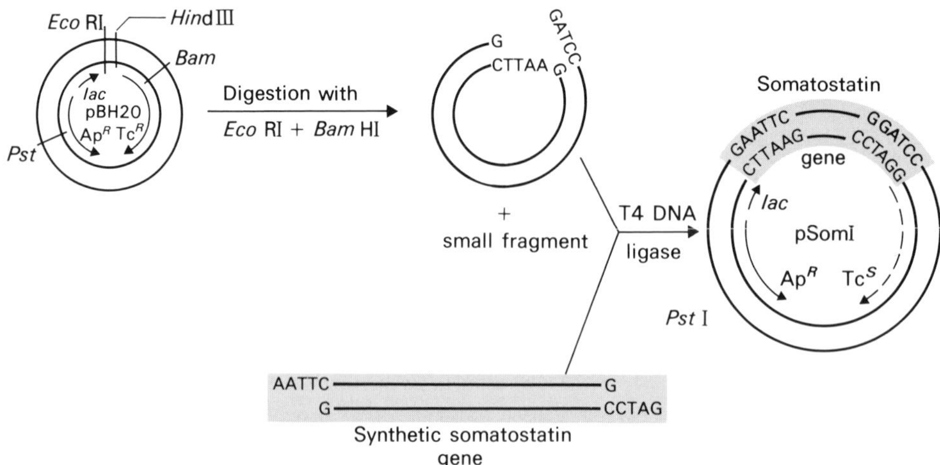

Fig. 3.11 The construction of pSom I by insertion of the synthetic somatostatin gene next to the *lac* control region of pBH20. Note that, for clarity, the step involving alkaline phosphatase (see text) has been omitted.

tosidase. The β-galactosidase structural gene has an *Eco* RI site near the COOH-terminus and the available data on the amino acid sequence of this protein suggested that it would be possible to insert the synthetic gene into this site and still maintain the proper reading frame. In order to do this, two new plasmids pSom II and pSom II-3 were created.

The formation of pSom II (Fig. 3.12)

Plasmid pSom I was digested with *Eco* RI and *Pst* I and the larger fragment, which contains the synthetic somatostatin gene, purified by gel electrophoresis. This fragment has lost the *lac* control region and part of the Ap^R gene. The Ap^R gene was restored by ligating this large fragment to the small fragment produced by digesting pBR322 with *Eco* RI and *Pst* I. The ligated mixture was used to transform *E. coli* to ampicillin resistance and the plasmid from one Ap^R clone selected and called pSom II.

The formation of pSom II-3 (Fig. 3.13)

Bacteriophage λ*plac* 5 which carries the *lac* control region and the entire β-galactosidase gene was digested with *Eco* RI and the mixture ligated with *Eco* RI-treated pSom II. The ligated mixture was used to transform *E. coli* and selection was made

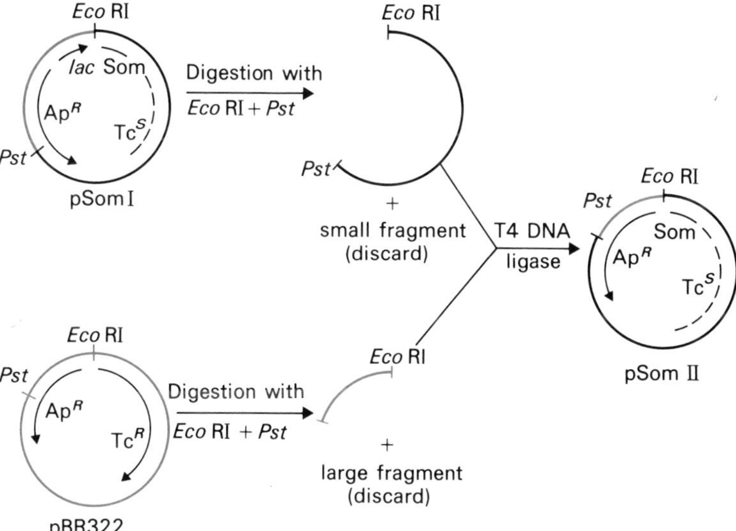

Fig. 3.12 Formation of pSom II. For clarity, only one strand of the plasmid DNA is shown.

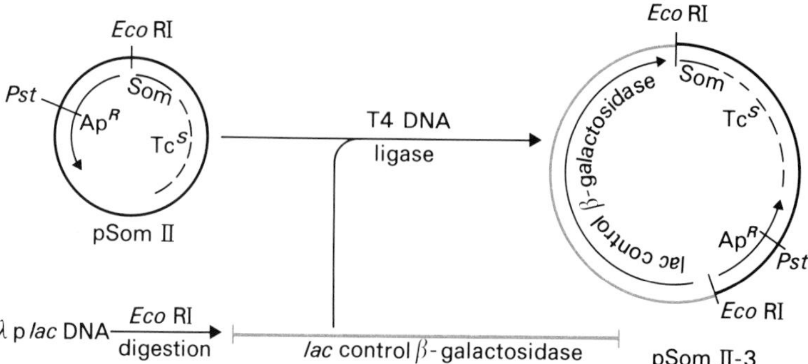

Fig. 3.13 Formation of pSom II-3 by insertion of *lac* region from λ*plac* into pSom II. For clarity, only one strand of DNA is shown.

for blue colonies on medium containing Xgal and ampicillin. Approximately 2% of the transformants were blue and analysis of the plasmid from them showed the presence of a 4.4 Mdal. fragment identical to that found by digesting λ*plac*5 with *Eco* RI.

Two orientations of the *Eco* RI *lac* fragment of λ*plac*5 are possible but only one of these would maintain the proper reading frame into the somatostatin gene. When a number of independent clones were examined, approximately 50% produced detectable somatostatin radioimmune activity and all

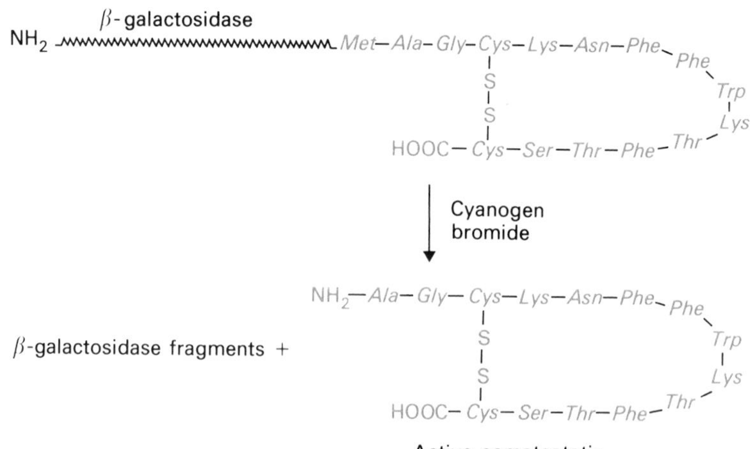

Fig. 3.14 Cleavage of the chimaeric protein by cyanogen bromide to yield active somatostatin. The somatostatin can readily be purified from cyanogen bromide and the fragments of β-galactosidase.

of these had the desired orientation of the *lac* operon. The non-producing clones were found to have the opposite orientation.

It should be noted that no somatostatin radioimmune activity was detected prior to cyanogen-bromide cleavage of the total cellular protein. Since the antiserum used in the radioimmune assay requires a free NH_2- terminal alanine, no activity was expected prior to cleavage. Methionine residues are the site of cyanogen bromide cleavage and it is for this reason that a methionine codon was included in the synthetic somatostatin gene (Figs 3.7 and 3.14).

The use of protein chemistry to generate native proteins

The somatostatin work just described demonstrated the feasibility of the synthetic gene approach for bacterial production of hormones. It also showed how the desired polypeptide can be released from a hybrid protein. Almost identical methods were used for insulin but these deserve some further comment.

The human insulin gene encodes a protein, preproinsulin, which has 58 more amino acid residues than insulin itself. During secretion of preproinsulin from the cells which produce it, 23 amino acids are removed from the N-terminus to generate proinsulin (see Chapter 7). The proinsulin assumes a three-dimensional structure similar to that adopted by insulin itself. Three disulphide bonds are formed during the folding process and these stabilize the molecule. A proteolytic event then cleaves out a stretch of 35 amino acids, the so-called C chain, from the middle of the proinsulin molecule (Fig. 3.15). The N-terminal portion of proinsulin becomes the B chain of insulin and the

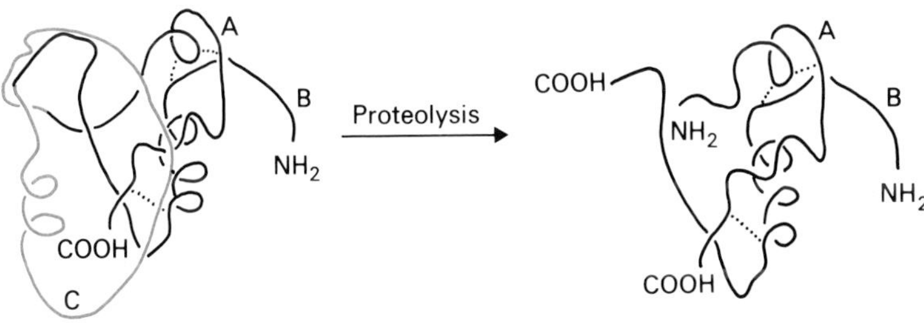

Fig. 3.15 The three-dimensional structure of human proinsulin and insulin. The three disulphide bonds which stabilize the correctly folded structures are shown as dotted lines.

C-terminal portion becomes the A chain. The A and B chains remain cross-linked in a stable structure by the disulphide bridges.

For the production of insulin in *E. coli*, chemically synthesized genes for the insulin A and B chains were cloned separately and attached in phase to the gene for β-galactosidase, just as with somatostatin. Thus two bacterial strains were constructed: one producing a fused protein carrying the human insulin A chain and the other a similar fused protein carrying the B chain (Goeddel *et al.* 1979b). The insulin chains were clipped from the β-galactosidase precursor by treatment with cyanogen bromide, purified and chemically linked *in vitro*. An alternative approach has been to clone the entire proinsulin gene which is then synthesized as a hybrid protein. The proinsulin is chemically cleaved from the *E. coli* carrier protein with cyanogen bromide and purified. Since the primary structure of human proinsulin has in it all the information needed to assume the proper configuration, proinsulin synthesized by bacteria can be stabilized *in vitro* by formation of the disulphide bonds. The proinsulin formed in this way is subjected to proteolysis to generate human insulin. To facilitate the conversion of the prohormone to the native hormone Wetzel *et al.* (1981) constructed a gene encoding an analogue of human proinsulin in which the normal 35 amino acid C peptide was replaced with a 'mini-C' peptide of 6 amino acids (Arg-Arg-Gly-Ser-Lys-Arg).

Protein chemistry has been used in a different way to obtain production of active β-endorphin, a polypeptide of 31 amino acid residues, from a cloned cDNA (Shine *et al.* 1980). Once again a hybrid β-galactosidase/β-endorphin protein was produced initially. Since β-endorphin does contain an internal methionine residue, use of cyanogen bromide cleavage was ruled out. The steps used to release the active β-endorphin from the hybrid protein depended upon the following points.

1 An arginine residue precedes the β-endorphin sequence. This can be a site for proteolytic cleavage by trypsin.

2 There are no internal arginine residues in the hormone.

3 The lysine residues in the hormone could be protected from attack by trypsin by reaction with citraconic anhydride *in vitro*.

Thus after treatment of the hybrid protein with citraconic anhydride, the β-endorphin having modified lysine residues was released by cleavage with trypsin. Native β-endorphin was then obtained by removal of the citraconic groups at pH 3.0.

Improved vectors derived from pBR322

As noted earlier, insertional inactivation frequently is used as a means of detecting the formation of recombinant molecules. The fact that cloning in pBR322 using *Eco* RI does not result in insertional inactivation is unfortunate for *Eco* RI is one of the most widely used enzymes. Bolivar (1978) has constructed two different pBR322-derived plasmids containing unique *Eco* RI sites in selectable markers. One of these plasmids, pBR324, carries the colicin E1 structural and immunity genes derived from pMB9. The position of a unique *Eco* RI site and a unique *Sma* I site in the colicin E1 structural gene provides a relatively easy selection for *Eco* RI and *Sma* I endonuclease-generated DNA fragments. The second plasmid, pBR325, carries the chloramphenicol-resistance gene derived from phage P1Cm. This plasmid has a unique *Eco* RI site in the Cm^R gene. Recovery of cells harbouring *Eco* RI-derived recombinant DNA molecules is facilitated by virtue of their $Ap^RTc^RCm^S$ phenotype.

Another derivative of pBR322 which now is widely used as a cloning vector is pAT153. Twigg and Sherratt (1980) observed that when a particular *Hae* II fragment was removed from Col E1 the plasmid copy number in *E. coli* increased five- to sevenfold. The corresponding *Hae* II fragment was removed from pBR322 to produce pAT153 (Fig. 3.16). Cells containing pAT153 have only an increase in plasmid content of 1.5 to 3 times compared with those containing pBR322. Although this increase is smaller than for the Col E1 deletion derivatives, pAT153 is still a useful cloning vector because of greater levels of plasmid and plasmid-specified gene products. In terms of biological containment pAT153 has a great advantage over pBR322. Although pBR322 is

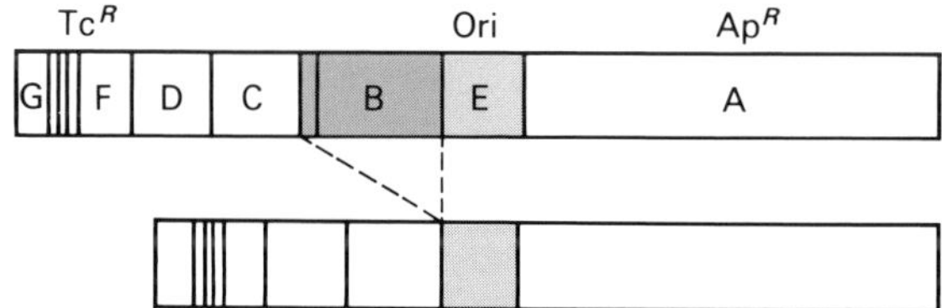

Fig. 3.16 The origin of pAT153. Plasmid pBR322 is cut into 11 fragments on complete digestion with *Hae* II. These fragments are assigned letters in alphabetical order on the basis of size. To construct pAT153, pBR322 DNA was partially digested with *Hae* II and re-ligated. According to Twigg and Sherratt (1980) only fragment B was removed but it is now known that the small fragment immediately to the left of B was also lost.

not self-transmissible, it can be mobilized at a frequency of 10^{-1} from cells containing a conjugative plasmid plus plasmid Col K. However, the *Hae* II fragment removed from pBR322 during the formation of pAT153 contains a DNA sequence (called *nic* or *bom*) essential for conjugal transfer. As a consequence pAT153 cannot be mobilized and this provides a means of biological containment.

Hayashi (1980) has constructed a plasmid called pKH47 by inserting about a 100 base-pair poly(dA):poly(dT) duplex segment into the unique *Pvu* II restriction enzyme site of pBR322. This insertion permits the separation of complementary strands of the linearized, denatured plasmid DNA by affinity

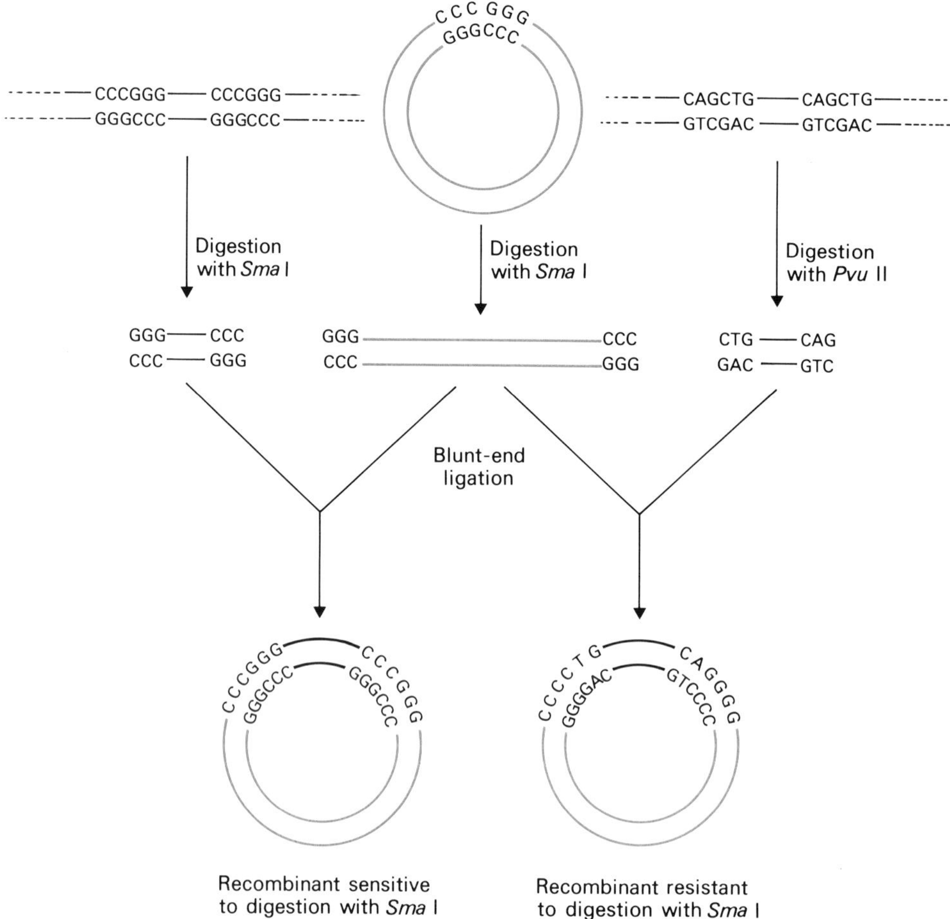

Fig. 3.17 Generation of endonuclease-sensitive and endonuclease-resistant chimaeras following blunt-end ligation.

chromatography on oligo(dA)- or oligo(dT)-cellulose columns. Once separated, the strands can be sequenced readily by the dideoxy chain terminator method (see p. 98). Since pKH47 retains all of the properties of pBR322 except a cleavage site for *Pvu* II it is a particularly useful vector for cloning DNA fragments whose sequence is required.

Many restriction enzymes generate 'blunt-ended' DNA fragments which can be cloned into pBR322 at any of its unique blunt-ended restriction sites, e.g. those for *Pvu* II or *Bal* I. However, unless the vector and the DNA fragment are cut with the same restriction endonuclease, ligation results in destruction of the recognition sites used to construct the recombinant and cloned fragments cannot be recovered precisely. Prentki and Krisch (1982) have constructed a derivative of pBR322 which is superior for the cloning of the blunt-ended DNA fragments. Their vector, called pHP34, has a unique *Sma* I site immediately flanked by two *Eco* RI sites. Blunt-ended DNA fragments cloned into the *Sma* I site of pHP34 can be recovered by digestion with *Eco* RI provided they do not contain any internal *Eco* RI sites.

If a blunt-ended fragment produced by cleavage with *Sma* I is ligated to a *Sma* I cut vector such as pHP34, in the resulting chimaera the fragment will be bounded by two *Sma* I sites (see Fig. 3.17). However, if the blunt-ended fragment is produced by a different enzyme, e.g. *Pvu* II, the resulting recombinant will contain no *Sma* I sites and this can be put to good use. After ligation the DNA can be treated with *Sma* I *prior* to transformation. This treatment will destroy most recircularized vector molecules thus permitting enrichment of recombinants.

Direct selection vectors

Most plasmid vectors contain at least two selectable markers. One marker is kept intact and is used to select transformants. The transformants then are screened to detect inactivation of the second marker indicating the insertion of foreign DNA. It would be more convenient if recombinant plasmids could be selected directly after transformation and consequently a number of workers have constructed direct selection vectors.

Schumann (1979) has described a plasmid (pKN80) which carries a fragment of phage Mu DNA encoding a killing function which is expressed efficiently upon transformation of pKN80 into Mu-sensitive bacteria. Cloning of DNA fragments at the unique *Hpa* I site of pKN80 results in insertional inactivation of the killing function. Roberts *et al.* (1980) have constructed plasmid, pTR262, which carries a Tc^R gene under the control of the phage lambda P_R promoter. Their selection procedure depends on inactivation of the lambda repressor gene, also on pTR262, by insertion of foreign DNA fragments into *Hin*d III or *Bcl* II sites contained within the repressor gene.

Two direct-selection vectors make use of the fact that sensitivity to some antibacterial agents is dominant over resistance. Plasmid pNO1523 carries both an Ap^R gene and a dominant gene which will confer Sm-sensitivity on a Sm^R host (Dean 1981). The plasmid contains unique *Sma* I and *Hpa* I restriction sites in the Sm-sensitivity gene. Insertion of a DNA segment into either of these sites inactivates the Sm^S gene. Transformation of the ligated DNA into a Sm^R host followed by a selection on medium containing Ap and Sm results in growth of only those bacteria containing recombinant plasmids. Hennecke *et al.* (1982) have described a similar vector carrying a Cm^R marker and a dominant p-fluorophenylalanine-sensitivity (pfp^S) marker. The latter gene contains a unique *Pst* I site. Direct selection is made by transforming pfp^R cells and plating on minimal medium containing pfp and Cm.

Low copy number plasmid vectors

High copy number plasmid vectors such as pBR322 are widely used because they are easily purified and, via a gene dosage effect, they can direct the synthesis of high levels of cloned gene products. However there are some genes which cannot be cloned on high copy number vectors because their presence seriously disturbs the normal physiology of the cell. Examples include genes which encode surface structural proteins, e.g. *omp*A (Beck & Bremer 1980), or proteins that regulate basic cellular metabolism, e.g. *pol*A (Murray & Kelley 1979). One strategy for cloning such genes is to use low copy number vectors. Two suitable low copy number vectors are phage lambda in its lysogenic state (see Chapter 4) and plasmid pSC101. Although pSC101 was the first plasmid to be used for cloning *in vitro*, as indicated on p. 53, it is seldom used because it carries only a single marker and insertional inactivation cannot be used to screen for recombinant clones.

Two sets of vectors have been constructed which retain the low copy number of pSC101 but have two more antibiotic-resistance markers and unique cleavage sites for many restriction endonucleases (Hashimoto-Gotoh *et al.* 1981, Stoker *et al.* 1982).

Runaway plasmid vectors

One reason for cloning a gene on a multicopy plasmid is to increase greatly its expression and hence facilitate purification of the protein it encodes. However, as indicated above, some genes cannot be cloned on high copy number vectors because excess gene product is lethal to the cell. The use of low copy number vectors avoids cell killing but this may be self-defeating since expression of the cloned gene will be reduced. A solution to this problem is to use runaway plasmid vectors.

The first runaway plasmid vectors were described by Uhlin *et al.* (1979). At 30 °C the plasmid vector is present in a moderate number of copies per cell. Above 35 °C all control of plasmid replication is lost and the number of plasmid copies per cell increases continuously. Cell growth and protein synthesis continue at the normal rates for 2–3 hours at the higher temperature. During this period products from genes on the plasmid are overproduced. Eventually inhibition of cell growth occurs and the cells lose viability but at this stage plasmid DNA may account for 50% of the DNA in the cell. The two runaway vectors described by Uhlin *et al.* (1979), now called pBEU1 and pBEU2, each carry a unique *Bam* HI site and a single antibiotic resistance marker.

More recently improved runaway vectors have been constructed. Yasuda and Takagi (1983) have described a runaway vector carrying a Km^R marker and having unique recognition sites for 5 restriction endonucleases and Uhlin *et al.* (1983) have described one encoding resistance to Ap and Tc, the latter marker with unique sites to permit insertional inactivation. Larsen *et al.* (1984) have constructed cloning vectors which are present in one copy per chromosome at temperatures below 37 °C and which display uncontrolled replication at 42 °C. In addition they carry a partitioning function (*par*, see p. 149) which stabilizes the plasmid at low temperatures when grown in the absence of selection pressure.

Chapter 4 Bacteriophage and Cosmid Vectors for *E. coli*

Essential features of bacteriophage lambda†

Bacteriophage λ is a genetically complex but very extensively studied virus of *E. coli*. Because it has been the object of so much molecular genetical research it was natural that, right

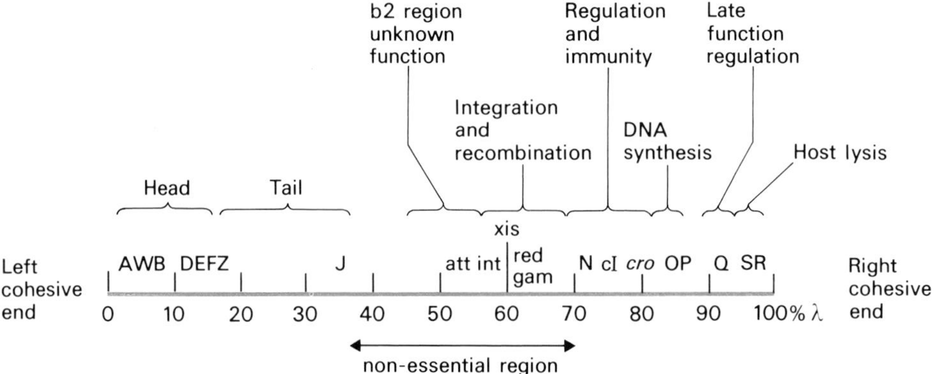

Fig. 4.1 Map of the λ chromosome, showing the physical position of some genes on the full length DNA of wild-type bacteriophage λ. Clusters of functionally related genes are indicated.

from the beginnings of gene manipulation, it should have been investigated and developed as a vector. The DNA of phage λ, in the form in which it is isolated from the phage particle, is a linear duplex molecule of about 48.5 kb pairs. The entire DNA

† Only those features of phage λ that are essential to understanding its use as a vector are discussed here. For a more general account the reader should consult a virology text such as Primrose and Dimmock (1980).

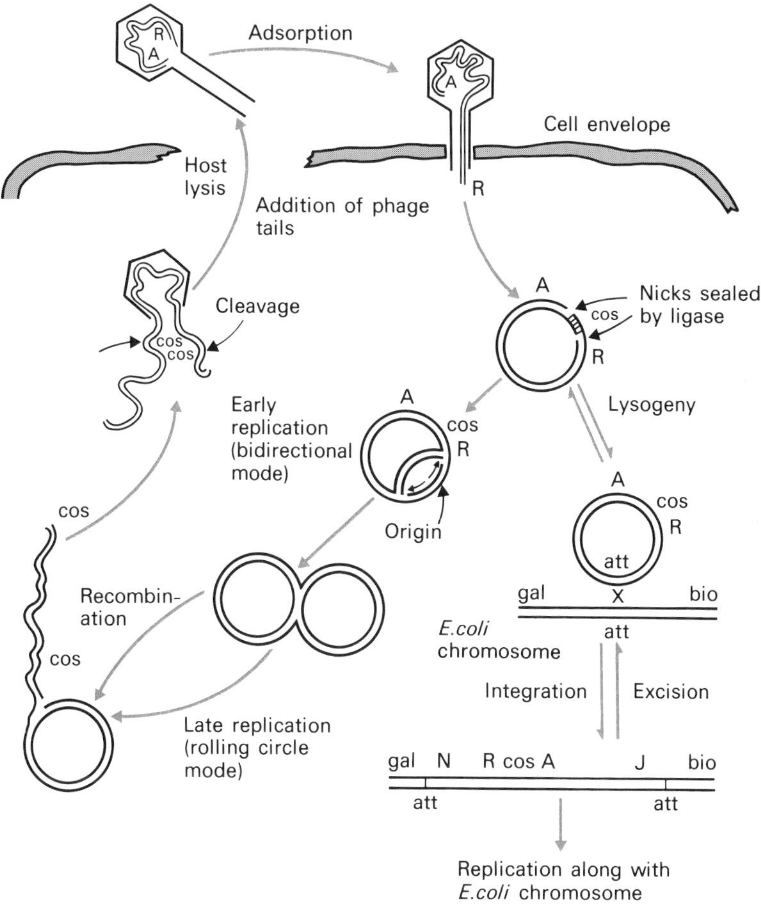

Fig. 4.2 Replication of phage λ DNA in lytic and lysogenic cycles. (After Kornberg 1980.)

sequence has been determined (Sanger *et al.* 1982). At each end are short single-stranded 5′-projections of 12 nucleotides which are complementary in sequence and by which the DNA adopts a circular structure when it is injected into its host cell, i.e. λ DNA naturally has cohesive termini which associate to form the *cos site*.

Functionally related genes of phage λ are clustered together on the map, except for the two positive regulatory genes *N* and *Q*. Genes on the left of the conventional linear map (Fig. 4.1) code for head and tail proteins of the phage particle. Genes of the central region are concerned with recombination (e.g. *red*) and the process of lysogenization in which the circularized chromosome is inserted into its host chromosome and stably replicated along with it as a prophage. Much of this central region, including these genes, is not essential for phage growth and can be deleted or replaced without seriously impairing the infectious growth cycle. Its dispensability is crucially important, as will become apparent later, in the construction of vector derivatives of the phage. To the right of the central region are genes concerned with gene regulation and prophage immunity to superinfection (*N*, *cro*, *c*I), followed by DNA synthesis (*O*, *P*), late function regulation (*Q*) and host cell lysis (*S*, *R*). Figure 4.2 illustrates the λ life-cycle.

Promoters and control circuits

As we shall see, it is possible to insert foreign DNA into the chromosome of phage λ derivatives and in some cases foreign genes can be expressed efficiently via λ promoters. We must therefore briefly consider the promoters and control circuits affecting λ gene expression.

In the lytic cycle, λ transcription occurs in three temporal stages; early, middle and late. Basically, early gene transcription establishes the lytic cycle (in competition with lysogeny); middle gene products replicate and recombine the DNA, and late gene products package this DNA into mature phage particles. Following infection of a sensitive host, early transcription proceeds from major promoters situated immediately to the left (P_L) and right (P_R) of the repressor gene (*c*I) (Fig. 4.3). This transcription is subject to repression by the product of the *c*I gene and in a lysogen this repression is the basis of immunity to superinfecting λ. Early in infection transcripts from P_L and P_R stop at termination sites t_L and t_{R_1}. The site t_{R_2} stops any transcripts that escape beyond t_{R_1}. Lambda switches from early to middle stage transcription by anti-termination. The *N* gene product, expressed from P_L, directs this switch. It interacts with RNA polymerase and, antagonizing the action of host termination protein ρ, permits it to ignore the stop signals so that P_L and P_R transcripts extend into genes such as *red* and *O* and *P* necessary for the middle stage. The early and middle transcripts and patterns of expression therefore overlap. The *cro* product, when sufficient has accumulated, prevents transcription from P_L and P_R. The gene Q is expressed from the distal portion of the extended P_R transcript and is responsible for the middle to late switch. This also operates by anti-termination. The *Q* product specifically anti-terminates the short P'_R transcript, extending it into the

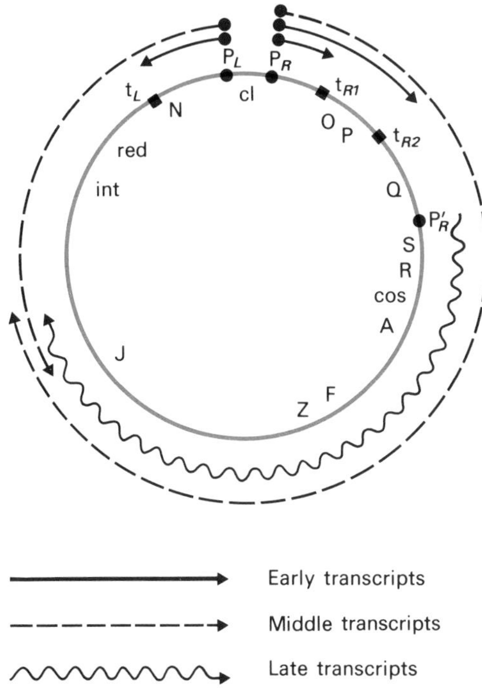

Fig. 4.3 Major promoters and transcriptional termination sites of phage λ. See text for details.

late genes, across the cohered *cos* region, so that many mature phage particles are ultimately produced.

Both *N* and *Q* play positive regulatory roles essential for phage growth and plaque formation; but an N^- phage *can* produce a small plaque if the termination site t_{R_2} is removed by a small deletion termed *nin* (N independent) as in λN^- *nin*.

Vector DNA

Wild-type λ DNA contains several target sites for most of the commonly used restriction endonucleases and so is not itself suitable as a vector. Derivatives of the wild-type phage have therefore been produced which either have a single target site at which foreign DNA can be inserted (*insertional* vectors), or have a pair of sites defining a fragment which can be removed and replaced by foreign DNA (*replacement* vectors). Since phage λ can accommodate only about 5% more than its normal complement of DNA, vector derivatives are constructed with deletions to increase the space within the genome. The shortest λ DNA molecules that produce plaques of nearly normal size are 25% deleted. Apparently if too much non-essential DNA is deleted from the genome it cannot be packaged into phage particles efficiently. This can be turned to advantage, for if the replaceable fragment of a replacement type vector either is removed by physical separation, or is effectively destroyed by treatment with a second restriction endonuclease that cuts it alone, then the deleted vector genome can give rise to plaques only if a new DNA segment is inserted into it. This amounts to positive selection for recombinant phage carrying foreign DNA.

Many vector derivatives, of both the insertional and replacement types, have been produced by several groups of researchers (e.g. Thomas *et al.* 1974, Murray & Murray 1975, Blattner *et al.* 1977, Leder *et al.* 1977). Most of these vectors have been constructed for use with *Eco* RI, *Bam* HI or *Hin*d III, but their application can be extended to other endonucleases by the use of linker molecules.

The λWES.λB′ phage is a useful vector that illustrates several important points. For details of the construction of this phage the reader is referred to papers of Thomas *et al.* (1974) and Leder *et al.* (1977). The DNA map of this vector is shown in Fig. 4.4. We can see that the phage has been constructed with three amber mutations in genes *W*, *E* and *S*. These reduce the likelihood of recombinants escaping from the laboratory environment since appropriate amber suppressor strains are very uncommon in nature. The fragment designated C in wild-type λ has been deleted by restriction and religation *in vitro*. In addition, the two most righthand *Eco* RI sites have been eliminated and a *nin* deletion introduced. The deletions create space for insertion of foreign DNA. The B′ fragment is the replaceable fragment. (The B fragment has inadvertently been inverted during construction of the vector and is designated B′.) In use, the vector DNA is digested with *Eco* RI, then the B′ fragment may be removed by preparative gel electrophoresis or other physical methods. Alternatively, this fragment can be destroyed by treatment of the *Eco* RI digest with *Sst* I. The *Eco* RI treated foreign DNA is then added to a mixture of vector arms, the mixture is ligated and used to transfect an appropriate amber suppressor strain of *E. coli* so that viable recombinant phage are recovered. Joining of the two DNA arms without insertion of foreign DNA results in a molecule that is too short (9.8% (B′ fragment) + 11.3% (C fragment) + 6.1% (*nin* deletion) = 27.2% less than λ^+) to produce viable phage even though it contains all of the genes necessary for lytic growth.

Figure 4.4 also shows the map of one of a set of 16 vector phages constructed by Blattner *et al.* (1977). These have been

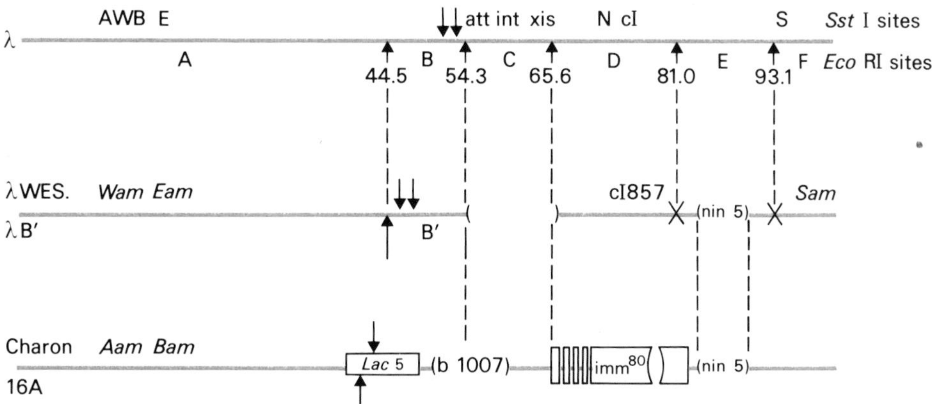

Fig 4.4 Physical map of λDNA and two vector derivatives, λWES.λB′ and charon 16A. Boxes indicate substitutions, but the lengths of substituted DNA are not exactly to scale. The λ regions are aligned. *Lac* 5 is a substitution from the *lac* region of *E. coli*. The box labelled *imm* 80 is a portion of phage ϕ80 DNA containing its immunity region. The region including four small boxes is derived from ϕ80 and is partially homologous to λ. Parentheses indicate deletions. Downward and upward arrows are *Sst* I and *Eco* RI restriction sites respectively. Numbers under *Eco* RI sites indicate the positions of the sites as percentages of the wild-type genome length.

aptly called Charon phages by their originators, after the old ferryman of Greek mythology who conveyed the spirits of the dead across the River Styx. Some of the Charon phages, like the one illustrated, have had amber mutations introduced in genes *A* and *B* in order to enhance biological containment. Charon 16A is an insertional vector with a single *Eco* RI site located in the gene for β-galactosidase (*lac* Z) which is included in the *lac* 5 DNA substitution. This is useful because there is a convenient colour test for the production of β-galactosidase. When the chromogenic substrate Xgal (see p. 62) is included in the plating medium, phage carrying *lac* 5 give dark blue plaques on Lac$^-$ indicator bacteria. Potential success with Charon 16A cloning is detected by insertional inactivation of the Lac function which results in colourless plaques.

Another useful screening method employing insertional inactivation has been exploited by Murray *et al.* (1977). Insertion of foreign DNA at the single target within the immunity region of one of their vector molecules destroys the ability of the phage to produce a functional repressor so that recombinants give clear plaques which are readily distinguished from the turbid plaques of parental phages formed by simple rejoining of the two fragments of the vector

DNA. This insertional inactivation of the *c*I gene also forms the basis of a powerful system for *selecting* recombinant formation in such phages as λ gt10 (Nathans & Hogness 1983, Young & Davis 1983) and λ NM1149 (Murray 1983). When the non-recombinant phages are plated on the *hfl* A (*h*igh *f*requency of *l*ysogeny) mutant of *E. coli* no plaques are formed. This is because lysogens, which are of course immune to superinfection, are created at such a high frequency in this host. However, recombinant phage have an inactive *c*I repressor gene, cannot form lysogens and therefore do form plaques.

Improved phage λ replacement vectors

As with plasmid vectors, improved phage λ vector derivatives have been developed in many laboratories, mostly with the aim of increasing the capacity for foreign DNA fragments generated by any one of several restriction enzymes (reviewed by Murray 1983). The maximum capacity can only be attained with vectors of the replacement type, so that there has also been an accompanying incentive to devise methods for positively selecting recombinant formation without the need for prior removal of the stuffer fragment. Even when steps are taken to remove the stuffer fragment by physical purification of vector arms, small contaminating amounts may remain, so that genetic selection for recombinant formation remains desirable. Two innovative means of achieving this are exploitation of the *A3* gene of phage T5 and the Spi^- phenotype.

Davison *et al.* (1979) have devised a vector in which the λB stuffer fragment of λWES.λB′ has been replaced by two identical 1.8 kb fragments of phage T5. These fragments carry the A3 gene which is known to prevent growth of T5 itself, or this adapted λ vector on *E. coli* carrying the plasmid Col Ib. Two fragments were found to be necessary in this new vector, called λgt WES.T5622, because a single 1.8 kb replacement of the λB fragment was too short to give viable phage. Positive selection for recombinant formation is imposed by plating on an *E. coli* host carrying plasmid Col Ib.

Wild-type λ cannot grow on *E. coli* strains lysogenic for phage P2; in other words the λ phage is Spi^+ (sensitive to P2 inhibition). It has been shown that the products of λ genes *red* and *gam*, which lie in the region 64–69% on the physical map, are responsible for the inhibition of growth in a P2 lysogen (Herskowitz 1974, Sprague *et al.* 1978, Murray 1983). Hence vectors have been derived (e.g. λL47 and λ1059) in which the stuffer fragment includes the region 64–69% so that recombinants in which this has been replaced by foreign DNA are phenotypically Spi$^-$ and can be positively selected by plating on a P2 lysogen (Karn *et al.* 1980, Loenen & Brammar 1980).

Deletion of the *gam* gene has other consequences. The *gam* product is necessary for the normal switch in λ DNA replication from the bi-directional mode to the rolling circle mode (see Fig. 4.2). *Gam*$^-$ phage cannot generate the concatemeric linear DNA which is normally the substrate for packaging into phage heads. However, *gam*$^-$ phage do form plaques because the *rec* and *red* recombination systems act on circular DNA molecules to form multimers which can be packaged. *Gam*$^-$ *red*$^-$ phage are totally dependent upon *rec*-mediated exchange for plaque formation on *rec*$^+$ bacteria. Lambda DNA is a poor substrate for this *rec* mediated exchange. Therefore, such phage make vanishingly small plaques unless they contain one or more short DNA sequences called *chi* (*c*ross-over *h*ot-spot *i*nstigator) sites, which stimulate *rec*-mediated exchange. The early replacement vector λWES.λB′ generates *red*$^-$ *gam*$^+$ clones. But many of the new replacement vectors with a large capacity (e.g. λL47, λ1059) generate *red*$^-$ *gam*$^-$ clones. These vectors have, therefore, been constructed so as to include a *chi* site within the non-replaceable part of the phage.

The most recent generation of lambda vectors combine a large capacity for foreign DNA, close to the theoretical limit of 23 kb, together with features that allow simple and efficient library construction (see Chapter 5). The replacement vectors EMBL3 and EMBL4 (Frischauf *et al.* 1983) have convenient polylinker (see p. 96) sequences flanking the replaceable fragment. Phages with inserts can be selected by their Spi$^-$ phenotype, and are *chi*$^+$. Derivatives of EMBL3 containing amber mutations are available (EMBL3 *Sam*, EMBL3 *Aam Bam*, EMBL3 *Aam Sam*). The inclusion of amber mutations in phage λ vectors not only increases biological containment, but can be used in a selective system for isolating DNA sequences linked to suppressor genes (see Chapter 13, Fig. 13.2), and in recombinational screening (see Chapter 6).

Expression of genes cloned in λ vectors

It is sometimes the aim of a gene manipulator to promote the expression of a gene which has been cloned so as to amplify the synthesis of a desirable gene product. There is much interest in improving the production of bacterial enzymes that are useful reagents in nucleic acid biochemistry itself, e.g. DNA ligase,

DNA polymerase and restriction endonucleases. Panasenko *et al.* (1977) have described a recombinant phage, constructed *in vitro*, carrying the *E. coli* DNA ligase gene which, after induction of the recombinant lysogen, results in a five hundredfold overproduction of the enzyme so that it represents 5% of the total cellular protein of *E. coli*. Dramatic amplification depends *inter alia* upon both increasing the gene dosage and ensuring efficient transcription. The gene dosage is increased as a result of phage DNA replication within the host; the level of transcription may be improved by suitable choice of vector and subsequent manipulation of the recombinant phage.

A great deal of our knowledge about expression of genes cloned in phage λ comes from the studies of N. E. Murray, W. J. Brammar and their colleagues on a model system in which genes from the *trp* operon of *E. coli* are inserted in the phage genome either by manipulation *in vitro* or by genetic methods *in vivo* (Hopkins *et al.* 1976, Moir & Brammar 1976). The following discussion is based on their work.

First, we must distinguish between cases where the inserted DNA does or does not include the bacterial *trp* promoter. If the insert *does* include its own promoter, the yield of *trp* enzymes can be enhanced simply by delaying cell lysis so that the number of gene copies is increased and the time available for expression is extended. This was originally achieved by making the vector S^-. Moir and Brammar obtained better amplification of gene products by including mutations in gene *Q* or *N*. In Q^- phage all the late functions, including that of *S*, are blocked, and in addition, packaging of the replicated DNA is prevented which even further extends the availability of the DNA. An N^- phage is also defective in late functions and although it replicates more slowly than N^+Q^- phage, yields of enzyme achieved were at least as great. In such infected cells anthranilate synthetase, the product of the *trp E* gene, comprised more than 25% of the total soluble protein.

λ *trp* phage *lacking* the *trp* promoter have been constructed so that *trp* expression is initiated at the promoter P_L of the leftward operon of phage λ. This operon has two useful features:

1 P_L is a powerful promoter;

2 the anti-termination effect of gene *N* expression permits transcription through sequences which might otherwise prevent expression of a distant inserted gene.

Once again, cell lysis and DNA packaging were prevented by mutations in *Q* and *S* and additionally the cro^- mutation was introduced so as to derepress transcription from P_L. Cells infected with such a phage may contain as much as 10% of the soluble protein as anthranilate synthetase. However, these phage were difficult to construct and propagate so an alternative approach was adopted. The *cro* gene lies within the immunity region and its product is immune-specific. The *cro* product of the heteroimmune phage 434 will not interact with P_L of λ. Hybrid phage containing P_L from λ but *cro* and P_R from 434 are therefore phenotypically Cro^- as far as leftward transcription is concerned. Infection with such a λ*trp* derivative, which also carried the S^- mutation, gave cells in which 25% of the soluble protein was anthranilate synthetase. Derivatives of this type which are *Nam* can be grown on a non-suppressing host providing they also carry the *nin* deletion of t_{R_2}. Thus amplification can be modulated by controlling the suppression of the *Nam* gene. This is a useful property since extreme overproduction of a product relaxes the selection that can be imposed on a recombinant clone and may lead to problems of instability.

This elegant exploitation of the genetics of phage λ demonstrates the advantages of a well characterized genetic system and shows that useful amplification of a variety of gene products may be achieved with λ vectors, even in cases where a strong promoter recognized by the host RNA polymerase does not accompany the inserted gene. Lessons learnt from this model system have been applied to the amplification of *E. coli* DNA polymerase I and T4 DNA ligase in induced lysogens of λ recombinants carrying these genes (Kelley *et al.* 1977, Murray *et al.* 1979, Wilson & Murray 1979).

Packaging phage λ DNA *in vitro*

So far, we have considered only one way of introducing manipulated phage DNA into the host bacterium, i.e. by transfection of competent bacteria (see Chapter 1). Using freshly prepared λ DNA that has not been subjected to any gene manipulation procedures, transfection will result in typically about 10^5 plaques per microgram of DNA. In a gene manipulation experiment in which the vector DNA is restricted, etc., and then ligated with foreign DNA, this figure is reduced to about 10^4–10^3 plaques per microgram of vector DNA. Even with perfectly efficient nucleic acid biochemistry some of this reduction is inevitable. It is a consequence of the random association of fragments in the ligation reaction which produces molecules with a variety of fragment combinations, many of which are inviable. Yet, in some contexts, 10^6 or more recombinants are required. The scale of such experiments can be kept within a reasonable limit (less than 100 μg vector DNA) by packaging the recombinant DNA into mature phage particles *in vitro*.

Placing the recombinant DNA in a phage coat allows it to be introduced into the host bacteria by the normal processes of phage infection, i.e phage adsorption followed by DNA injection. Depending upon the details of the experimental design, packaging *in vitro* yields about 10^6 plaques per microgram of vector DNA after the ligation reaction.

Figure 4.5 shows some of the events occurring during the packaging process that take place within the host during normal phage growth and which we now require to perform *in vitro*. Phage DNA in concatemeric form, produced by a rolling circle replication mechanism (see Fig. 4.2), is the substrate for the packaging reaction. In the presence of phage head precursor (the product of gene *E* is the major capsid protein) and the product of gene *A*, the concatemeric DNA is cleaved into monomers and encapsidated. Nicks are introduced in opposite strands of the DNA, 12 nucleotide pairs apart at each *cos* site, to produce the linear monomer with its cohesive termini. The product of gene *D* is then incorporated into what now becomes a completed phage head. The products of genes *W* and *FII*, among others, then unite the head with a separately assembled tail structure to form the mature particle.

The principle of packaging *in vitro* is to supply the ligated recombinant DNA with high concentrations of phage head precursor, packaging proteins and phage tails. Practically this is most efficiently performed in a very concentrated mixed lysate of two induced lysogens, one of which is blocked at the prehead stage by an amber mutation in gene *D* and therefore accumulates this precursor while the other is prevented from forming any head structure by an amber mutation in gene *E* (Hohn & Murray 1977). In the mixed lysate, genetic complementation occurs and exogenous DNA is packaged. Although concatemeric DNA is the substrate for packaging (covalently joined concatemers are of course produced in the ligation reaction by association of the natural cohesive ends of λ), the in-vitro system *will* package added monomeric DNA.

There are two potential problems associated with packaging *in vitro*. First, endogenous DNA derived from the induced prophages of the lysogens used to prepare the packaging lysate can itself be packaged. This can be overcome by choosing the appropriate genotype for these prophages, i.e. excision upon

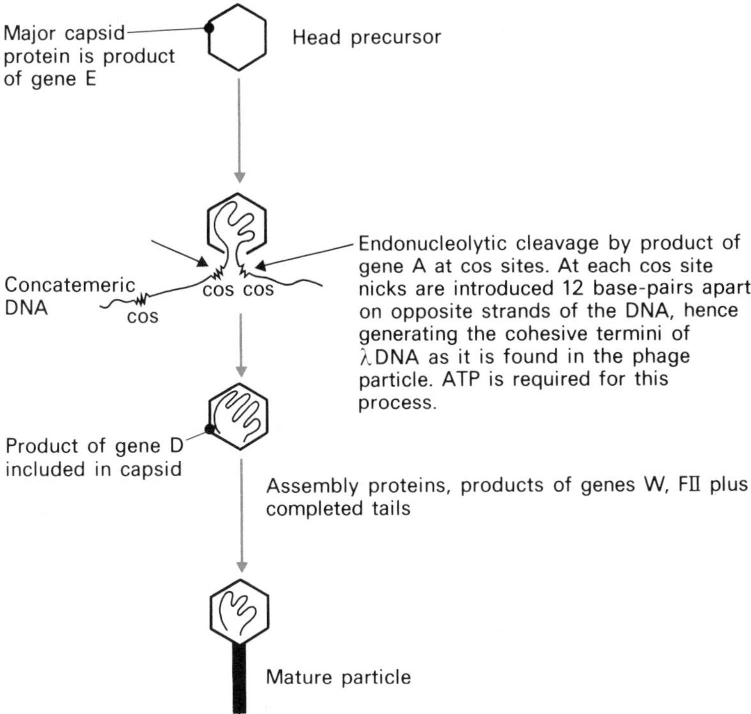
Major capsid protein is product of gene E
Head precursor
Concatemeric DNA
cos
cos cos
Endonucleolytic cleavage by product of gene A at cos sites. At each cos site nicks are introduced 12 base-pairs apart on opposite strands of the DNA, hence generating the cohesive termini of λDNA as it is found in the phage particle. ATP is required for this process.
Product of gene D included in capsid
Assembly proteins, products of genes W, FII plus completed tails
Mature particle

Fig. 4.5 Simplified scheme showing packaging of phage λDNA into phage particles.

induction is inhibited by the *b*2 deletion (Gottesman & Yarmolinsky 1968) and *imm 434* immunity will prevent plaque formation if an *imm 434* lysogenic bacterium is used for plating the complex reaction mixture. Additionally, if the vector does not contain any amber mutation an Su^- indicator can be used so that endogenous DNA will not give rise to plaques. The second potential problem arises from recombination in the lysate between exogenous DNA and induced prophage markers. If troublesome, this can be overcome by using recombination-deficient (i.e. red^-, rec^-) lysogens and by UV-irradiating the cells used to prepare the lysate, so eliminating the biological activity of the endogenous DNA (Hohn & Murray 1977).

Cosmid vectors

As we have seen, concatemers of unit-length λ DNA molecules can be efficiently packaged if the *cos* sites—substrates for the packaging dependent cleavage—are 37–52 kb apart (75–105% the size of λ^+ DNA). In fact, only a small region in the proximity of the *cos* site is required for recognition by the packaging system (Hohn 1975).

Plasmids have been constructed which contain a fragment of λ DNA including the *cos* site (Collins & Brüning 1978, Collins & Hohn 1979). These plasmids have been termed *cosmids* and can be used as gene cloning vectors in conjunction with the in-vitro packaging system. Figure 4.6 shows a gene cloning scheme employing a cosmid. Packaging the cosmid recombinants into phage coats imposes a desirable selection upon their size. With

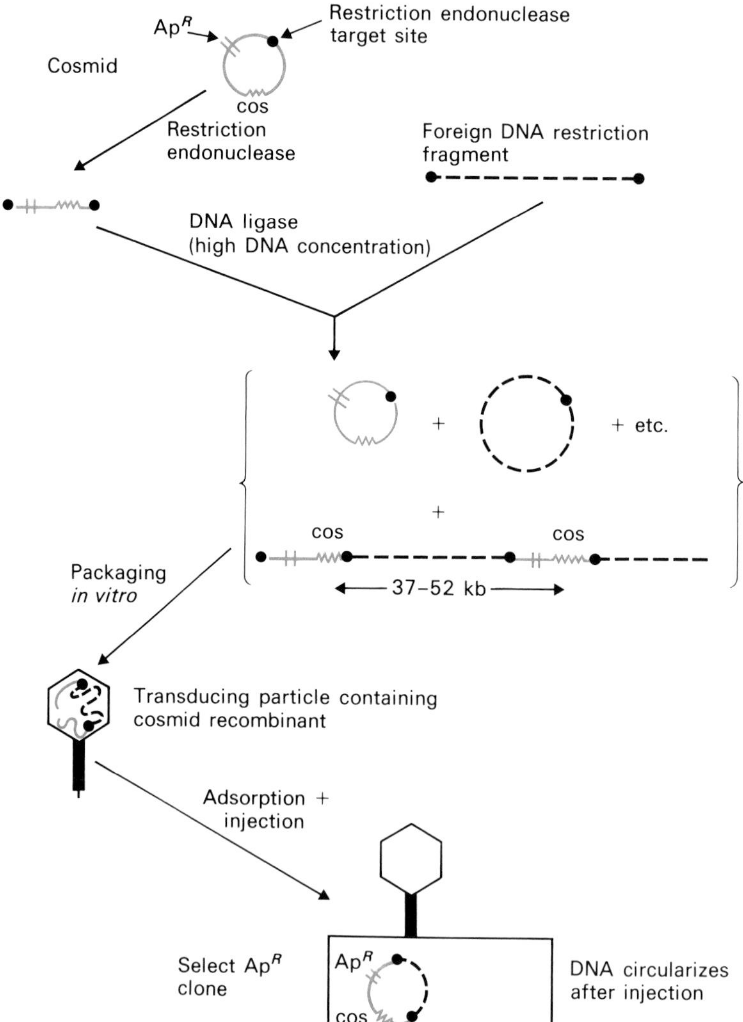

Fig. 4.6 Simple scheme for cloning in a cosmid vector. See text for details.

a cosmid vector of 5 kb we demand the insertion of 32–47 kb of foreign DNA—much more than a phage λ vector can accommodate. Note that after packaging *in vitro*, the particle is used to infect a suitable host. The recombinant cosmid DNA is injected and circularizes like phage DNA but replicates as a normal plasmid without the expression of any phage functions. Transformed cells are selected on the basis of a vector drug resistance marker.

Used in this way, cosmids provide an efficient means of cloning large pieces of foreign DNA. In addition, the system produces a very low background of non-recombinant clones without recourse to selection by insertional inactivation or pretreatment of the linearized vector DNA with alkaline phosphatase (see p. 33). Because of their capacity for large fragments of DNA, cosmids are particularly attractive vectors for constructing libraries of eukaryotic genome fragments. Partial digestion with restriction endonuclease provides suitably large fragments. However, there is a potential problem associated with the use of partial digests in this way. This is due to the possibility of two or more genome fragments joining together in the ligation reaction hence creating a clone containing fragments that were not initially adjacent in the genome. This would give an incorrect picture of their chromosomal organization. The problem can be overcome by size-fractionation of the partial digest.

Even with sized foreign DNA, in practice cosmid clones may be produced that contain non-contiguous DNA fragments ligated to form a single insert. The problem can be solved by dephosphorylating the foreign DNA fragments so as to prevent their ligation together. This method is very sensitive to the exact ratio of target to vector DNAs (Collins & Bruning 1978) because vector-to-vector ligation can occur. Furthermore, recombinants with a duplicated vector are unstable and break down in the host by recombination, resulting in the propagation of non-recombinant cosmid vector.

Such difficulties have been overcome in a cosmid cloning procedure devised by Ish-Horowicz and Burke (1981). By appropriate treatment of the cosmid vector pJB8 (Fig. 4.7) lefthand and righthand vector ends are purified which are incapable of self-ligation but which accept dephosphorylated foreign DNA. Thus the method eliminates the need to size the foreign DNA fragments and prevents formation of clones containing short foreign DNA or multiple vector sequences.

An alternative solution to these problems has been devised by Bates and Swift (1983) who have constructed cosmid c2XB.

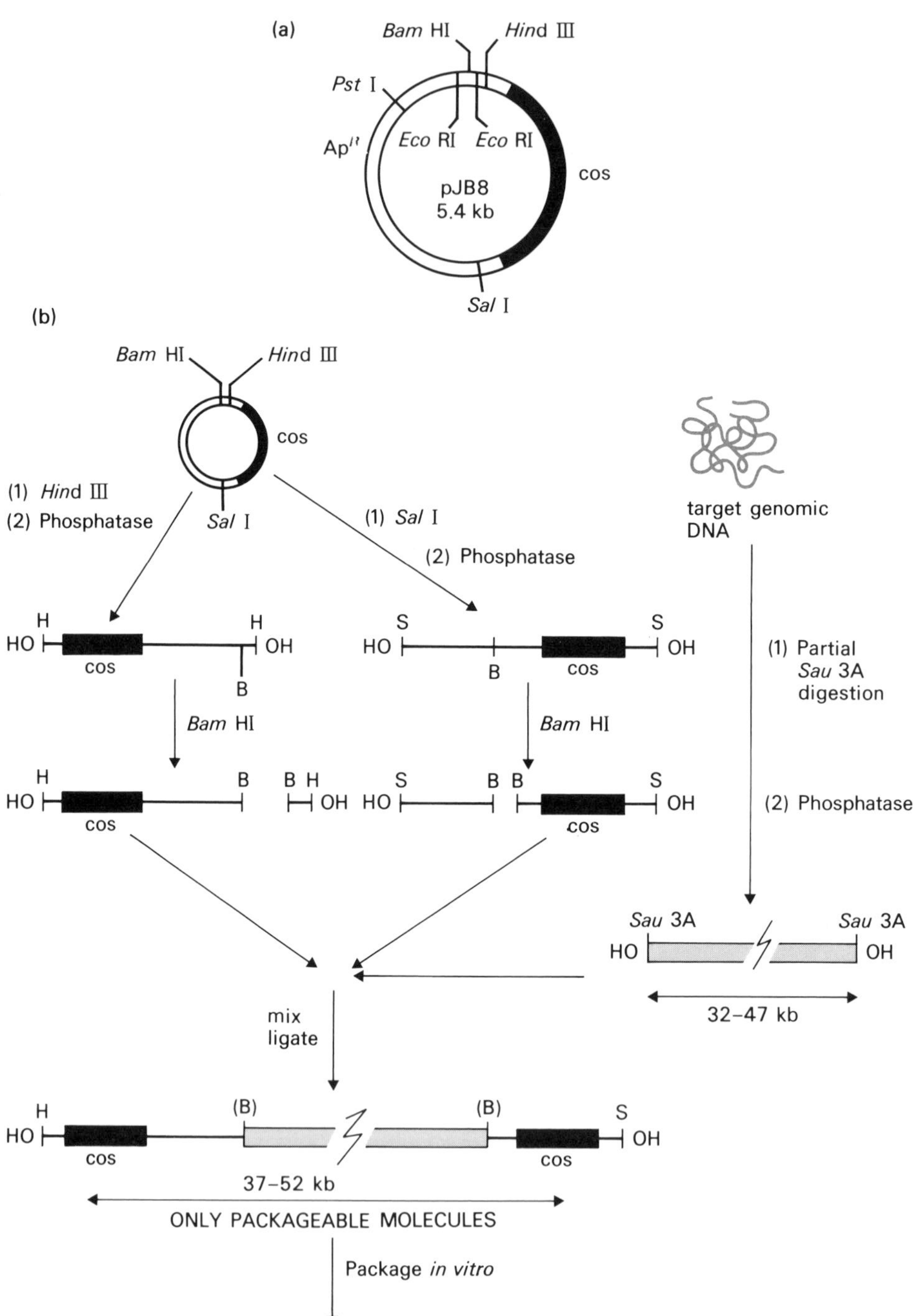

Fig. 4.7 Cosmid cloning scheme of Ish-Horowicz and Burke. (a) Map of cosmid pJB8. (b) Application to the construction of a genomic library of fragments obtained by partial digestion with *Sau* 3A. This restriction endonuclease has a tetranucleotide recognition site and generates fragments with the same cohesive termini as *Bam* HI (see p. 27).

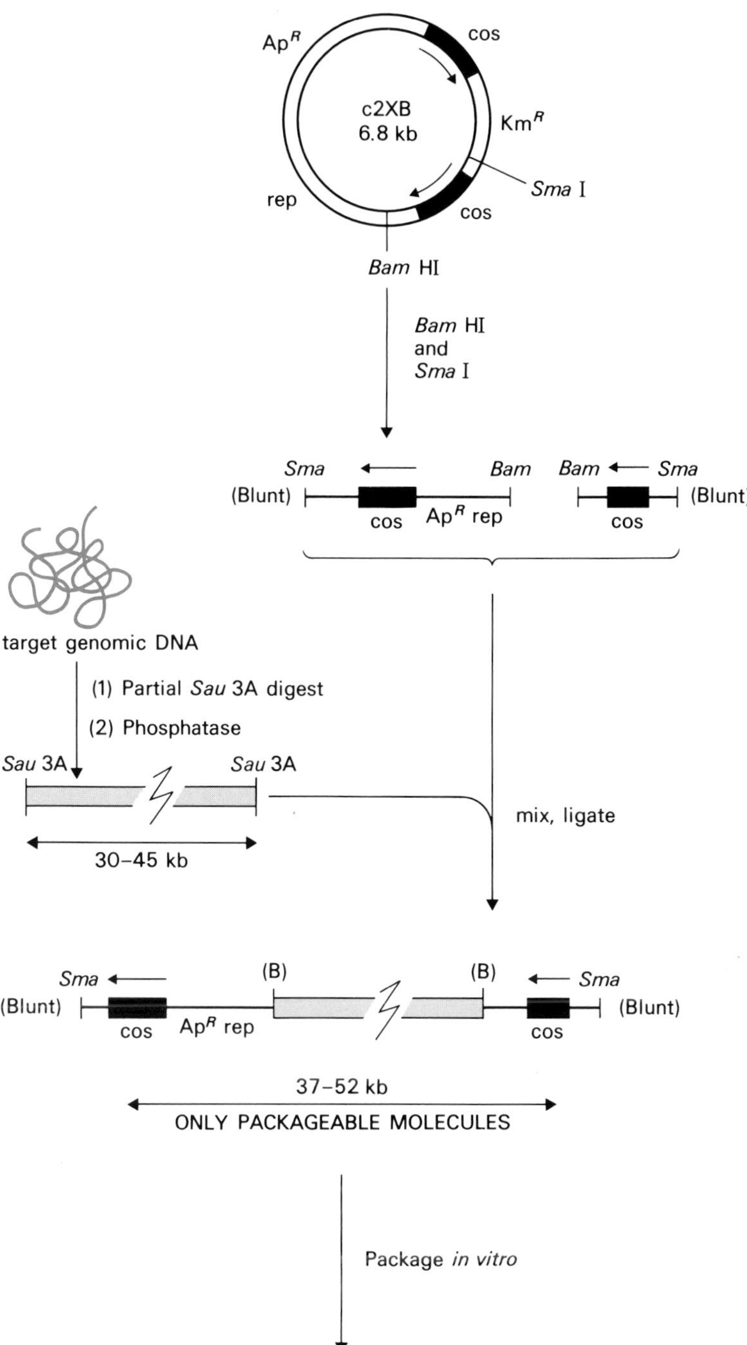

Fig. 4.8 Cosmid cloning scheme of Bates and Swift. The cosmid c2XB contains two *cos* sites, separated by a site for the restriction endonuclease *Sma* I which creates blunt ends. These blunt ends ligate only very inefficiently under the conditions used and effectively prevent the formation of recombinants containing multiple copies of the vector.

This plasmid carries a *Bam* HI insertion site and two *cos* sites separated by a blunt-end restriction site (Fig. 4.8). The creation of these blunt ends, which ligate only very inefficiently under the conditions used, effectively prevents vector self-ligation in the ligation reaction.

Phasmid vectors

A second combination of plasmid and phage λ sequences has been devised to exploit the virtues of each type of vector. This combination consists of a plasmid vector carrying a λ attachment (λ*att*) site. The plasmid may insert into a phage λ genome by means of the site-specific recombination mechanism of the phage which is normally responsible for recombinational insertion of the phage into the bacterial chromosome during lysogen formation. This reversible recombinational insertion of plasmid into the phage is referred to as 'lifting' the plasmid and generates a phage genome containing one or more plasmid molecules (depending upon the length of the plasmid). These novel genetic combinations are called *phasmids* (Brenner *et al.* 1982). They contain functional origins of replication of the plasmid and of λ, and may be propagated as a plasmid or as a phage in appropriate *E. coli* strains. Reversal of the lifting process releases the plasmid vector.

Phasmids may be used in a variety of ways; for instance, DNA may be cloned in the plasmid vector in a conventional way and then the recombinant plasmid can be lifted onto the phage. Phage particles are easy to store, they have an effectively infinite shelf-life, and screening phage plaques by molecular hybridization often gives cleaner results than screening bacterial colonies (see Chapter 6). Alternatively, a phasmid may be used as a phage cloning vector, from which subsequently a recombinant plasmid may be released.

DNA CLONING WITH SINGLE-STRANDED DNA VECTORS

M13, f1 and fd are filamentous coliphages containing a circular single-stranded DNA molecule. These coliphages are being developed as cloning vectors for they have a number of advantages over other vectors, including the usual two classes of vector for *E. coli*, plasmids and phage lambda. However, in order to appreciate their advantages it is essential to have a basic understanding of the biology of filamentous phages.

The biology of the filamentous coliphages

The phage particles have dimensions 900 nm × 9 nm and contain a single-stranded circular DNA molecule which is 6407 (M13) or 6408 (fd) nucleotides long. The complete nucleotide sequences of fd and M13 are available and they are 97% homologous. The differences consist mainly of isolated nucleotides here and there, mostly affecting the redundant bases of codons, with no blocks of sequence divergence. Partial sequencing of f1 DNA indicates that it is very similar to M13 DNA.

The filamentous phages only infect strains of enteric bacteria harbouring F pili. The adsorption site appears to be the end of the F pilus but exactly how the phage genome gets from the end of F pilus to the inside of the cell is not known. Replication of phage DNA does not result in host cell lysis. Rather, infected cells continue to grow and divide, albeit at a rate slower than uninfected cells, and extrude virus particles. Up to 1000 phage particles may be released into the medium per cell per generation.

The single-stranded phage DNA enters the cell by a process in which decapsidation and replication are tightly coupled. The capsid proteins enter the cytoplasmic membrane as the viral DNA passes into the cell while being converted to a double-stranded replicative form (RF). The RF multiplies rapidly until about 100 RF molecules are formed inside the cell. Replication of the RF then becomes asymmetric due to the accumulation of a viral-encoded single-stranded specific DNA binding protein. This protein binds to the viral strand and prevents synthesis of the complementary strand. From this point on only viral single strands are synthesized. These progeny single strands are released from the cell as filamentous particles following morphogenesis at the cell membrane. As the DNA passes through the membrane the DNA binding protein is stripped off and replaced with capsid protein.

Why use single-stranded vectors?

For many experiments with cloned DNA, e.g. heteroduplex analysis, isolation of complementary RNA and, in particular, the newer methods of DNA sequencing, single-stranded DNA is required. Techniques such as isopycnic centrifugation in the presence of alkali or ribopolymers are laborious and not particularly effective means of preparative separation of the two strands of a DNA helix. The use of single-stranded DNA vectors is an attractive means of combining, in one simple experiment, cloning, amplification, and strand separation of an originally double-stranded DNA fragment.

As single-stranded vectors the filamentous phages have a number of advantages. First, the phage DNA is replicated via a double-stranded circular DNA (RF) intermediate. This RF can be purified and manipulated *in vitro* just like a plasmid. Second, both RF and single-stranded DNA will transfect competent *E. coli* cells to yield either plaques or infected colonies, depending on the assay method. Third, the size of the phage particle is governed by the size of the viral DNA and therefore there are no packaging constraints. Indeed, viral DNA up to six times the length of M13 DNA has been packaged (Messing *et al.* 1981). Finally, with these phages it is very easy to determine the orientation of an insert. Although the relative orientation can be determined from restriction analysis of RF there is an easier method (Barnes 1980). If two clones carry the insert in opposite directions, the single-stranded DNA from them will hybridize and this can be detected by agarose gel electrophoresis. Phage from as little as 0.1 ml of culture supernate can be used in assays of this sort making mass screening of cultures very easy.

In summary, as vectors, filamentous phages possess all the advantages of plasmids while producing particles containing single-stranded DNA in an easily obtainable form.

Development of filamentous phage vectors

Unlike λ the filamentous coliphages do not have any non-essential genes which can be used as cloning sites. However, in M13 there is a 507 base pair intergenic region, from position 5498–6005 of the DNA sequence, which contains the origins of DNA replication for both the viral and the complementary strands. In most of the vectors developed so far foreign DNA has been inserted at this site although it is possible to clone at the carboxy-terminal end of gene IV (Boeke *et al.* 1979). The wild-type phages are not very promising as vectors because they contain very few unique sites within the intergenic region: *Asu* I in the case of fd, and *Asu* I and *Ava* I in the case of M13. However, a site does not have to be unique to be useful as the example below shows.

The first example of M13 cloning made use of one of ten *Bsu* I sites in the genome, two of which are in the intergenic region (Messing *et al.* 1977). For cloning, M13 RF was partially digested with *Bsu* I and linear full-length molecules isolated by agarose gel electrophoresis. These linear monomers were blunt end ligated to a *Hin*d II restriction fragment comprising the *E. coli lac* regulatory region and the genetic information for the α-peptide of β-galactosidase. The complete ligation mixture was used to transform a strain of *E. coli* with a deletion of the β-galactosidase α-fragment and recombinant phage detected by intragenic complementation on media containing IPTG and Xgal. The IPTG is a gratuitous inducer of β-galactosidase and Xgal a chromogenic substrate; where complementation occurs a blue colour is produced. One of the blue plaques was selected and the virus in it designated M13 mp1.

Insertion of DNA fragments into the *lac* region of M13 mp1 destroys its ability to form blue plaques making detection of recombinants easy. However, the *lac* region only contains unique sites for *Ava* II, *Bgl* I and *Pvu* I and three sites for *Pvu* II and there are no sites anywhere on the complete genome for the commonly used enzymes such as *Eco* RI or *Hin*d III. To remedy this defect Gronenborn and Messing (1978) introduced an *Eco* RI site into the *lac* region of mp1 and the way they did so is particularly interesting. From DNA sequence data and restriction mapping (Fig. 4.9) it was known that a single base change (guanine residue 13 → adenine) in the codon for the 5th amino acid residue of the β-galactosidase α-fragment would create an *Eco* RI site. This, in turn, would lead to an aspartate residue being replaced by an asparagine residue which would have no significant effect on the complementation properties of the α-peptide.

Methylation of guanine has been shown to cause it to mispair with uracil. Therefore single-stranded DNA from M13 mp1 particles was treated with the methylating agent N-methyl-N-nitrosourea and then transformed into cells and allowed to undergo several cycles of replication. The ccc RF DNA was isolated from these cells and digested with *Eco* RI. Linear molecules of genome length were separated from undigested molecules by agarose gel electrophoresis, excised from the gel and recircularized by way of their cohesive ends. The ligated molecules were transformed into cells and the resulting virus particles isolated. Following this procedure three individual clones with unique *Eco* RI restriction sites at different positions in the phage genome were isolated. Two of these M13 mp2 and M13 mp3, were the result of the conversion of *Eco* RI* sequences in the *lac* fragment of M13 mp1 corresponding to the positions of amino acids 5 and 119 respectively. The *Eco* RI site of the third mutant was elsewhere in the genome.

The introduction of an *Eco* RI site into the *lac* region of M13

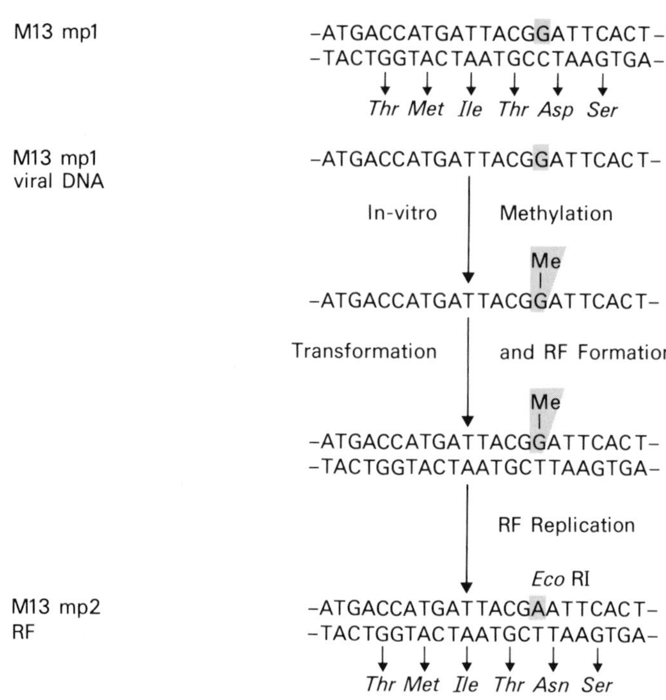

Fig. 4.9 In-vitro mutagenesis of the *lac* region of M13 mp1 to produce M13 mp2. Note that the methylguanine pairs with thymine but during replication this thymine will pair with adenine resulting in a G:C base pair being replaced by an A:T base pair.

mp1, which is itself resistant to cleavage by *Eco* RI, creates a unique site to clone DNA fragments. Gronenborn and Messing (1978) have shown that insertion of *Eco* RI-derived fragments into M13 mp2 leads to inactivation or reduction of the β-galactosidase activity. Furthermore, expression of functions coded by the inserted DNA can be controlled by the *lac* regulatory region. However, to improve the versatility of M13 as a vector, Messing *et al.* (1981) constructed M13 mp7 which has a multipurpose cloning site in the *lac* region.

The actual construction of M13 mp7 is too complex to describe in detail here but a summary is provided in Fig. 4.10. Starting with M13 mp2, Rothstein *et al.* (1979) removed the single *Bam* HI site in gene III. Then a synthetic linker containing a *Bam* HI site was inserted at the *Eco* RI site to generate mWJ43. This phage still gives blue plaques on media containing IPTG and Xgal. Into this *Bam* HI site Messing *et al.* (1981) introduced yet another oligonucleotide linker, this time one containing *Pst* I and *Sal* I sites, to create M13 mp71. Since the reading frame of the *lac* region of M13 mp71 is still unaltered, a functional α-

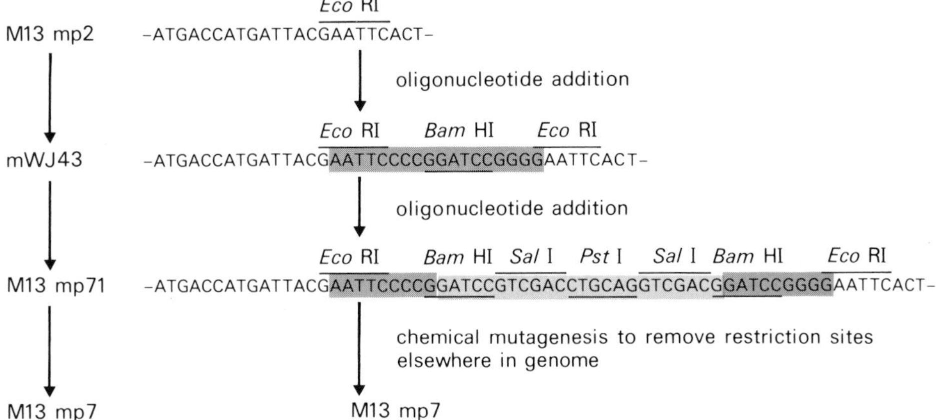

Fig. 4.10 The derivation of M13 mp7. Only the base sequence at the beginning of the *lac* region is shown.

peptide of β-galactosidase is still produced. Although in this phage only the *Pst* I site is unique, the sites for *Sal* I, *Bam* HI and *Eco* RI are also usable for cloning because the ensuing loss of small inserts still results in a functional *lac* sequence. The sequence 5′-GTCGAC-3′ is cleaved by endonucleases *Sal* I, *Acc* I and *Hinc* II. Unfortunately, due to ambiguities in their recognition sequence the latter two endonucleases each cleave one additional site, both located in gene II. Consequently, these two sites were removed by chemical mutagenesis to generate M13 mp7.

The M13 mp7 vector has a symmetrical multiple restriction site, or *polylinker* region. This has the limitation that DNA fragments with dissimilar ends cannot be inserted because treatment of the polylinker site with a pair of restriction enzymes (e.g. *Sal* I and *Eco* RI) will generate a vector with ends derived from the outer pair of restriction sites (*Eco* RI). In order to overcome this problem Messing and his co-workers have constructed new derivatives, M13 mp8, mp10, mp11 (Messing & Vieira 1982), mp18 and mp19 (Norrander *et al.* 1983), which have unpaired restriction sites in non-symmetrical polylinker regions (see Appendix 6). These vectors have the advantage that DNA fragments with dissimilar ends can be cloned, and the orientation of the insert is fixed. M13 mp8 and mp9 have similar polylinker regions in opposite orientations so that a foreign DNA fragment can be inserted either way round. The M13 mp10/mp11 pair, and the mp18/mp19 pair, also have their polylinker sites in opposite orientations. In practice, after such vectors have been digested with a pair of restriction enzymes it is often convenient to isolate the large vector band from an agarose gel, hence discarding the small restriction fragment. When ligation reactions are performed in the presence of the foreign DNA fragment, simple reclosure of the non-recombinant vector cannot occur so that only white, recombinant plaques are obtained.

In this section we have concentrated on the vectors developed by Messing and his co-workers. There are several reasons for this. First, the clever use of mutagenesis to insert and remove restriction sites. Second, the construction of vectors with polylinkers. The convenience afforded by such polylinkers in gene manipulation experiments has been widely appreciated. Plasmid vectors have been constructed which also incorporate such polylinker sites. These are the pUC plasmids (Vieira & Messing 1982, Norrander *et al.* 1983). This principle has also been extended to the specialized vector πVX (see Chapter 6) and to the phage lambda vectors EMBL3 and EMBL4 (see p. 81). Finally, the M13 mp vectors have revolutionized large-scale DNA sequencing. This latter aspect is covered in the next section.

Sequencing single-stranded DNA

In order to fully appreciate the usefulness of single-stranded vectors it is necessary to understand the chain terminator DNA sequencing procedure (Sanger *et al.* 1977). This procedure capitalizes on two properties of DNA polymerase. First, its ability to synthesize faithfully a complementary copy of a single-stranded DNA template. Second, its ability to use 2′3′ dideoxynucleoside triphosphates as substrates. Once the analogue is incorporated the 3′ end lacks a hydroxyl group and no longer is a substrate for chain elongation; thus the growing DNA chain is terminated.

In practice, the Klenow fragment of DNA polymerase I, which lacks the $5' \rightarrow 3'$ exonuclease activity of the intact enzyme, is used to synthesize a complementary copy of the single-stranded target sequence. Initiation of DNA synthesis requires a primer and usually this is a chemically synthesized oligonucleotide which is annealed close by.

DNA synthesis is carried out in the presence of the four deoxynucleoside triphosphates, one or more of which is labelled with ^{32}P, and in four separate incubation mixes containing a low concentration of one each of the four dideoxynucleoside triphosphate analogues. Therefore, in each reaction there is a population of partially synthesized radioactive DNA molecules, each having a common 5′ end, but

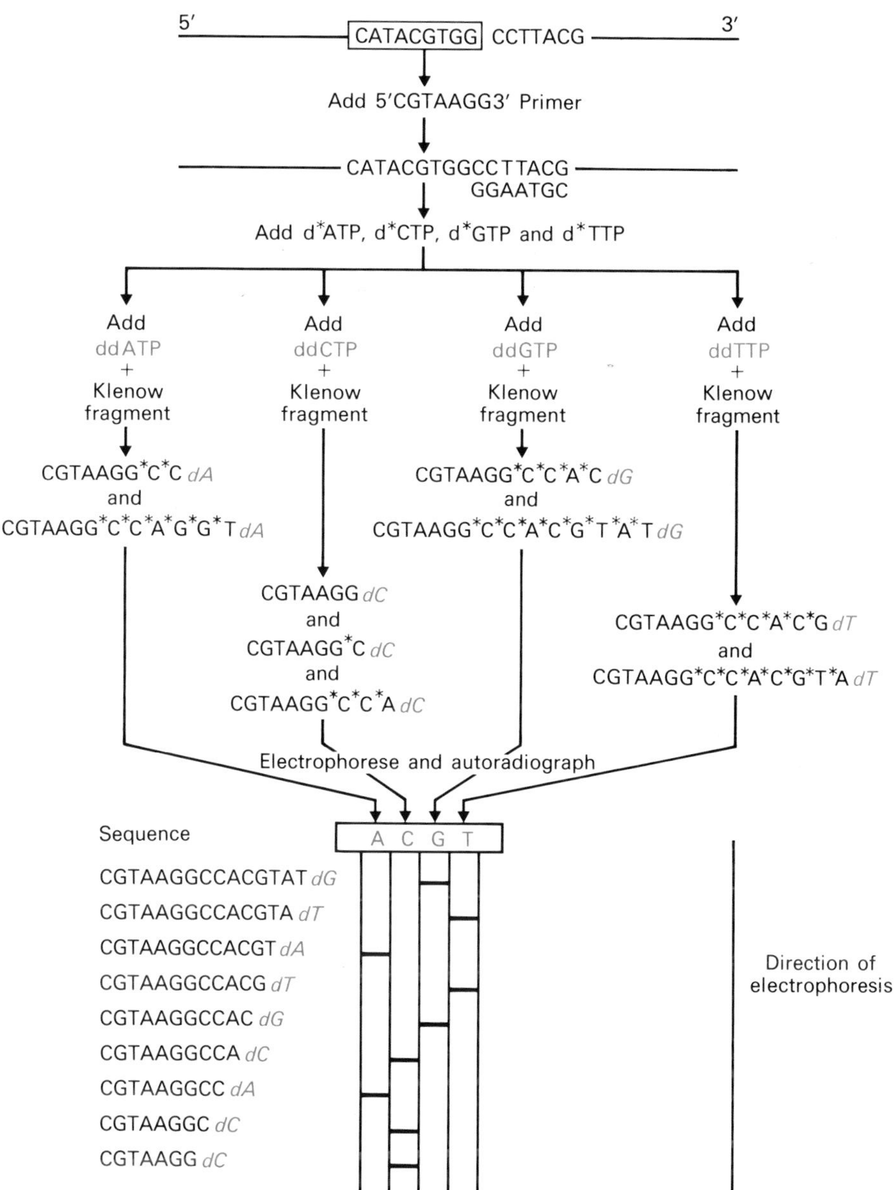

Fig. 4.11 DNA sequencing with dideoxynucleoside triphosphates as chain terminators. In this figure asterisks indicate the presence of ^{32}P and the prefix 'd' indicates the presence of a dideoxynucleoside. At the top of the figure the DNA to be sequenced is enclosed within the box. Note that unless the primer is also labelled with a radioisotope the smallest band with the sequence CGTAAGG*dC*will not be detected by autoradiography as no labelled bases were incorporated.

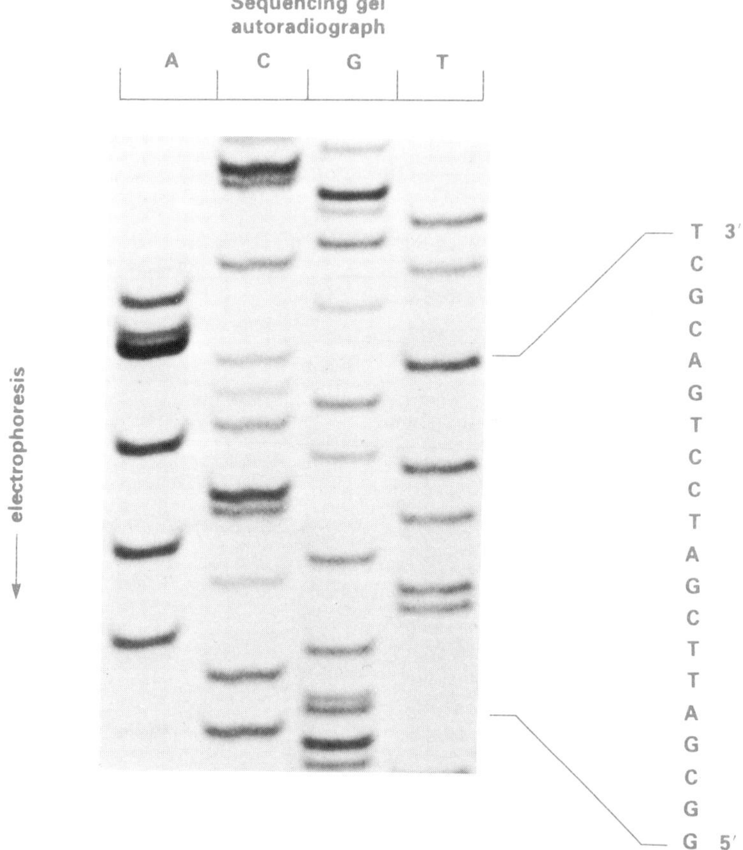

Fig. 4.12 Autoradiograph of a sequencing gel obtained with the chain terminator DNA sequencing method.

each varying in length to a base-specific 3′ end. After a suitable incubation period the DNA in each mixture is denatured, electrophoresed side by side, and the radioactive bands of single-stranded DNA detected by autoradiography. The sequence can then be read off directly from the autoradiograph as shown in Figs 4.11 and 4.12.

The DNA to be sequenced is cloned into one of the clustered cloning sites in the *lac* region of M13 mp-series of vectors. Recombinants are detected by the formation of white plaques on media containing IPTG and Xgal (Fig. 4.13). Virus is isolated from these white plaques, stocks prepared and the single-stranded viral DNA extracted for sequencing. The real beauty of M13 mp-series is that cloning into the same, specific region of the genome obviates the need for isolation of many different primers since a single primer can be used for all inserts. The

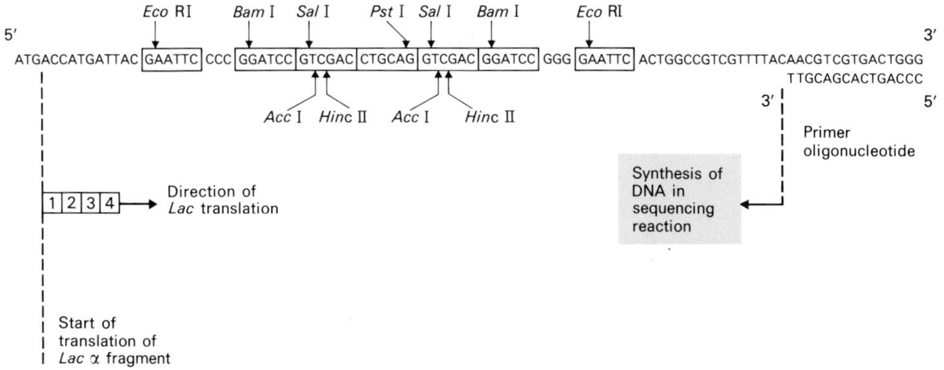

Fig. 4.13 Sequence of M13 mp7 DNA in the vicinity of the multi-purpose cloning region. The upper sequence is that of M13 mp7 from the ATG start codon of the β-galactosidase α-fragment, through the multi-purpose cloning region and back into the β-galactosidase gene. The horizontal bars indicate the recognition sites for the enzymes shown. The short sequence at the right-hand is that of the primer used to initiate DNA synthesis across the cloned insert. The numbered boxes correspond to the amino acids of the β-galactosidase fragment.

original primer was a short restriction fragment which was cloned in pBR322 and which was complementary to a region of the *lac* Z gene immediately adjacent to the righthand *Eco* RI insertion site (Anderson *et al.* 1980, Heidecker *et al.* 1980). Messing *et al.* (1981) have developed a more suitable 15-base synthetic oligonucleotide which primes just to the right of the polylinker. This has subsequently been found to have low homology with a second site in M13, so that an improved 15-base oligonucleotide primer has been synthesized. This has no homology at any secondary site (Norrander *et al.* 1983).

Oligonucleotide-directed mutagenesis: Point mutations and insertion/deletion formation

Oligonucleotide-directed mutagenesis is the most specific and general form of mutagenesis available. Any chosen nucleotide in a long DNA fragment can be specifically changed to any of the three other possible nucleotides. The method involves priming in-vitro DNA synthesis with a chemically synthesized oligonucleotide (7–20 nucleotides long) that carries a base mismatch with the complementary wild-type sequence. As shown in Fig. 4.14 the method requires that the DNA to be mutated is available in single-stranded form and cloning in M13 is particularly suitable for providing this. However, DNA cloned in a plasmid and obtained in duplex form can also

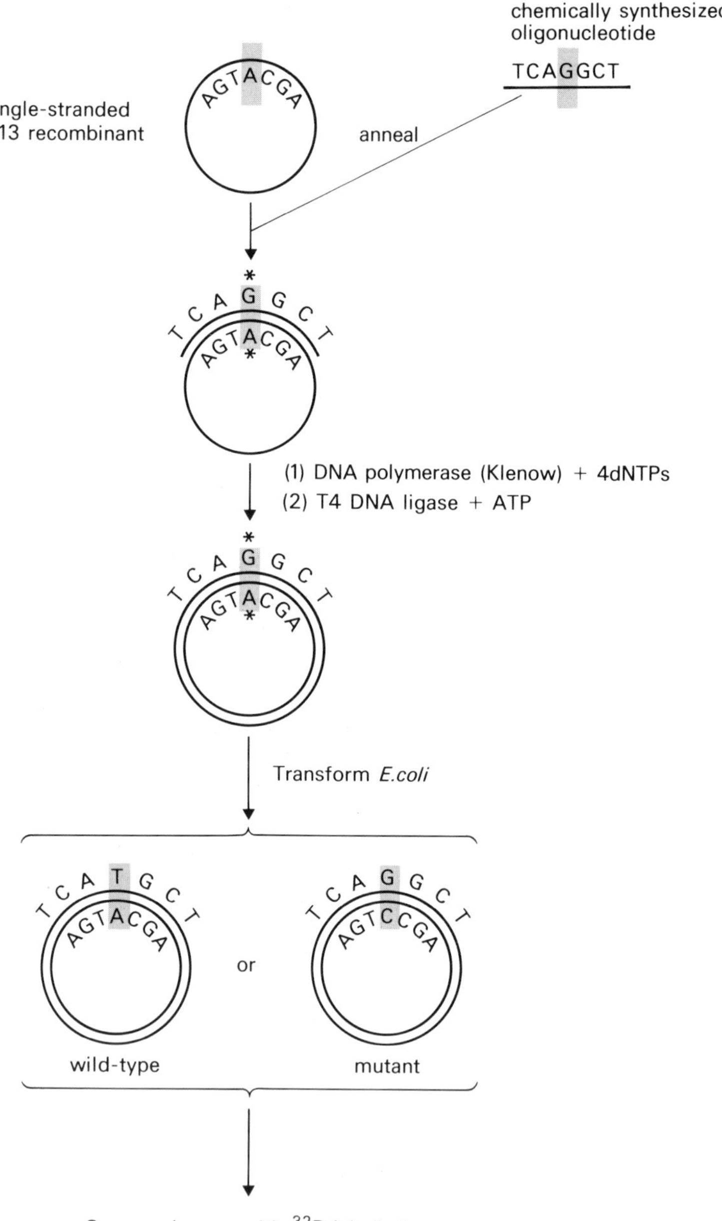

Fig. 4.14 Oligonucleotide-directed mutagenesis. Asterisks indicate mismatched bases.

be converted to a partially single-stranded molecule that is suitable (Dalbadie-McFarland *et al.* 1982). The synthetic oligonucleotide primes DNA synthesis and is itself incorporated into the resulting heteroduplex molecule. After transformation of host *E. coli*, this heteroduplex gives rise to homoduplexes whose sequences are either that of the original wild-type DNA or that containing the mutated base (Gillam *et al.* 1980, Dalbadie-McFarland *et al.* 1982, Zoller & Smith 1983). The frequency with which mutated clones arise, compared with wild-type clones, may be low. In order to pick out mutants, the clones can be screened by nucleic acid hybridization (see Chapter 6) with ^{32}P-labelled oligonucleotide as probe. Under suitable conditions of stringency, i.e. temperature and cation concentrations, a positive signal will be obtained only with mutant clones. This allows ready detection of the desired mutant (Wallace *et al.* 1981, Traboni *et al.* 1983). In order to check that the procedure has not introduced other, adventitious, changes in the sequence it is prudent to check the sequence of the mutant directly by DNA sequencing.

The particular value of site-directed mutagenesis lies in the ability to study the structure–function relationship of a DNA sequence; for example, mutant proteins can be encoded with very specific changes in particular amino acid residues which may be directly involved in catalysis or determining substrate specificity (see p. 292).

A variation of the procedure outlined above involves oligonucleotides containing inserted or deleted sequences. As long as stable hybrids are formed with single-stranded wild-type DNA, priming of in-vitro DNA synthesis can occur giving rise ultimately to clones corresponding to the inserted or deleted sequence (Wallace *et al.* 1980, Norrander *et al.* 1983).

Chapter 5 Cloning Strategies and Gene Libraries

Cloning strategies

Any DNA cloning procedure has four essential parts: a method for generating DNA fragments; reactions which join foreign DNA to the vector; a means of introducing the artificial recombinant into a host cell in which it can replicate; and a method of selecting or screening for a clone of recipient cells that has acquired the recombinant (Fig. 5.1). In previous chapters DNA cutting and joining reactions have been described, and the properties of several phage and plasmid vectors have been discussed together with the factors governing the choice between the various cutting and joining methods and different vector molecules. These choices will depend upon what type of clones are wanted, cDNA or genomic DNA clones. We also have to consider whether we wish to increase the frequency of the desired sequence in our starting material by a prior enrichment procedure. Alternatively we can compare gene libraries in which the desired sequence has not been enriched. The latter is sometimes called a *shotgun* approach.

Genomic DNA libraries

As an example, let us suppose that we wish to clone a single-copy gene from the human genome. We might simply digest total human DNA with a restriction endonuclease such as *Eco* RI, insert the fragments into a suitable phage λ vector and then attempt to isolate the desired clone. How many recombinants would we have to screen in order to isolate the right one? Assuming *Eco* RI gives, on average, fragments about 4 kb long, and given that the human haploid genome is 2.8×10^6 kb, we can see that over 7×10^5 independent recombinants must be prepared and screened in order to have a reasonable chance of including the desired sequence. In other words we have to obtain a very large number of recombinants which together contain a complete collection of all (or nearly all) of the DNA

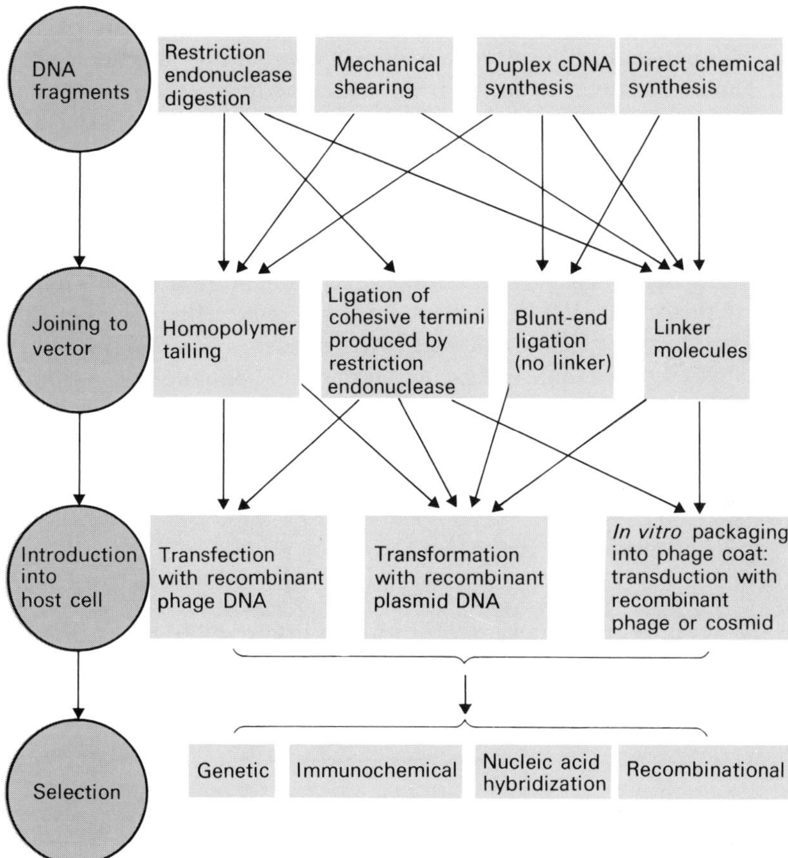

Fig. 5.1 Generalized scheme for DNA cloning in *E. coli*. Favoured routes are shown by arrows.

sequences in the entire human genome. Such a collection from which we withdraw the desired clone is called a *gene library* or *gene bank*.

There are two problems with the above approach. First, the gene may be cut internally one or more times by *Eco* RI so that it is not obtained as a single fragment. This is likely if the gene is large. Also, it may be desired to obtain extensive regions flanking the gene or whole gene clusters. Fragments averaging about 4 kb are likely to be inconveniently short. Alternatively, the gene may be contained on an *Eco* RI fragment that is larger than the vector can accept. In this case the appropriate gene would not be cloned at all.

These problems can be overcome by cloning *random* DNA fragments of a large size (~20 kb). Since the DNA is randomly fragmented, there will be no systematic exclusion of any sequence. Furthermore, clones will overlap one another, giving an opportunity to 'walk' from one clone to an adjacent one (see Chapter 13). Because of the larger size of each cloned DNA, fewer clones are required for a complete or nearly complete library. How many clones are required? Let n be the size of the genome relative to a single cloned fragment. Thus for the human genome, 2.8×10^6 kb, and for a cloned fragment size of 20 kb, $n = 1.4 \times 10^5$. The number of independent recombinants required in the library must be greater than n, because sampling variation will lead to the inclusion of some sequences several times, and the exclusion of other sequences in a library of just n recombinants. Clarke and Carbon (1976) have derived a formula which relates the probability (P) of including any DNA sequence in a random library of N independent recombinants:

$$N = \frac{ln\,(1 - P)}{ln\left(1 - \frac{1}{n}\right)}.$$

Therefore, to achieve a 95% probability ($P = 0.95$) of including any particular sequence in a random human genomic DNA library of 20 kb fragment size:

$$N = \frac{ln\,(1 - 0.95)}{ln\left(1 - \frac{1}{1.4 \times 10^5}\right)}$$

$$= 4.2 \times 10^5.$$

Notice that a considerably higher number of recombinants is required to achieve a 99% probability, for here $N = 6.5 \times 10^5$.

How can appropriately sized, random fragments be produced? A variety of methods are available. Random breakage by mechanical shearing is appropriate, but a much more commonly used procedure involves restriction endonucleases. In the strategy devised by Maniatis *et al.* (1978) (Fig. 5.2) the target DNA is restricted with a mixture of *two* restriction enzymes. These enzymes have tetranucleotide recognition sites, which therefore occur frequently in the target DNA and in a *limit* double-digest would produce fragments averaging less than 1 kb. The restriction reaction is carried out only to a partial extent, so that the bulk of fragments are relatively large, in the range 10–30 kb. These are effectively a random set of overlapping fragments. These can be fractionated by velocity centrifugation on a sucrose gradient or by preparative gel electrophoresis, so as to give a random

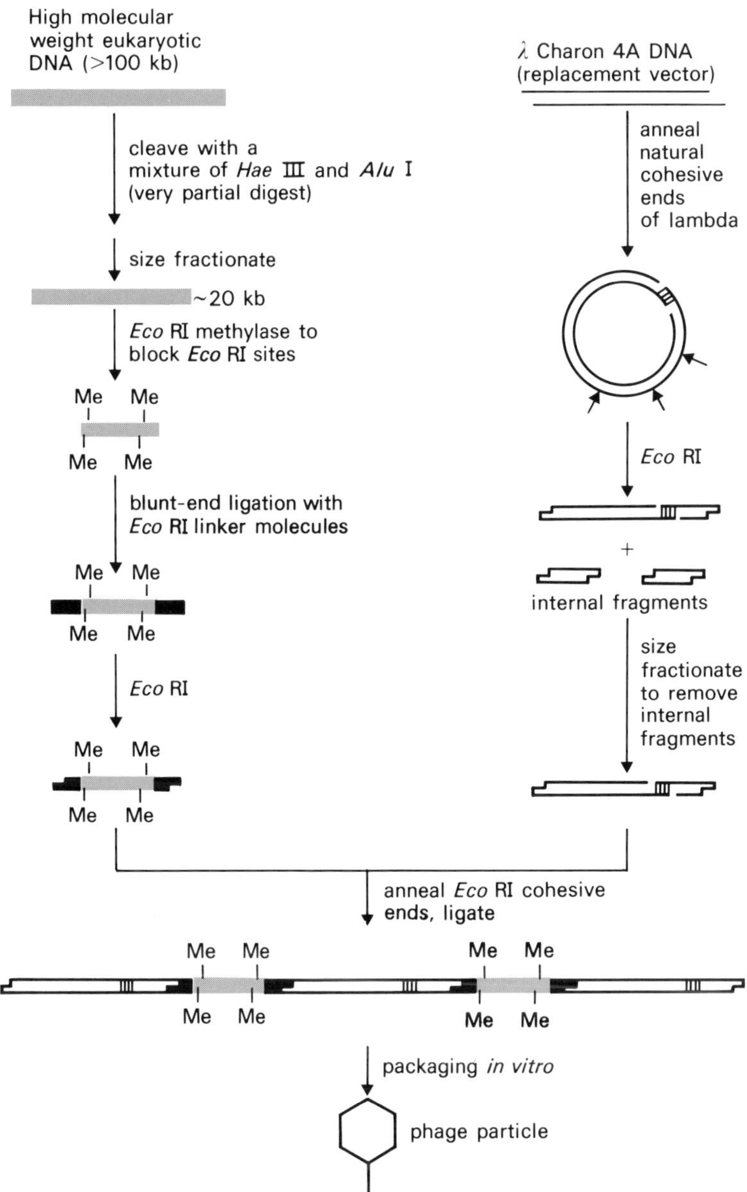

Fig. 5.2 Maniatis' strategy for producing a representative gene library.

population of fragments of about 20 kb which are suitable for insertion into the phage lambda vector. Packaging *in vitro* ensures that an appropriately large number of independent recombinants can be recovered, which will give an almost completely representative library.

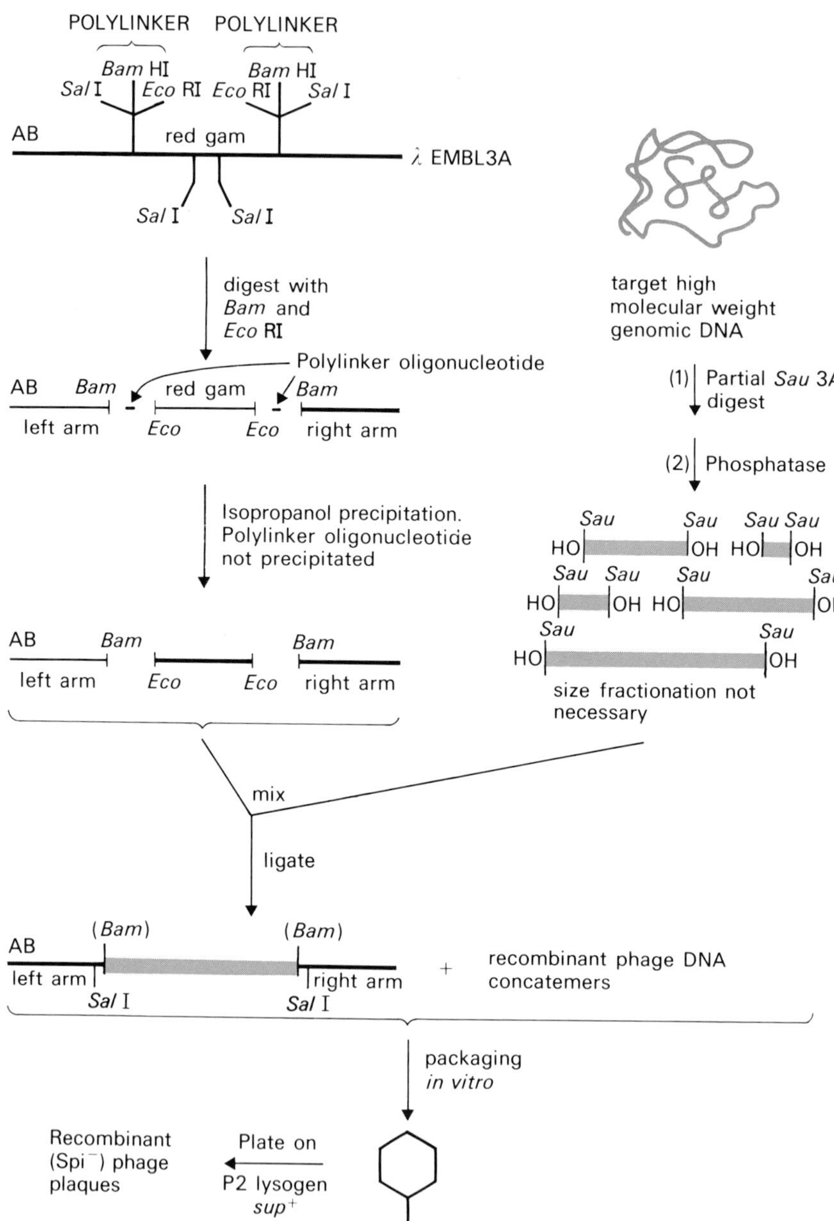

Fig. 5.3 Creation of a genomic DNA library using the phage lambda vector EMBL3A.

High molecular weight genomic DNA is partially digested with *Sau* 3A. The fragments are treated with phosphatase to remove their 5′-phosphate groups. The vector is digested with *Bam* HI and *Eco* RI which cut within the polylinker sites. The tiny *Bam* HI/*Eco* RI polylinker fragments are discarded in the isopropanol precipitation.

In Maniatis' strategy, the use of two different restriction endonucleases with completely unrelated recognition sites, *Hae* III and *Alu* I, assists in obtaining fragmentation that is nearly random. These enzymes both produce blunt ends, and the cloning strategy requires linkers (Fig. 5.2). A convenient simplification can be achieved by using a *single* restriction endonuclease which cuts frequently, such as *Sau* 3A. This will create a partial digest that is slightly less close to random than that achieved with a pair of enzymes. However, it has the great advantage that the *Sau* 3A fragments can be readily inserted into high-capacity phage lambda vectors, such as λL47 or λEMBL3 (see Chapter 4) which have been digested with *Bam* HI (Fig. 5.3). This is because *Sau* 3A and *Bam* HI create the same cohesive ends (see p. 27). These partial digestion methods, coupled with packaging the phage lambda recombinants, have been the most widely employed strategies for creating genomic DNA libraries.

In place of the phage lambda vectors cosmid vectors may be chosen. These also have the high efficiency afforded by packaging *in vitro* and have an even higher capacity than any phage lambda vector. However, there are two drawbacks in practice. First, most workers find that screening libraries of phage lambda recombinants by plaque hybridization gives cleaner results than screening libraries of bacteria containing cosmid recombinants by colony hybridization (see Chapter 6). Plaques usually give less of a background hybridization than do colonies. Second, it may be desired to retain and store an amplified genomic library. With phage, the initial recombinant DNA population is packaged and plated out. It can be screened at this stage. Alternatively, the plates containing the recombinant plaques can be washed to give an *amplified* library of recombinant phage. The amplified library can then be stored

---

The vector arms are then ligated with the partially digested genomic DNA. The phosphatase treatment prevents the genomic DNA fragments from ligating together. Non-recombinant vector cannot reform because the small polylinker fragments have been discarded. The only packageable molecules are recombinant phages. These are obtained as plaques on a P2 lysogen of sup^+ *E. coli*. The Spi^- selection ensures recovery of phage lacking *red* and *gam* genes. A sup^+ host is necessary because, in this example, the vector carries amber mutations in genes A and B. These mutations increase biological containment, and can be applied to selection procedures such as recombinational selection (see Chapter 6), or tagging DNA with a sup^+ gene (see Chapter 13). Ultimately, the foreign DNA can be excised from the vector by virtue of the *Sal* I sites in the polylinker.

almost indefinitely; phage have a long shelf-life. The amplification is so great that samples of this amplified library could be plated out and screened with different probes on hundreds of occasions. With bacterial colonies containing cosmids it is also possible to store an amplified library (Hanahan & Meselson 1980), but bacterial populations cannot be stored as readily as phage populations. There is often an unacceptable loss of viability when the bacteria are stored.

A word of caution is necessary when considering the use of any amplified library. This is the possibility of *distortion*. Not all recombinants in a population will propagate equally well, e.g. variations in target DNA size or sequence may affect replication of a recombinant phage, plasmid or cosmid. Therefore, when a library is put through an amplification step particular recombinants may be increased in frequency, decreased in frequency or lost altogether. Development of modern vectors and cloning strategies has simplified library construction to the point where many workers now prefer to create a new library for each screening, rather than risk using a previously amplified one.

The ease with which random libraries can be created and screened, and the possibility of chromosome walking, means that the shotgun approach has now become very widely adopted. However, as an alternative approach we could obtain a partially purified DNA fraction which is enriched in the desired sequence. The task of screening would then be diminished correspondingly. This approach is now rather outdated but it was necessary at a time before packaging *in vitro* had been developed, because then it was not possible readily to produce enough independent clones for a complete library. One method that has been employed very successfully for such enrichment is chromatography upon the medium RPC-5 which consists of a quaternary ammonium salt, the extractant, supported upon a matrix of plastic beads. Originally developed for high-resolution reversed phase chromatography of tRNAs, it will fractionate milligram quantities of DNA fragments generated by restriction endonucleases (Hardies & Wells 1976). Leder and his co-workers (Tilghman *et al.* 1977) loaded an *Eco* RI digest of total mouse genomic DNA onto an RPC-5 column, and upon elution with a concentration gradient of sodium acetate, obtained a series of DNA fractions which were assayed for their ability to hybridize with mouse globin cDNA. A fraction was identified as being substantially enriched in a DNA fragment bearing β-globin sequences. The discrimination on the RPC-5 column chromatography is only slightly dependent upon fragment size, so that additional enrichment (to about 500-fold) could be obtained by combining it with preparative gel electrophoresis. The final enriched fraction was ligated into the λWES.λB vector, and out of about 4300 plaques obtained by transfection, 3 positive clones were detected.

Multiple-copy genes, even without prior enrichment, do not, of course, require the screening of so many clones as single-copy genes. In fact, the clustered 5*s* DNA copies and the extrachromosomal rDNA copies of the frog *Xenopus* are present in so many copies that they can be isolated in pure form on the basis of their buoyant density difference from bulk DNA in centrifugation gradients of caesium salts. Thus these genes can be purified without gene cloning, although with one significant difference. Studies of the non-cloned sequences deal with average properties of the material. It is only by cloning individual copies that microheterogeneity within genes or spacers, down to the nucleotide sequence level, can be investigated.

cDNA cloning

Cloned eukaryotic cDNAs have their own special uses which derive from the fact that they lack the intron sequences which are usually present in the corresponding genomic DNA. Introns are non-coding sequences which often occur within eukaryotic gene sequences. They can be situated within the coding sequence itself, where they then interrupt the co-linear relationship of the gene with the encoded polypeptide. They may also occur in the 5′ or 3′ untranslated regions of the gene, but in any event they are copied into RNA by RNA polymerase when it transcribes the gene. The initial, primary transcript is a precursor to mRNA. It goes through a series of processing events in the nucleus before appearing in the cytoplasm as mature mRNA. These events include the removal of intron sequences by a process called splicing. When cDNA is derived from mRNA it therefore lacks intron sequences. Since removal of eukaryotic intron transcripts by splicing does not occur in bacteria, eukaryotic cDNA clones find application where bacterial expression of the foreign DNA is necessary, either as a prerequisite for detecting the clone (see Chapter 6), or because the polypeptide product is the primary objective. Also, where the sequence of the genomic DNA is known, the position of intron/exon boundaries can be assigned by comparison with the cDNA sequence.

A second situation where cDNA cloning is carried out

involves the analysis of temporally regulated gene expression in development, or tissue-specific gene expression. By using the 'plus and minus' screening procedure (see Chapter 6) it is possible to screen a cDNA clone library to identify cDNA clones derived from mRNA molecules present in one cell type but absent in another cell type.

As with genomic DNA, it may be appropriate to isolate cDNA from purified mRNA. Alternatively a cDNA clone library may be prepared and screened for particular sequences. Before proceeding further it is necessary to consider the nature of mRNA populations in tissues. In many tissues and cultured cells mRNAs are present at widely different *abundances*, i.e. some mRNA types are present in large numbers per cell, others may be present at just a few copies per cell. Table 5.1 gives some representative examples. Notice that in the chicken oviduct one mRNA type is superabundant. This is the mRNA encoding ovalbumin, the major egg-white protein. Therefore, this mRNA population is naturally so enriched in ovalbumin mRNA that cloning the ovalbumin cDNA presents no problem in screening. The clones could be identified by screening a small number of recombinants; the hybrid released translation procedure would be appropriate (see Chapter 6).

Another appropriate strategy for obtaining abundant cDNAs is to clone the cDNA directly in an M13 vector such as M13 mp8. A set of clones can then be sequenced immediately and identified on the basis of the polypeptide that each encodes. A successful demonstration of this *shotgun sequencing* strategy is given by Putney *et al.* (1983) who determined DNA sequences of 178 randomly chosen muscle cDNA recombinants. Complete amino acid sequences were available for 19 abundant muscle-related proteins. Altogether, they were able to identify

Table 5.1 Abundance classes of typical mRNA populations.

Source	Number of different mRNAs	Abundance (molecules/cell)
Mouse liver cytoplasmic poly(A)$^+$	9	12 000
	700	300
	11 500	15
Chick oviduct polysomal poly(A)$^+$	1	100 000
	7	4000
	12 500	5

References: Mouse (Young *et al.* 1976); chick oviduct (Axel *et al.* 1976).

clones corresponding to 13 of these 19 proteins, including interesting protein variants.

For cDNA clones in the low abundance class it is usual to construct a cDNA library. The formula of Clarke and Carbon (see p. 104) can be adapted to estimate the number of recombinants required for a reasonably complete library. Typically, 10^5 clones will be sufficient for low abundance mRNAs from most cell types. Once again the high efficiency obtained by packaging *in vitro* makes phage λ vectors attractive for obtaining large numbers of cDNA clones. Insertional vectors such as λgt10, λNM1149 (see Chapter 4) or λgt11 (see Chapter 6) are particularly well suited for such cDNA cloning.

Is it worth enriching for a particular mRNA or cDNA before cloning? Only in special circumstances is a ready purification possible and attractive. In general, the most commonly used primary screening technique involves colony or plaque hybridization with radioactive or immunochemical probes. This is applicable to very large libraries and the effort involved is largely independent of the number of recombinants to be screened. There is, therefore, usually little to be gained by attempting to enrich the starting mRNA or cDNA in order to reduce the number of recombinants to be screened. The isolation of clones by such screening procedures has become a commonplace technical feat which effectively performs a purification that is impossible by any other means.

Chapter 6 Recombinant Selection and Screening

The task of isolating a desired recombinant from a population of bacteria or phage depends very much upon the cloning strategy that has been adopted; for instance, when a cDNA derived from a purified or abundant mRNA is to be cloned, the task is relatively simple—only a small number of clones need to be screened. Isolating a particular single-copy gene sequence from a complete mammalian genomic library requires techniques in which hundreds of thousands of recombinants can be screened.

The purpose of this chapter is to outline important methods that have been developed for the selection and screening of recombinant clones. It is not an exhaustive account; variations of these themes will no doubt occur to the thoughtful reader, and particular experimental systems will sometimes present special opportunities that can be turned to advantage.

Genetic methods

(i) *Selection for presence of vector*

When combined with microbiological techniques, genetic selection is a very powerful tool since it can be applied to large populations. All useful vector molecules carry a selectable genetic marker or property. Plasmid and cosmid vectors carry drug resistance or nutritional markers, and in the case of phage vectors plaque formation is itself the selected property. Genetic selection for presence of the vector is a prerequisite stage in obtaining the recombinant population. As we have seen this can be refined to distinguish recombinant molecules and non-recombinant, parental vector. Insertional inactivation of a drug resistance marker, or of a gene such as β-galactosidase for which there is a colour test, are examples of this (see p. 62). With certain replacement type lambda vectors, and with cosmid vectors, size selection by the phage particle selects recombinant formation.

(ii) *Selection of inserted sequences*

If an inserted foreign gene in the desired recombinant is expressed, then genetic selection may provide the simplest method for isolating clones containing the gene. Cloned *E. coli* DNA fragments carrying biosynthetic genes can be identified by complementation of nonrevertible auxotrophic mutations in the host *E. coli* strain. A related example comes from the work of Cameron *et al.* (1975) who have cloned the *E. coli* DNA ligase gene in a phage λgt.λB vector. They exploited the inability of λ*red*$^-$ phage (the vector is *red*$^-$ by deletion of the C fragment) to form plaques on *E. coli lig* ts at the permissive temperature, whereas λ*red*$^-$ phage will form plaques on *E. coli* Lig$^+$. Recombinant phage carrying the wild-type ligase function could therefore be selected simply by their ability to form plaques through complementation of the host deficiency when plated on *E. coli lig* ts.

It has been found that certain eukaryotic genes are expressed in *E. coli* and can complement auxotrophic mutations in the host bacterium. Ratzkin and Carbon (1977) inserted fragments of yeast DNA, obtained by mechanical shearing, into the plasmid Col E1 using a homopolymer tailing procedure. They transformed *E. coli his*B mutants with recombinant plasmid and, by selecting for complementation, isolated clones carrying an expressed yeast *his* gene.

A similar approach has even been applied successfully to cloned mouse sequences. Chang *et al.* (1978) constructed a population of recombinant plasmids containing cDNA that was derived from an unfractionated mouse cell mRNA preparation in which dihydrofolate reductase (DHFR) mRNA was present. Mouse DHFR is much less sensitive to inhibition by the drug trimethoprim than is *E. coli* DHFR, so that by selecting transformants in medium containing the drug, clones were isolated in which resistance was conferred by synthesis of the mouse enzyme. This was an early example of expression of a mammalian structural gene in *E. coli*. The factors affecting expression of heterologous genes are complex, and an efficient selection procedure was required in order to identify clones actually synthesizing mouse DHFR amongst those containing non-expressed DHFR cDNA (see Chapter 7).

Immunochemical methods

Immunochemical detection of clones synthesizing a foreign protein has also been successful in cases where the inserted gene sequence is expressed. A particular advantage of the method is that genes which do not confer any selectable property on the host can be detected, but it does of course require that specific antibody is available.

A number of laboratories have developed similar immunochemical detection methods (Skalka & Shapiro 1976, H. A. Ehrlich *et al.* 1978). The method of Broome and Gilbert (1978) has been applied most widely. It depends upon three points:

1 an immune serum contains several IgG types that bind to different determinants on the antigen molecule;

2 antibody molecules absorb very strongly to plastics such as polyvinyl, from which they are not removed by washing;

3 IgG antibody can be readily radiolabelled with ^{125}I by iodination *in vitro*.

These properties are elegantly exploited in the following way. First, transformed cells are plated on agar in a conventional Petri dish. A replica plate must also be prepared because subsequent procedures kill these colonies. The bacterial colonies are then lysed in one of a number of ways—by exposure to chloroform vapour, by spraying with an aerosol of virulent phage, or by using a host bacterium that carries a thermo-inducible prophage. This releases the antigen from positive colonies. A sheet of polyvinyl that has been coated with the appropriate antibody (unlabelled) is applied to the surface of the plate whereupon the antigen complexes with the bound IgG. The sheet is removed and exposed to ^{125}I-labelled IgG. The ^{125}I-IgG can react with the bound antigen via antigenic determinants at sites other than those involved in the initial binding of antigen to the IgG-coated sheet, as shown in Fig. 6.1. Positively reacting colonies are detected by washing the sheet and making an autoradiographic image. The required clones can then be recovered from the replica plate. The method can be applied with only minor modification to plates bearing phage plaques instead of transformed colonies.

Two further aspects of the immunochemical method deserve mention. First, detection of altered protein molecules is possible providing that the alteration does not prevent cross-reaction with antibody. Thus Villa-Komaroff *et al.* (1978) have isolated *E. coli* clones containing cDNA sequences from rat preproinsulin mRNA. The cDNA was inserted by dG-dC homopolymer tailing at the *Pst* I site of pBR322. Using anti-insulin antibody, they isolated a clone which expressed a fused protein composed of the N-terminal region of β-lactamase (from pBR322) and a region of the proinsulin protein linked through a stretch of six glycine residues encoded by $d(G)_{18}$ of

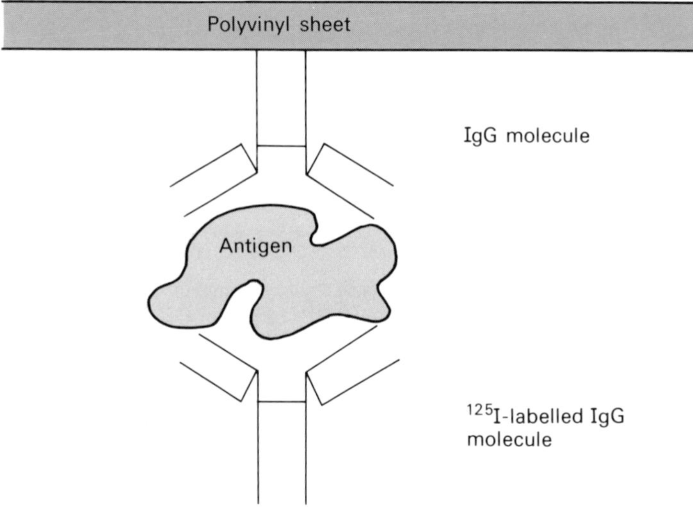

Fig. 6.1 Antigen–antibody complex formation in the immunochemical detection method of Broome Gilbert. (See text for details.)

the joint (see p. 132). Second, the two-site detection method employed by Broome and Gilbert is particularly suited to the detection of novel genetic constructions; for instance, by coating polyvinyl discs with IgG prepared from an immune serum directed against one protein, and detecting the immobilized antigen with ^{125}I-antibodies directed against another protein, only hybrid polypeptide molecules, synthesized as a result of DNA recombination, would produce an autoradiographic response.

A very efficient exploitation of the immunochemical detection method involves the phage lambda expression vector λgt11 (Young & Davis 1983). This vector carries the *E. coli lac* Z gene. A unique *Eco* RI site is located within the β-galactosidase coding region. Recombinant libraries can be constructed in which eukaryotic cDNA has been inserted, by means of linkers, into the *Eco* RI site. In such recombinants the β-galactosidase is insertionally inactivated and, depending upon the translational phase at the fusion junction, hybrid proteins are expressed. In a population of cDNA recombinants in which duplex cDNA has been synthesized by hairpin self-priming (see Chapter 2) we can expect a proportion of recombinants containing any particular cDNA to be in phase. The vector can accept up to 8.3 kb of insert DNA and complete cDNA libraries containing large numbers of independent recombinants can be constructed readily because of the efficiency endowed by packaging *in vitro*.

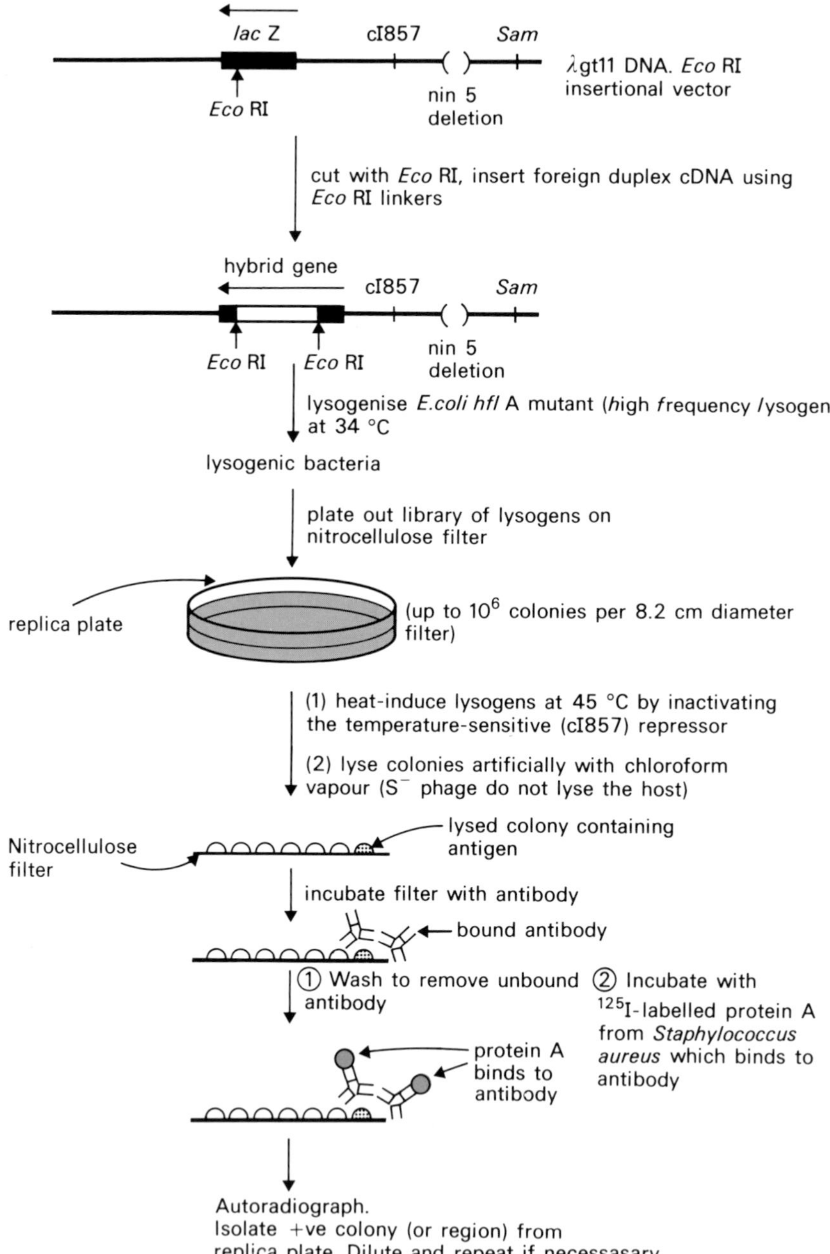

Fig. 6.2 Immunochemical screening applied to the expression vector λgt11 and its derivatives. The duplex cDNA is inserted within the *lac* Z gene. In a proportion of recombinants the insertion will be in the correct translational reading phase so as to direct the synthesis of a hybrid protein which will be detected by reaction with antibody raised

Immunochemical screening of the library is achieved by lysogeny of the phage library in *E. coli hfl* A (a *h*igh *f*requency *l*ysogeny mutant). The lysogens produce detectable amounts of hybrid protein upon induction (Fig. 6.2). Up to 10^6 colonies can be screened on a single 8.2 cm diameter filter.

Nucleic acid hybridization methods

Various recombinant detection methods employing hybridization with DNA isolated and purified from the transformed cells have been developed. However, these have been almost entirely superseded by the method of Grunstein and Hogness (1975) who have developed a screening procedure to detect DNA sequences in transformed colonies by hybridization *in situ* with radioactive 'probe' RNA. Their procedure can rapidly determine which colony amongst thousands contains the required sequence. A modification of the method allows screening of colonies plated at a very high density (Hanahan & Meselson 1980).

The colonies to be screened are first replica plated onto a nitrocellulose filter disc that has been placed on the surface of an agar plate prior to inoculation (Fig. 6.3). A reference set of these colonies on the master plate is retained. The filter bearing the colonies is removed and treated with alkali so that the bacterial colonies are lysed and their DNAs are denatured. The filter is then treated with proteinase K to remove protein and leave denatured DNA bound to the nitrocellulose, for which it has a high affinity, in the form of a 'DNA-print' of the colonies. The DNA is fixed firmly by baking the filter at 80 °C. The defining, labelled RNA is hybridized to this DNA and the result of this hybridization is monitored by autoradiography. A colony whose DNA-print gives a positive autoradiographic result can then be picked from the reference plate.

Variations of this procedure can be applied to phage plaques

---

against the required protein. The functions of the *c*I and S genes are discussed in Chapter 4. The *hfl* A mutation of *E. coli* results in a very high frequency of lysogenization by phage lambda. These lysogens express detectable amounts of hybrid protein when they are induced by raising the temperature so as to inactivate the temperature-sensitive *c*I repressor carrying the c1857 mutation. The frequency of lysogenization is high on the *hfl* A strain, but some non-lysogens will be present on the filter. The procedure can be improved by incorporating a drug resistance marker, e.g. ampicillin resistance (Kemp *et al.* 1983) or kanamycin resistance, borne on transposon Tn5, into the vector. Lysogenic bacteria can then be selected in the presence of the drug.

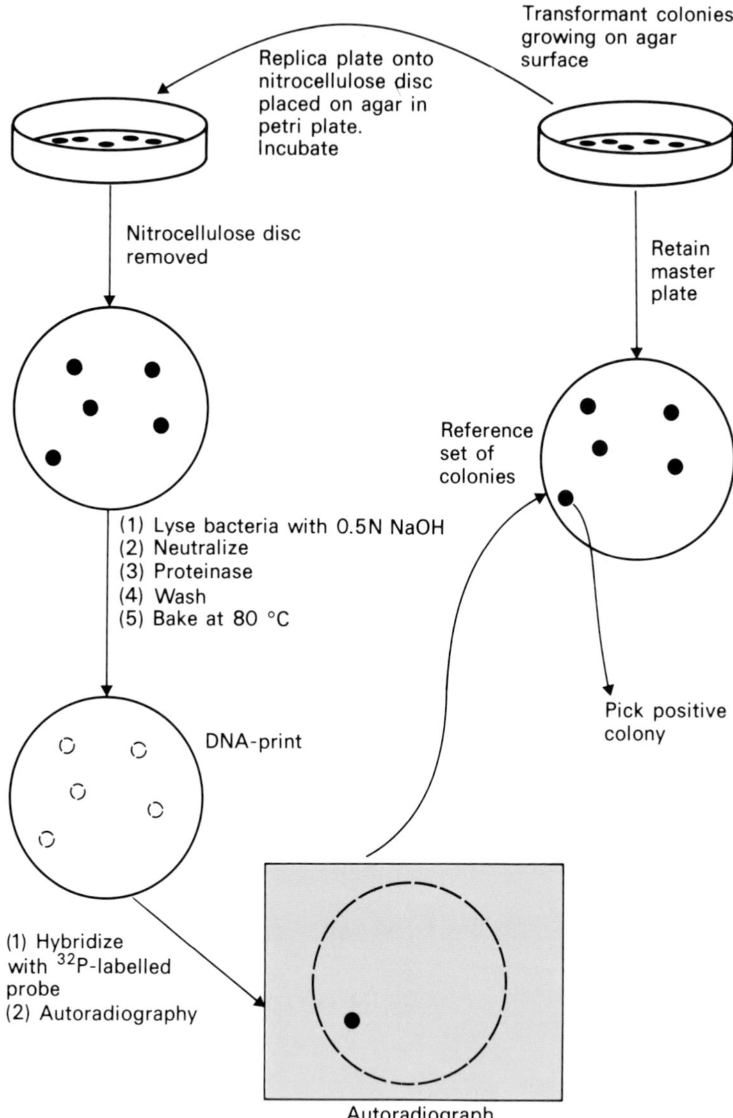
Transformant colonies growing on agar surface
Replica plate onto nitrocellulose disc placed on agar in petri plate. Incubate
Nitrocellulose disc removed
Retain master plate
Reference set of colonies
(1) Lyse bacteria with 0.5N NaOH
(2) Neutralize
(3) Proteinase
(4) Wash
(5) Bake at 80 °C
DNA-print
Pick positive colony
(1) Hybridize with 32P-labelled probe
(2) Autoradiography
Autoradiograph

Fig. 6.3 Grunstein-Hogness method for detection of recombinant clones by colony hybridization.

(Jones & Murray 1975, Kramer *et al.* 1976). Benton and Davis (1977) devised a method in which the nitrocellulose filter is applied to the upper surface of agar plates, making direct contact between plaques and filter. The plaques contain a considerable amount of unpackaged recombinant DNA which binds to the filter and can be denatured, fixed, hybridized, etc.

This method has the advantage that several identical DNA prints can easily be made from a single phage plate: this allows the screening to be performed in duplicate, and hence with increased reliability, and also allows a single set of recombinants to be screened with two or more probes.

The great advantage of the hybridization method is its generality. It does not require expression of the inserted sequences and can be applied to any sequence provided a suitable radioactive probe is available. As originally described, the method employed ^{32}P-labelled RNA, such as ribosomal

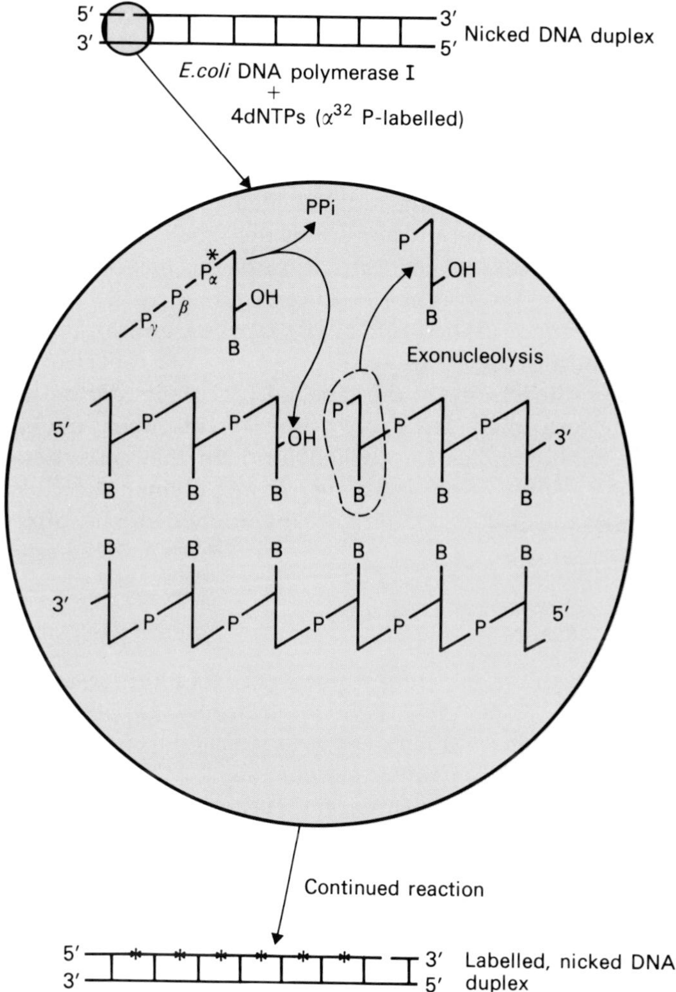

Fig. 6.4 ^{32}P-labelling of duplex DNA by nick-translation. Asterisks indicate radiolabelled phosphate groups.

RNA. However with slight modification it can be applied to ^{32}P-labelled DNA probes. Thus, for instance, a 'library' of human DNA fragments cloned in the Charon 4A phage vector was screened for clones containing globin sequences using ^{32}P-labelled denatured DNA of plasmids containing human globin cDNA sequences as probe (Lawn *et al.* 1978).

The DNA probe used in hybridizations of this kind is most conveniently labelled by a process known as nick-translation (Rigby *et al.* 1977). This somewhat confusing term does not relate to protein synthesis (translation), but rather to the translation or movement of a nick along a duplex DNA molecule (Fig. 6.4). Nicks can be introduced at widely separated sites in DNA by very limited treatment with DNase I. At such a nick, which exposes a free 3′-OH group, DNA polymerase I of *E. coli* will incorporate nucleotides successively. Concomitant hydrolysis of the 5′-terminus by the 5′→3′ exonucleolytic activity of polymerase I releases 5′-mononucleotides. If the four deoxynucleoside triphosphates are radiolabelled (e.g. α^{32}P-dNTPs), the reaction progressively incorporates the label into a duplex that is unchanged except for translation of the nick along the molecule. Because the original nicking occurs virtually at random, a DNA preparation is effectively labelled uniformly to a degree depending upon the extent of nucleotide replacement and specific radioactivity of the labelled precursors. Often the reaction is performed with only one of the four deoxynucleoside triphosphates in labelled form. In that case, non-uniform base composition, especially homopolymer stretches, will lead to a non-uniform pattern of labelling which may affect subsequent applications.

The probe problem

Screening procedures which rely on nucleic acid hybridization are, as we have seen, general in application and powerful. Using these procedures it is now possible easily to isolate any gene sequence from virtually any organism, *if a probe is available*. The problem of gene isolation then is a problem of obtaining a suitable probe. There are several solutions to this problem.

1 In certain specialized cell types or tissues particular mRNAs are abundant or super-abundant. The corresponding cDNA clones can be isolated by screening small numbers of recombinants directly by such methods as shotgun sequencing or HRT (see below). These cDNA probes can then be used to isolate genomic sequences.

2 Nucleic acid sequences encoding certain proteins have been sufficiently conserved in evolution such that cross-species nucleic acid hybridization is possible. Examples of effective heterologous probes include histone sequences [sea urchin *vs.* *Xenopus* (Old, R. W. *et al.* 1982)], and actin sequences [*Dictyostelium vs. Xenopus* (Cross *et al.* 1984)]. This means that the particular biological advantages of one experimental system can be exploited to isolate a gene sequence, which may then provide a probe for corresponding genes in other organisms.

3 An oligonucleotide, which needs to be only about 15–20 nucleotides long (see Chapter 13), and which corresponds to a part of the sequence encoding the protein in question, can be synthesized chemically. This requires that short, oligopeptide, stretches of amino acid sequence must be known. The DNA sequence (or its complement) encoding the protein can then be deduced from the genetic code. However, a problem arises here because of the degeneracy of the code. Most amino acids are encoded by more than one codon. For this reason, oligopeptide sequences known to contain tryptophan or methionine residues are particularly valuable, because these two amino acids have single codons, and the number of possibilities is thereby reduced. Thus, for example, the oligopeptide His—Phe—Pro—Phe—Met may be identified and chosen to provide a probe sequence. This oligopeptide could be encoded by the following sequences:

$$5'\text{- }\ \mathrm{CA}{}^{\mathrm{T}}_{\mathrm{C}}\mathrm{TT}{}^{\mathrm{T}}_{\mathrm{C}}\mathrm{CC}\begin{smallmatrix}\mathrm{T}\\ \mathrm{C}\\ \mathrm{A}\\ \mathrm{G}\end{smallmatrix}\mathrm{TT}{}^{\mathrm{T}}_{\mathrm{C}}\mathrm{ATG}\quad 3'$$

These 32 different sequences do not have to be synthesized individually. It is possible to perform a mixed addition reaction for each polymerization step where the sequence is degenerate. Therefore only one, mixed, probe is prepared and radiolabelled.

Plus and minus screening

This is a variant of the nucleic acid hybridization method that is particularly suitable for isolating tissue-specific or developmentally regulated cDNA sequences or clones derived from mRNAs that are induced by particular treatments.

Let us consider, for example, the isolation of cDNAs derived from mRNAs which are abundant in the gastrula embryo of the frog *Xenopus* but which are absent, or present at low abundance, in the egg. A cDNA clone library is prepared from gastrula mRNA. Replica filters carrying identical sets of recombinant clones are then prepared (Fig. 6.5). One of these filters is then probed with ^{32}P-labelled mRNA (or cDNA) from gastrula

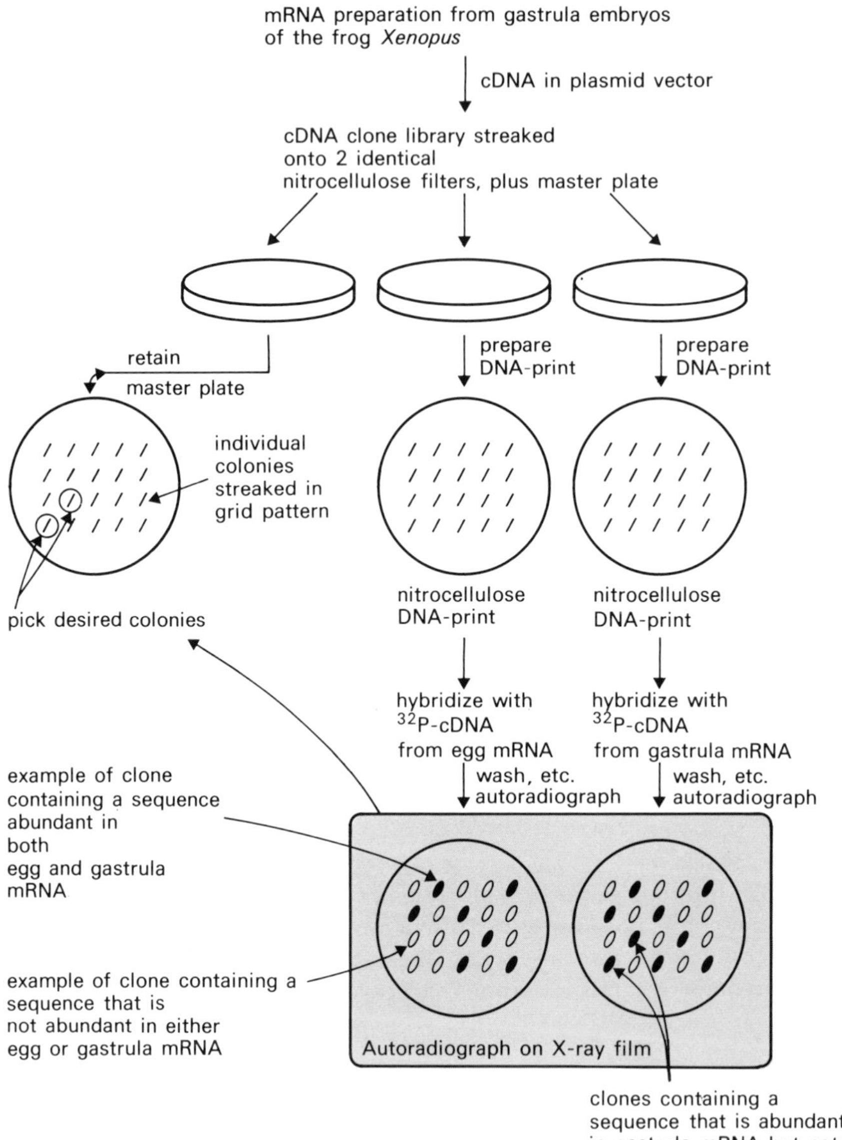

Fig. 6.5 Plus and minus screening. See text for details.

embryos, and one with ^{32}P-labelled mRNA (or cDNA) from the egg. Some colonies will give a positive signal with both probes, these represent cDNAs derived from mRNA types that are abundant at both stages of development. Some colonies will not give a positive signal with either probe, these correspond to mRNA types present at undetectably low abundance in both tissues. This is a feature of using probes derived from mRNA populations: only abundant or moderately abundant sequences in the probe carry a significant proportion of the label and are effective in hybridization. Importantly, some colonies give a positive signal with the gastrula probe, but not with the egg probe. These should, therefore, correspond to the required sequences.

Such plus and minus screening has been applied to the analysis of the development of *Xenopus* by Dworkin and Dawid (1980) and to the slime mould *Dictyostelium* by Williams and Lloyd (1979). In both instances it was estimated that a 5-fold difference in mRNA abundance could be detected in this procedure.

Recombinational probe

This ingenious and powerful method is based upon homologous recombination in the *E. coli* host (Seed 1983). The probe sequence here is inserted into a specially constructed plasmid vector, πVX. This is a very small plasmid of 902 bp, derived from the Col E1 replicon, which contains a convenient polylinker sequence and a suppressor tRNA gene, *sup* F (Fig. 6.6). Genomic phage lambda libraries are propagated on recombination-proficient *E. coli* containing the probe-πVX recombinant plasmid. Phage carrying sequences homologous to the probe acquire an integrated copy of the plasmid by homologous recombination. Phage bearing integrated probe-πVX can then be recovered and isolated by growth under appropriate selective conditions. This is most easily achieved by using a phage λ vector carrying an amber mutation suppressible by *sup* F (e.g. EMBL3 derivatives, see Chapter 4). By finally plating on a non-suppressing *E. coli* only those phage that have integrated a *sup* F gene, by virtue of homology with the probe, can form plaques.

This method can be applied readily to very large numbers ($>5 \times 10^6$) of phage in a genomic library, and has the advantage of being very quick, providing that the probe-πVX recombinant plasmid has been constructed at a prior stage. The shortest probe segment giving high recombination has been found to be about 60 bp long. A certain amount of sequence divergence can be tolerated in the homologous recombination event, but it has been shown that the probe does discriminate between a perfectly homologous sequence and one with 8.2% sequence divergence (Seed 1983).

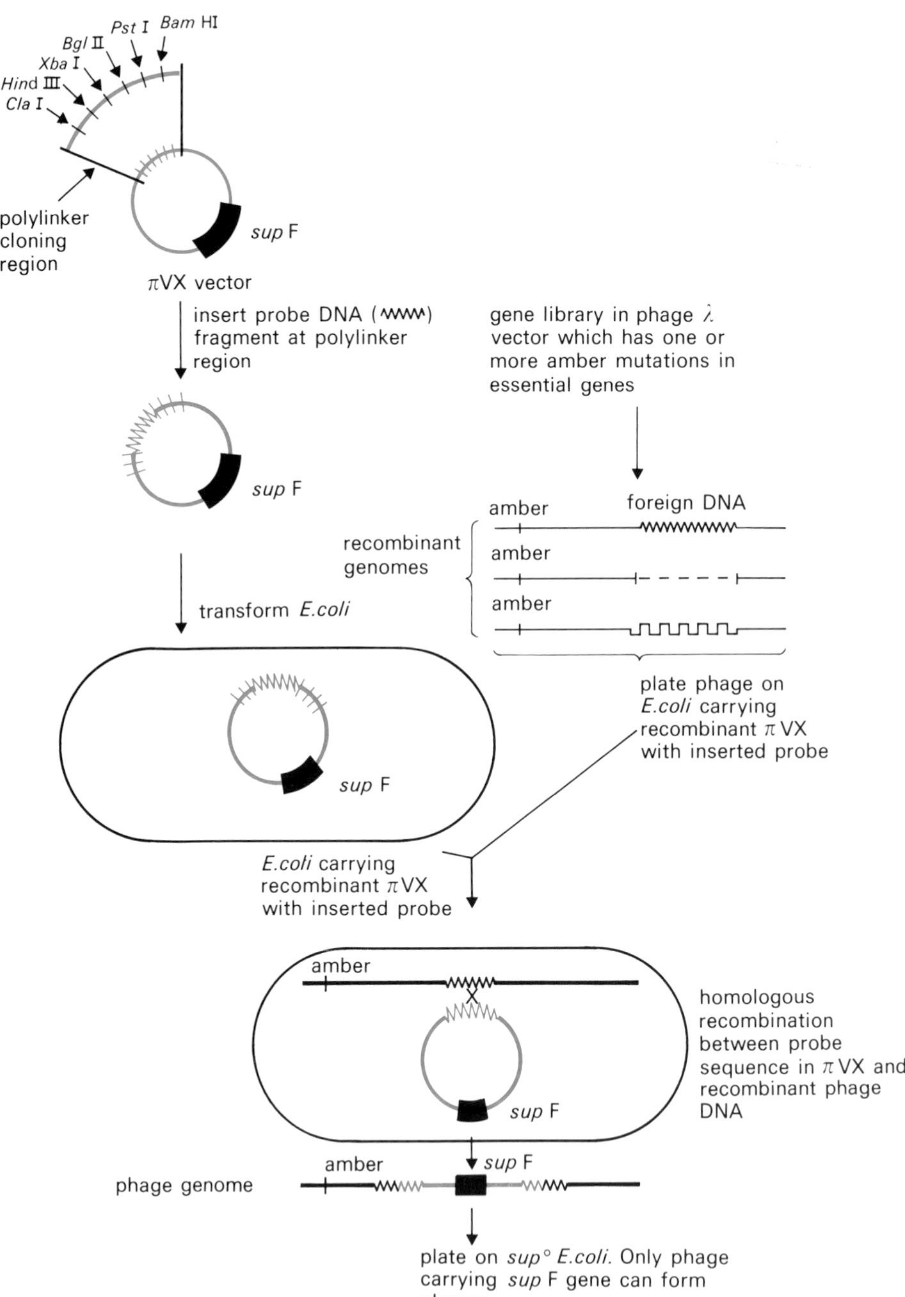

Fig. 6.6 Recombinational probing. See text for details.

Hybrid released translation (HRT) and hybrid arrested translation (HART)

These methods enable a cloned DNA to be correlated with the protein(s) which it encodes. Hybrid *released* translation is a direct method in which cloned DNA is bound to a cellulose nitrate filter and hybridized with an unfractionated preparation of mRNA or even total cellular RNA. The filter is then washed and hybridized messenger is eluted by heating in low salt buffer or in a buffer containing formamide. The recovered mRNA is then translated in a cell-free translation system such as that derived from wheat germ or rabbit reticulocytes, and radiolabelled polypeptide products whose synthesis it directs are analysed on an appropriate gel electrophoresis system (Fig. 6.7).

Hybrid *arrested* translation is based upon the fact that an mRNA will not direct the synthesis of protein in a cell-free translation sytem when it is in a hybrid form with its DNA complement (Paterson *et al.* 1977). In a model experiment, rabbit globin mRNA, which is a mixture of mRNA species encoding α- and β-globin polypeptides, was mixed with denatured DNA of the recombinant plasmid pβG1 which contains a rabbit β-globin cDNA sequence. Conditions were chosen so that DNA–RNA hybridization was favoured whilst suppressing re-association of the plasmid DNA (i.e. high formamide concentration). The nucleic acids were recovered from the hybridization mixture, and added to a cell-free translation system prepared from wheat germ. ^{35}S-methionine was included so that synthesis of radioactive polypeptides could be analysed by polyacrylamide gel electrophoresis and autoradiography. It was apparent that the pre-hybridized mRNA was rendered inactive by hybridization with its DNA complement. Full translational activity of the β-globin mRNA was recovered when the pre-hybridized mRNA preparation was dissociated by a brief heating before addition to the cell-free system.

HRT and HART are powerful methods. Unfractionated mRNA preparations from most cell types direct the synthesis of several hundreds of distinct abundant polypeptides that can be resolved on 2-dimensional electrophoresis systems. It should be quite feasible to isolate recombinant cDNAs derived from such a mixed population of mRNAs, and correlate them with their corresponding polypeptides.

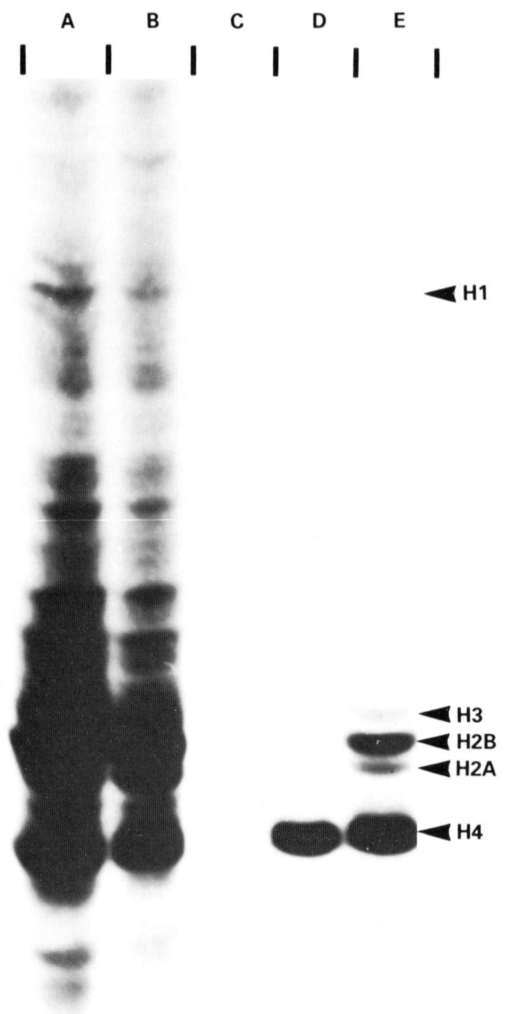

Fig. 6.7 *Hybrid Released Translation*. Application of HRT to two cloned DNAs containing histone sequences of *X. laevis*. The figure shows an autoradiograph of a polyacrylamide gel analysis of radiolabelled polypeptides synthesized in a wheat germ cell-free translation system with ^{3}H-lysine. The tracks correspond to the following RNA additions to the cell-free system: A and B, total RNA from *X. laevis* ovary; C, no added RNA; D, RNA released from a filter bearing histone H4 cDNA sequences cloned in pAT153; E, RNA released from a filter bearing H3, H2B, H2A and H4 sequences clustered in a genomic DNA fragment cloned in λWES. These filters had been hybridized with total ovary RNA. Positions of marker histones are indicated.

Chapter 7 Expression in *E. coli* of Cloned DNA Molecules

Following the demonstration (see p. 53) that a gene (Ap^R) originating from *S. aureus* can function in an unrelated bacterium, *Escherichia coli*, it was widely assumed that genes from any bacterium could be expressed in any other. This belief was based on the expectation that, in parallel with the universality of the genetic code, the other parts of the gene-to-phenotype biochemical pathway would also be universal. This idea was strengthened by the observation that genes from two lower eukaryotes, *Saccharomyces cerevisiae* (Struhl *et al.* 1976) and *Neurospora crassa* (Vapnek *et al.* 1977) are also expressed in *E. coli*. For that reason the genes from higher organisms were also expected to be expressed in bacteria. However, these favourable reports were followed by a whole series which indicated that many cloned genes were not expressed in their new genetic background. An explanation of these failures is provided by consideration of the steps involved in the gene-to-phenotype pathway.

Synthesis of a functional protein depends upon transcription of the appropriate gene, efficient translation of the mRNA and, in many cases, post-translational processing and compartmentalization of the nascent polypeptide. A failure to perform correctly any one of these processes can result in the failure of a given gene to be expressed. Transcription of a cloned insert requires the presence of a promoter recognized by the host RNA polymerase. Efficient translation requires that the mRNA bears a ribosome binding site. In the case of an *E. coli* mRNA a ribosome binding site includes the translational start codon (AUG or GUG) and a sequence that is complementary to bases on the 3′ end of 16*s* ribosomal RNA. Shine and Dalgarno (1975) first postulated the requirement for this homology, and various S-D sequences, as they are often called, have been found in almost all *E. coli* mRNAs examined. Identified S-D sequences vary in length from 3–9 bases and precede the translational start codon by 3–12 bases (Steitz 1979). A common post-translational modification of proteins involves cleavage of a *signal* sequence whose function is to direct the passage of the protein through the cell membrane (Blobel & Doberstein 1975, Inouye & Beckwith 1977). Another phenomenon, possibly related to the processing of proteins, is their degradation. It is known that the short polypeptides encoded by genes which have undergone nonsense mutations are rapidly degraded in *E. coli* while the wild-type proteins are stable. It can be envisaged that the foreign proteins would be rapidly degraded in the new host if their configuration or amino acid sequence did not protect them from intracellular proteases.

From the above discussion it is clear that the first requirement for expression in *E. coli* of a structural gene, inserted *in vitro* in a DNA molecule, is that the gene be placed under the control of an *E. coli* promoter. When this is done the protein product may be synthesized from its own N terminus but more often than not a fusion protein is produced. A few examples are given in Table 7.1 and some of them are discussed in more detail below. It should be noted that in some cases it may be possible to cleave the fusion protein *in vitro* to yield native protein and two examples have already been discussed: somatostatin and β-endorphin (see pp. 60 and 68).

Construction of vectors that give improved transcription of inserts

In order to place genes under the control of an *E. coli* promoter, Fuller (see Backmann *et al.* 1976) constructed a 'portable promoter' in the form of a DNA fragment known as *Eco* RI (UV5). This fragment carries the regulatory region of the *lac* operon and the UV5 mutation bounded by *Eco* RI generated cohesive ends. The UV5 mutation makes the system insensitive to catabolite repression. The fragment also carries the L8 mutation but this does not affect its use. The L8 mutation is only mentioned here for the sake of completeness and to avoid confusion when reading the literature. One of the *Eco* RI sites is located 65 base pairs downstream from the start point of transcription, a position corresponding to the ninth amino acid residue of the *lac* Z gene. Backmann *et al.* (1976) have constructed a plasmid in which this *Eco* RI (UV5) fragment is fused to the repressor gene (*c*I) of bacteriophage λ. Strains carrying this plasmid overproduced the repressor and transcription of the *c*I gene was largely under *lac* control.

Most of the expression vectors in current use are based on either the *lac* UV5 promoter, as described above, or on the

Table 7.1 Kinds of protein produced when cloned DNA fragments are linked to bacterial promoters.

Kind of protein	Selected examples	Reference
Fusion protein	Rat growth hormone Ovalbumin Fibroblast interferon	Seeburg *et al.* 1977 Mercerau-Puijalon *et al.* 1978 Houghton *et al.* 1980
Cleavable fusion protein	Somatostatin β-endorphin Human insulin	Itakura *et al.* 1977 Shine *et al.* 1980 Goeddel *et al.* 1979b
Native protein (Protein synthesized from its own N terminus)	SV40 t antigen Human growth hormone Mouse dihydrofolate reductase	Roberts *et al.* 1979a Goeddel *et al.* 1979a Chang *et al.* 1978

tryptophan (*trp*) promoter which controls the expression of the genes involved in tryptophan biosynthesis in *E. coli*. Examples of eukaryotic genes expressed in *E. coli* by means of one or other of these promoters are shown in Table 7.1. Two other controllable promoters which have been used are the λP_L promoter (see Chapter 4) and the *rec*A promoter (Feinstein *et al.* 1983, Shirakawa *et al.* 1984). The *rec*A gene of *E. coli* normally is repressed by the product of the *lex*A gene but is induced by substances such as nalidixic acid which damage DNA.

There are two tests which can be used to show that expression of a cloned gene is under the control of the selected *E. coli* promoter. First, expression of the foreign gene should be obtained when it is placed in the correct orientation relative to the promoter but not when it is in the wrong orientation. Second, expression of the gene should be enhanced by those environmental conditions which normally lead to activation of the promoter, e.g. addition of an inducer such as IPTG for *lac*-based expression vectors or starvation for tryptophan with *trp*-based vectors.

Positioning cloned inserts in the correct translational reading frame: Expression of fusion proteins

When expression of a foreign gene is dependent on its fusion to an *E. coli* gene it is important that the correct translational reading frame is maintained. There is only a one in three chance that two randomly selected fragments will be fused in phase. Thus for most gene fusions adjustments would have to be made at the gene junction. To avoid this laborious procedure Charnay *et al.* (1978) constructed a set of vectors, each having a single *Eco* RI restriction site, in which the cloned gene can be placed in each of the three possible reading frames relative to the translation initiation site of the *lac*Z gene. They started with a λ*lac* derivative (λΔZUV5) having a unique *Eco* RI site in the ninth codon of the β-galactosidase gene. The reading frame of this *Eco* RI site was defined arbitrarily as ϕ1. λΔZ vectors in which the inserted fragments can be positioned in phases ϕ2 and ϕ3 (λΔZ2 and λΔZ3) were constructed by two successive additions of 2G.C. base pairs to each of the *Eco* RI (UV5) fragments (Fig. 7.1).

The procedure for adding 2 base pairs to each end of the 203 base pair *Eco* RI (UV5) fragment is shown in Fig. 7.2. The fragment was treated with S1 nuclease in order to produce flush ends with G.C as the final base pair. Using T4 DNA ligase a synthetic *Eco* RI linker then was attached to each end of the fragment and this was followed by digestion with *Eco* RI endonuclease. This resulted in the production of a 207 base pair fragment called *Eco* RI 207 (UV5). Using the same procedure, *Eco* RI 207 (UV5) was converted to *Eco* RI 211 (UV5). Both the 207 and 211 base pair fragments were then inserted into a phage λ vector. Using methods which are irrelevant to this discussion

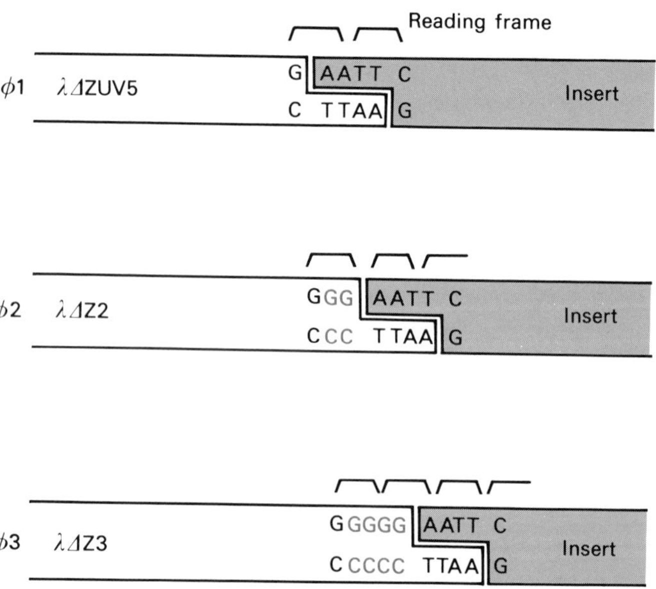

Fig. 7.1 The different reading frames with respect to the translation initiation site of the *lac* Z gene presented by the three λ vectors.

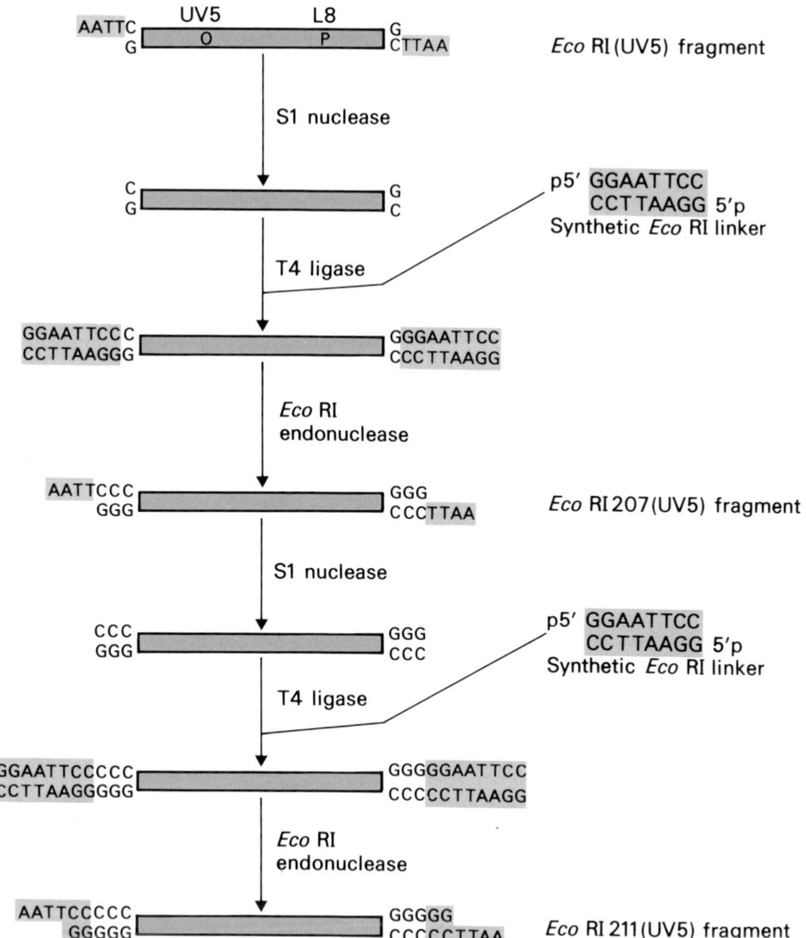

Fig. 7.2 Construction of the *Eco* RI 211 (UV5) fragment by two successive additions of two G-C base pairs to each end of the *Eco* RI (UV5) fragment.

the unwanted *Eco* RI sites upstream from the promoter on each fragment were removed to generate λΔZ2 and λΔZ3.

Tacon *et al.* (1980) using a slightly different methodology have constructed a similar set of plasmid vectors but in which the expression of the cloned inserts is under the control of the *trp* promoter. As with the *lac* vectors of Charnay *et al.* (1978) use of this set of vectors results in the production of a fused protein. These *trp*-based vectors are used widely and among the proteins expressed with them are influenza virus surface antigens (Emtage *et al.* 1980) and synthetic genes coding for poly(L-aspartyl-L-phenylalanine) (Doel *et al.* 1980) and urogastrone (Smith *et al.* 1982).

An alternative strategy for obtaining fusion proteins: Cloning in the *Pst* I site of pBR322

An alternative method of putting an insert under the control of an *E. coli* promoter without the concomitant problem of translational reading frame adjustment has been used by Villa-Komaroff *et al.* (1978). They cloned cDNA transcripts of rat preproinsulin mRNA at the unique *Pst* I site of pBR322 (see Fig. 3.5). This site lies in the Ap^R gene at a position corresponding to amino acid residues 183 and 184. Consequently, an insert at the *Pst* I site should result in the production of a fused gene product. To effect this insertion, use was made of the homopolymer tailing procedure (Fig. 7.3). The advantage of this method is that in each recombinant molecule the lengths of the repeating G-C base-paired joints may be different and at least some of them will be in the correct reading frame.

Manipulation of cloned genes to achieve expression of native proteins

So far we have described methods for placing a cloned DNA fragment under the control of an *E. coli* promoter. In many instances these methods result in the production of a hybrid protein carrying amino-terminal amino acids from β-lactamase, β-galactosidase or the *trp*E gene product. It would be more satisfactory if the gene-promoter fusion always produced a native protein.

Tacon *et al.* (1983) have described the construction of two plasmid vectors which facilitate the expression of native proteins. Both make use of the *trp* promoter and associated S-D sequence. In one of them, pWT551, the gene to be cloned is inserted at a *Hin*d III site but expression requires the presence of an initiator ATG or GTG codon at the start of the coding region. The second vector, pWT571, has an *Eco* RI site overlapping the initiation codon (ATG AATTC). Cloning a gene in the correct reading frame in this site will result in expression of a protein having the sequence fMet—Asn at the N-terminus. Alternatively, digestion with endonuclease *Eco* RI followed by S1 nuclease to remove the 5′-AATT extensions permits blunt end ligation to the initiation codon of a fragment encoding a protein lacking N-formylmethionine.

Secretion of foreign proteins

β-lactamase, the product of the Ap^R gene of pBR322, is a periplasmic protein. It is synthesized as a pre-protein with a 23 amino acid leader sequence which serves as a signal to direct the secretion of the protein to the periplasmic space. This signal sequence is removed as the protein traverses the membrane. Insertion of a foreign gene into the Ap^R gene should permit

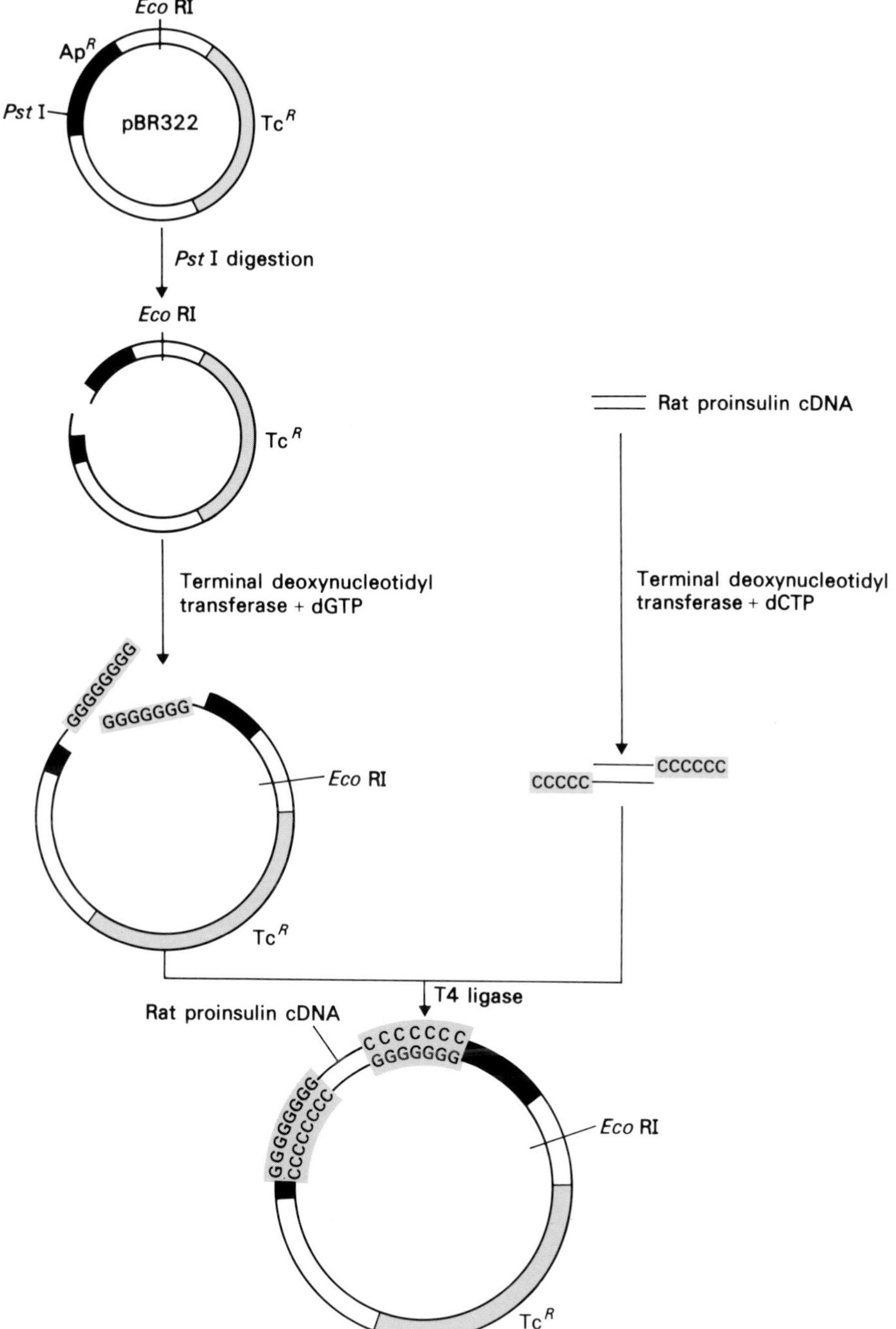

Fig. 7.3 Insertion of rat proinsulin cDNA at the *Pst* I site of pBR322.

expression of the foreign sequence as a fusion product which is transported to the periplasmic space. The insertion of the rat proinsulin gene at the *Pst* I site of pBR322 (see p. 132) provides a good test of this hypothesis. As expected, the β-lactamase-proinsulin fusion protein was found in the periplasmic space (Villa-Komaroff *et al.* 1978).

Talmadge and Gilbert (1980) have constructed a series of plasmids in which the unique *Pst* I site of pBR322 has been moved so that it lies within or near the pre-β-lactamase signal sequence. The genes for rat proinsulin and rat preproinsulin were cloned into the *Pst* I sites of these different plasmids resulting in the formation of hybrid β-lactamase (prokaryotic) and insulin (eukaryotic) signal sequences. By comparing the levels of insulin antigen in the *E. coli* periplasmic space, Talmadge *et al.* (1980) were able to deduce the structural requirements of a signal sequence and to show that a eukaryotic signal sequence can function in a prokaryote.

The experiments described above show that a protein which normally is secreted by a eukaryote can also be secreted by *E. coli*. The question remains, can a protein which normally is cytoplasmic be engineered in such a way that it is secreted? Analysis of the signal sequences from a number of secreted proteins indicated a number of features common to all of them. At the N-terminus they have a run of polar amino acid residues, including one which is positively charged, followed by a 'core' of 10–15 hydrophobic residues. However, more than just the signal sequence may be required for secretion. Koshland and Botstein (1982) isolated chain-terminating mutants of the Ap^R gene. Cells carrying these mutant genes synthesized truncated β-lactamases which did not appear in the periplasmic space as usual but remained in the cytoplasmic membrane. Thus at least the carboxy-terminus of β-lactamase plays an important role in the secretory process.

Further evidence for a role of the structural sequence of the protein in determining its cellular location comes from the work of Tommassen *et al.* (1983). They constructed a hybrid gene which encoded the signal sequence as well as 158 amino acids of β-lactamase and the complete structural sequence of the outer membrane PhoE protein. The fusion protein expressed by this gene was found in the outer membrane rather than the periplasm. It is not always possible to obtain secretion of a protein which normally is cytoplasmic and Ito *et al.* (1981) have shown that this is the case for β-galactosidase.

Stability of foreign proteins in *E. coli*

There are a number of examples of the instability of cloned gene products. Details of the degradation of somatostatin were given earlier (see p. 64). Human fibroblast pre-interferon and interferon also are known to be unstable (Taniguchi *et al.* 1980). Talmadge and Gilbert (1982) have shown that cellular location can affect stability. They measured the half-life of rat proinsulin in *E. coli* and found it to be 2 minutes when located in the cytoplasm but over 10 times greater in the periplasm.

Various strategies have been developed to cope with the instability of foreign proteins in *E. coli*. In the case of somatostatin, degradation was prevented by producing a fused protein consisting of somatostatin and β-galactosidase and subsequently cleaving off the somatostatin with cyanogen bromide. Although there are a number of examples in the literature of the stabilization of proteins in this way, the disadvantage is that the subsequent production of a native protein *in vitro* is not always possible (see p. 67). An alternative approach is to use certain mutants of *E. coli* which have a reduced complement of intracellular proteases. One particularly useful mutant lacks the *lon* protease, a DNA-binding protein with ATP-dependent proteolytic activity. A quite different strategy has been developed by Simon *et al.* (1983). It is based on the observation that during bacteriophage T4 infection of *E. coli* turnover of normal cellular proteins continues but there is a marked decrease in the degradation of abnormal proteins (e.g. fragments produced as a result of nonsense mutations). They cloned the T4 gene responsible for this effect and called it *pin* for *p*rotease *in*hibition. They found that labile eukaryotic proteins, e.g. fibroblast interferon, encoded by genes cloned in *E. coli* were stabilized in cells in which the T4 *pin* gene was expressed.

Detecting expression of cloned genes

When it is known that a cloned DNA fragment codes for a particular protein or RNA species then suitable tests can be devised to determine if that gene is expressed in its new host. Thus somatostatin or insulin can be detected by sensitive radioimmunoassays, 5*s* RNA by hybridization to suitable probes and enzymes either by appropriate assay procedures or by complementation of auxotrophic mutations in a suitable host (see p. 113). But how does one detect expression of cloned genes when the function of these genes is not known? The answer is to look for the synthesis of novel proteins or RNA species in cells carrying the recombinant molecules. However, with wild-type cells the detection of new proteins or RNA species is almost impossible because of the concurrent expression of the host genome. Three approaches have been devised to circumvent this problem. Two of them are in-vivo methods: expression of plasmid-borne genes can be detected by the use of *mini-cells* or by UV irradiation of the host cell. The latter method is known as the *maxi-cell* method and is also applicable to genes cloned in λ vectors. The third method is a coupled in-vitro transcription and translation system.

Mini-cells are small, spherical, anucleate cells which are produced continuously during the growth of certain mutant strains of bacteria. Because of their size difference, mini-cells and normal-sized cells can be separated easily on sucrose gradients. Mini-cells purified in this way from plasmid-free parents can be shown to contain normal amounts of protein and RNA but to lack DNA. *In vivo*, these mini-cells do not incorporate radioactive precursors into RNA or protein. By contrast, mini-cells produced from plasmid-carrying parents contain significant amounts of plasmid DNA. Plasmid-containing mini-cells are capable of RNA and protein synthesis and would appear to be ideal for detecting the expression of genes carried by recombinant plasmids.

In general, there is a correlation between the genotype of a plasmid and the polypeptides synthesized by mini-cells containing that plasmid. However, there are complications since the introduction of deletions or DNA insertions into plasmids can have unpredictable results; for example, a 1 megadalton deletion in a Col E1-derived plasmid prevented the synthesis in mini-cells of polypeptides of 56 000, 42 000, 30 000 and 28 000 daltons. The explanation of this result was apparent after the demonstration that the latter three polypeptides were degradation products of colicin E1 (Meagher *et al.* 1977b). The insertion into a kanamycin-resistance gene of a DNA fragment containing the *Eco* RI methylase gene caused the synthesis in mini-cells of *Eco* RI methylase and two other polypeptides. Again, this result could be explained since it was known that the kanamycin-resistance protein was inactivated by the insertion and that there was a portion of another gene on the inserted DNA fragment. These facts permit the prediction that two new polypeptides would be produced. When the DNA insertion is of unknown function, e.g. randomly cleaved eukaryotic DNA sequences, interpretation of the expression of the DNA fragments in mini-cells could still be complicated.

The maxi-cell method (Sancar *et al.* 1979) is based on two

observations. First, when irradiated with UV light, *E. coli rec*A *uvr*A cells stop DNA synthesis and chromosomal DNA is extensively degraded so that only a small amount remains several hours later. Second, if these cells contain a Col E1-like multicopy plasmid (e.g. pBR322), the plasmid molecules that do

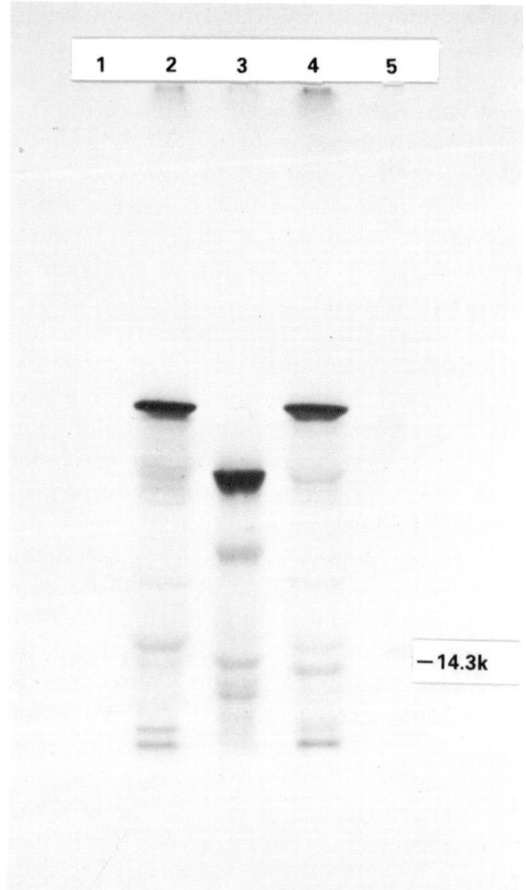

Fig. 7.4 Autoradiograph produced after transcription and translation of plasmid DNA *in vitro*. Plasmid DNA was transcribed and translated *in vitro* in the presence of ^{35}S methionine, the protein products separated by SDS-polyacrylamide electrophoresis and the individual proteins detected by autoradiography. Track 1, DNA-free control; track 2, plasmid pAT153; track 3, plasmid pACYC184,; track 4, plasmid pWT111; track 5, molecular weight marker protein. Tracks 2–4 were loaded with equivalent amounts of radioactivity (500 000 c.p.m.) and the gel was exposed to the photographic film for 3.5 h. The major band in tracks 2 and 4 is β-lactamase and in track 3 chloramphenicol acetyl transferase. Note that many truncated polypeptides are produced but that these are minor constituents.

not receive a UV hit continue to replicate with plasmid DNA levels increasing about 10-fold by 6 h post-irradiation. If a radioactive amino acid is added a few hours after irradiation, the vast majority of the proteins which are labelled are plasmid-encoded.

The expression of genes cloned in λ vectors can be studied by infection of bacteria previously irradiated so severely with UV light that their own gene expression is effectively eliminated by DNA damage. Under these conditions the products of the cloned genes are seen against a background of λ-specified proteins. If the expression of the DNA insert is independent of λ promoters then even this background can be eliminated by infecting UV-irradiated cells lysogenized with a non-inducible mutant of λ (Newman *et al.* 1979). Under these conditions sufficient λ repressor is present in the cells to prevent transcription of phage genes.

Bacterial cell-free systems for coupled transcription and translation of DNA templates were first described by Zubay and colleagues as early as 1967. More recently these have been modified (Pratt *et al.* 1981) to permit expression *in vitro* of genes contained on bacterial plasmids or bacteriophage genomes. Since DNA fragments generated by restriction endonuclease cleavage can be used to programme polypeptide synthesis, it is possible readily to assign polypeptides to small regions of the coding template. There are two other advantages to the system. First, incorporation of radioactive label into protein is far more efficient than is possible using in-vivo methods. This makes the system very sensitive and ^{35}S-labelled polypeptides can be identified by polyacrylamide gel electrophoresis after a few hours autoradiography (Fig. 7.4). Second, DNA from other prokaryotes can be expressed efficiently. This is particularly useful when cloning in organisms other than *E. coli* in which the *mini-cell* and *maxi-cell* techniques have not been developed.

MAXIMIZING THE EXPRESSION OF CLONED GENES

In the previous section we discussed those features of gene expression in *E. coli* which need to be considered if detectable synthesis of a cloned gene product is to be obtained. However, much of the interest surrounding in-vitro gene manipulation concerns the commercial applications, e.g. production of vaccines or human hormones in *E. coli*. Here detectable expression is not sufficient; rather it must be maximized. The key factors affecting the level of expression of a cloned gene are shown in Table 7.2, and, as will be seen in the following section, most of them exert their effect at the level of translation.

Table 7.2 Factors affecting the expression of cloned genes.

Promoter strength
Translational initiation sequences
Codon choice
Secondary structure of mRNA
Transcriptional termination
Plasmid copy number
Plasmid stability
Host cell physiology

Constructing the optimal promoter

A large number of promoters for *E. coli* RNA polymerase have been analysed and a recent compilation (Hawley & McClure 1983) gives the sequence of 168 of them. Clearly we would like to use this information to develop the most efficient promoter possible. Comparison of many promoters has led to the formulation of a consensus sequence which consists of the −35 region (5′-TTGACA-) and the −10 region or Pribnow box (5′-TATAAT), the transcription start point being assigned position +1. Of the four promoters used most widely in expression vectors none shows absolute identity with the consensus sequence (Fig. 7.5). In trying to identify the strongest promoter it is essential to have a measure of the relative efficiencies of the different candidates and a suitable system has been devised (Russell & Bennett 1982, de Boer *et al.* 1983a). The promoter to be tested is placed in front of a promoter-less galactokinase (*gal*K) gene carried on a plasmid and the level of galactokinase synthesized in a $GalK^-$ host used as a measure of promoter strength. Using the galactokinase system it was shown that promoter strength is directly proportional to the degree of similarity with the consensus sequence. Thus it is not surprising that compared with the *lac* promoter the two synthetic hybrid promoters *tac* I and *tac* II (Fig. 7.5) were eleven and eight times stronger respectively. Furthermore the distance between the −35 and −10 regions is important. Hawley and McClure have analysed a number of mutations which affect this spacing. In all cases the promoter was stronger if the spacing was moved closer to 17 base pairs and weaker if moved further away from 17 base pairs.

Because of the increased strength of the *tac* promoters relative to the *lac* and *trp* promoters, Amann *et al.* (1983) have cloned a *tac* promoter on a series of plasmid vectors that facilitate the expression of cloned genes. These vectors contain various cloning sites followed by transcription signals. In

		-35 REGION		-10 REGION	
CONSENSUS	• • •	TTGACA	1 2 3 4 5 6 7 8 9 10 11 12 13 14 15 16 17	TATAAT	• •
lac	GGC	TTTACA	CTTTATGCTTCCGGCTCG	TATATT	GT
trp	CTG	TTGACA	ATTAATCAT CGAACTAG	TTAACT	AG
λP_L	GTG	TTGACA	TAAATACCA CTGGCGGT	GATACT	GA
rec A	CAC	TTGATA	CTGTATGAA GCATACAG	TATAAT	TG
tac I	CTG	TTGACA	ATTAATCAT CGGCTCG	TATAAT	GT
tac II	CTG	TTGACA	ATTAATCAT CGAACTAG	TTTAAT	GT

Fig. 7.5 The base sequence of the −10 and −35 regions of four natural promoters and two hybrid promoters.

addition, Amann *et al.* (1983) have described plasmids that facilitate the conversion of the *lac* promoter to the *tac* promoter.

Maximal transcription of a cloned gene may require more than just a strong promoter. By fusing the *E. coli tyr*T promoter region to the galactokinase gene and using galactokinase activity as a measure of promoter strength Lamond and Travers (1983) were able to show that efficient expression from the *tyr*T promoter requires specific sequences upstream of the canonical promoter elements. However, it must be pointed out that the *tyr*T gene encodes a bacterial $tRNA^{tyr}$ and is subject to stringent control, i.e. expression is reduced during amino acid starvation. Indeed, Travers (1984) has compared the sequences of a number of promoters under stringent control and found conserved nucleotides immediately 3′ to the −10 region.

Optimizing translation initiation

As well as giving clues to the essential structural features of promoters, DNA sequence analysis has yielded considerable information about those factors which affect translational initiation. Details of these sequences can be found in the compilation of Stormo *et al.* (1982). In addition to the AUG initiation codon three more or less conserved structural features have been recognized. First and foremost is the S-D sequence which contains all or part of the polypurine sequence 5′ UAAGGAGGU 3′. Second, in polycistronic messengers at least, there may be a requirement for one or more termination codons in the ribosome binding site. Finally, the ribosome binding site of genes encoding highly expressed proteins, e.g. phage capsid proteins and ribosomal proteins, contains all or part of the sequence PuPuUUUPuPu.

Using a gene expression system which includes a portable Shine-Dalgarno region de Boer *et al.* (1983b) have examined the effect on translation of varying the sequence of the four bases that follow the SD region. The presence of four A residues or four T residues in this position gave the highest translational efficiency. Translational efficiency was 50% or 25% of maximum when the region contained, respectively, four C residues or four G residues.

The composition of the triplet immediately preceding the AUG start codon also affects the efficiency of translation. For translation of β-galactosidase mRNA the most favourable combinations of bases in this triplet are UAU and CUU. If UUC, UCA or AGG replaced UAU or CUU the level of expression was 20-fold less (Hui *et al.* 1984).

Based on this information the reader can be forgiven for thinking that it would be easy to construct a consensus ribosome binding site using oligonucleotide synthesis. The first problem is that for maximal expression of a cloned gene the ribosome binding site for that gene must be close to the promoter. The second problem is highlighted by experiments where the distance between the gene and the promoter was varied.

Gene-promoter separation: the effect of mRNA secondary structure

Cleavage of λ*plac* 5.1 DNA with a combination of *Eco* RI and *Alu* I restriction endonucleases produces a mixture of fragments one of which is 95 base pairs long and contains the *lac* promoter and β-galactosidase gene S-D sequence. Roberts *et al.* (1979b) constructed a series of recombinant plasmids in which this promoter-bearing fragment was located at varying distances in front of the λ*cro* gene. Nine different recombinant plasmids were selected, transformed into *E. coli* and the level of *cro* protein in each clone measured. The DNA sequence across each of the *lac–cro* fusion was also determined. The results obtained are summarized in Fig. 7.6. The most striking feature of these results was the enormous difference (>2000-fold) in the levels of *cro* protein produced by the different recombinants. Since the same promoter is being used in each plasmid it is uniform in each case. Iserentant and Fiers (1980) have constructed secondary structure models for the RNA transcripts produced by the plasmids of Roberts *et al.* (1979b). From an analysis of these structures they concluded that for good expression the initiation and, to a lesser extent, the S-D site

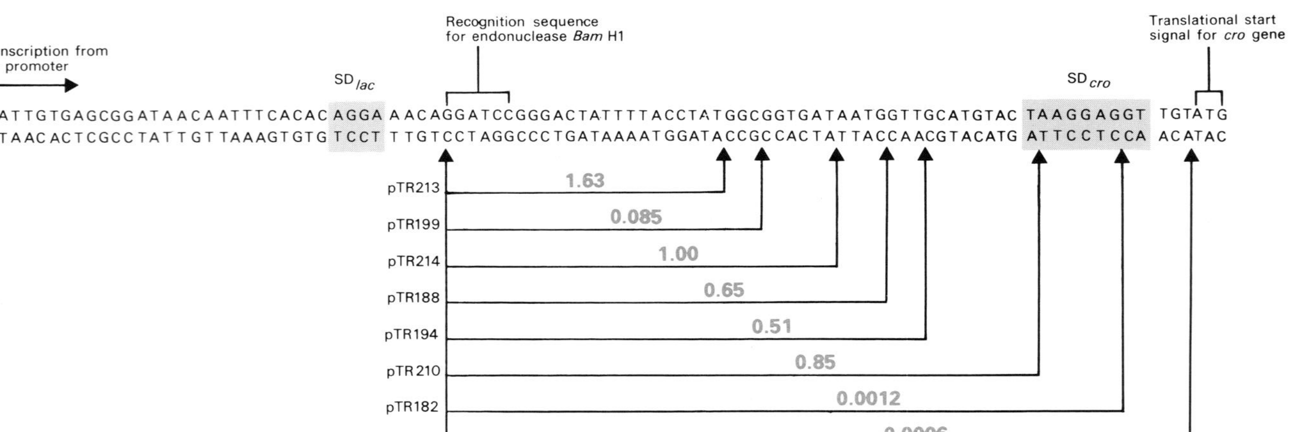

Fig. 7.6. Effect on *cro* protein production of gene promoter separation. Shown is a portion of the sequence of pTR161 extending from the *lac* promoter to the startpoint of translation of the *cro* gene. Also shown is the extent of the deletion in eight derivatives of pTR161. The figures on the brackets indicate the amount of *cro* protein synthesized relative to pTR161. The deletions were created after *Bam* HI digestion by the method outlined in Fig. 7.8.

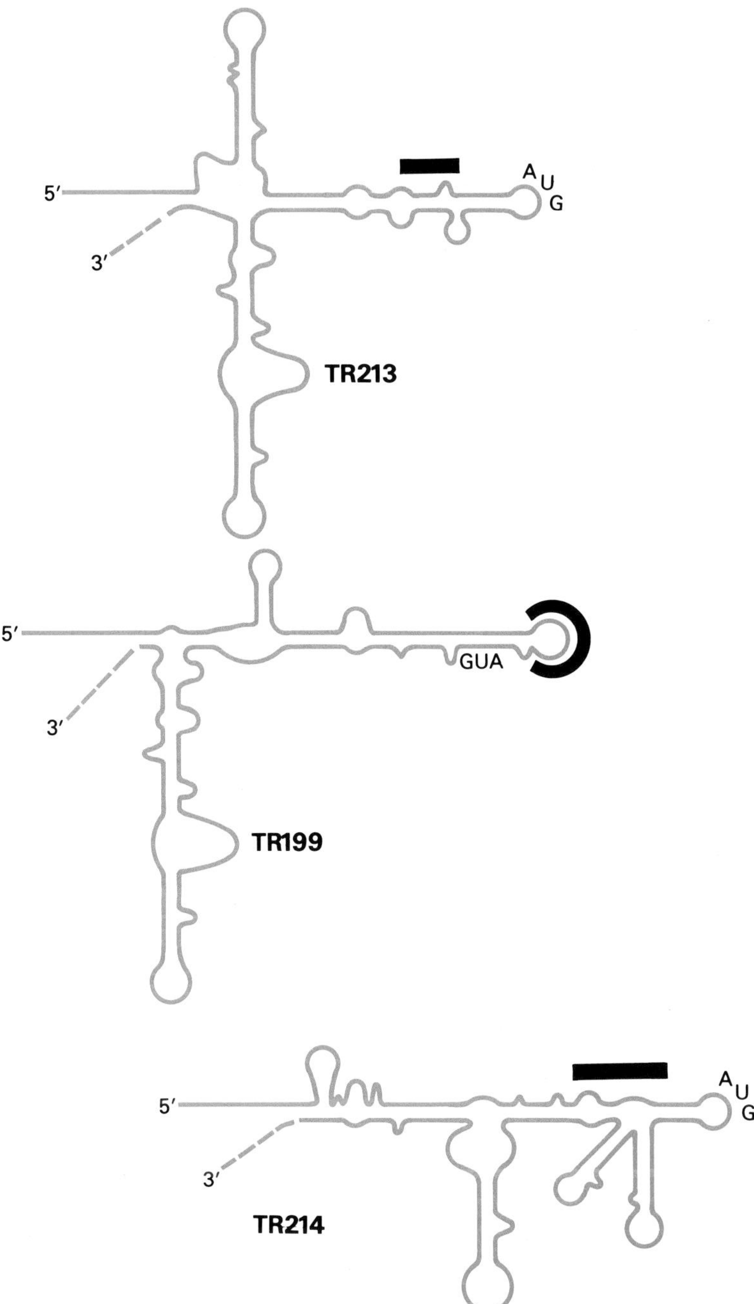

Fig. 7.7 Postulated secondary structure on the *cro* mRNA synthesized by three of the plasmids shown in Fig. 7.6. Note that for clone TR199, in which *cro* is poorly expressed, the initiation codon is base-paired in a stem structure. The solid bars indicate the location of the S-D sequence.

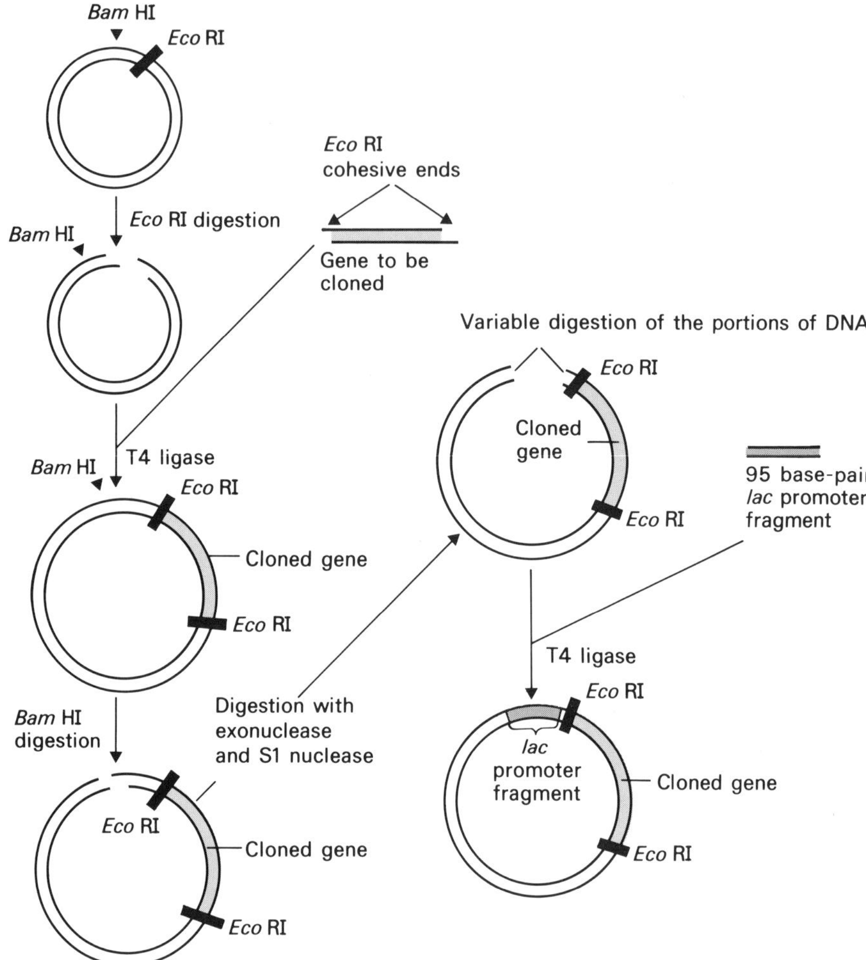

Fig. 7.8 General method for varying the distance between the *lac* promoter and a cloned gene. The example shown uses a hypothetical plasmid carrying unique closely spaced sites for *Eco* RI and *Bam* HI.

must be accessible (Fig. 7.7). Thus the initiation of translation involves interaction between an activated 30*s* ribosomal subunit and the 5′ terminal region of a mRNA *which is already folded* in a specific secondary structure. Confirmation that the sequence between the promoter and the ATG codon affects the level of gene expression has been provided by Chang *et al.* (1980), Shepard *et al.* (1982) and Gheysen *et al.* (1982). These latter workers were able to confirm the conclusions of Iserentant and Fiers (1980) that the varying levels of expression are a reflection of secondary structure of the mRNA. In addition,

Schottel *et al.* (1984) have shown that overlapping of the translation control region by translation initiated upstream, whether in phase or out of phase, greatly reduces the level of expression of the desired gene product.

Based on the above work a strategy for maximizing gene expression has been proposed: namely, a bank of clones is established in which the gene under study is placed at varying distances from the promoter. These recombinant clones then are screened and the one giving the highest yield is selected. Roberts *et al.* (1979b) have described a general method whereby this might be achieved (Fig. 7.8). The gene is cloned in a plasmid such that a unique restriction site is located within 100 base pairs of the 5′ end of the gene. The plasmid is opened at that site and varying amounts of DNA excised with exonuclease III and S1 nuclease. The promoter fragment is then inserted, in this case the 95bp *lac* promoter fragment, and the plasmid closed by T4 DNA ligase. This produces a set of plasmids bearing the promoter separated by varying distances from the cloned gene. However, there is a limitation with this method: its use is restricted to genes whose product can be assayed readily. This problem led Guarente *et al.* (1980) to devise a method to maximize expression of cloned genes in the absence of assays for their gene products.

The method in question makes use of a *lac* I–*lac* Z fusion in which the carboxyterminal end of the *lac* Z gene is fused in

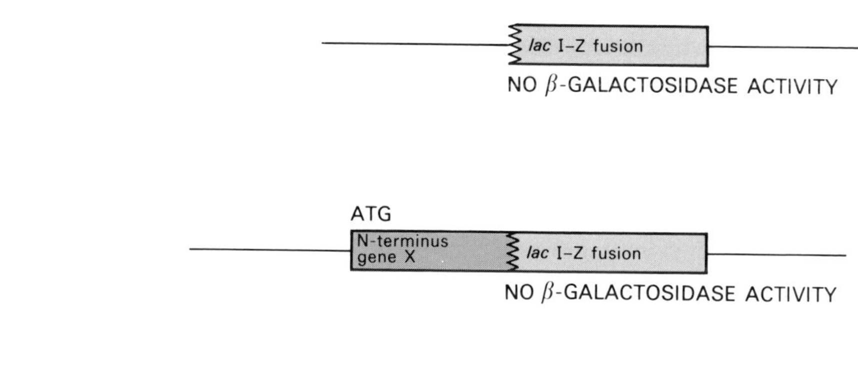

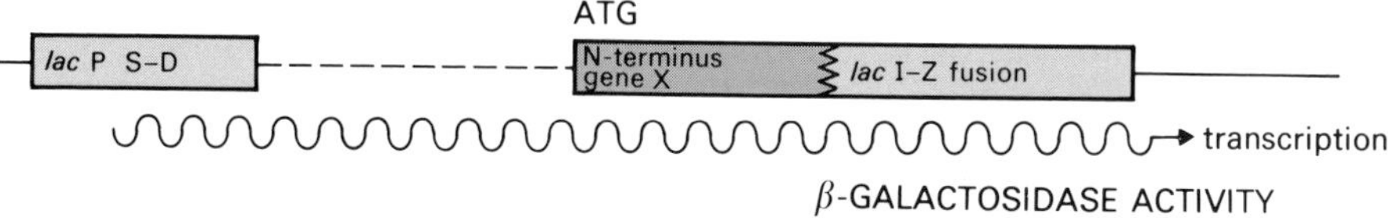

Fig. 7.9 Method for maximizing expression of cloned genes in the absence of assays for their gene products (see text for details).

phase to a fragment of the *lac* I gene. The *lac* Z fragment, which comprises most of the gene, encodes a protein fragment that has β-galactosidase activity. This activity is not affected by additional protein sequences at its amino-terminus. The *lac* I-Z fused gene is not expressed because the *lac* I promoter at the end of the I gene has been deleted. Now suppose that the gene being studied, and whose protein product is difficult to study, is gene X. A fragment of DNA bearing the amino-terminal end of gene X is then joined to the *lac* I-Z fusion (Fig. 7.9) so that the two are in the same translational reading frame. The *lac* promoter and SD sequence are then placed at varying distances in front of the ATG initiation codon, as before. Those promoter placements which favour efficient expression of gene X are identified by their β-galactosidase activity. Gene X is then reconstituted intact, with the portable promoter in place, by recombination *in vitro* or *in vivo*.

The effect of codon choice

The genetic code is degenerate and hence for most of the amino acids there is more than one codon. However, in all genes, whatever their origin, the selection of synonymous codons is distinctly non-random (for reviews see Gouy & Gautier 1982, Grosjean & Fiers 1982). The bias in codon usage has two components: correlation with tRNA availability in the cell and non-random choices between pyrimidine ending codons. Ikemura (1981a) measured the relative abundance of the 26 known tRNAs of *E. coli* and found a strong positive correlation between tRNA abundance and codon choice. Later, Ikemura (1981b) noted that the most highly expressed genes in *E. coli* contain mostly those codons corresponding to major tRNAs but few codons of minor tRNAs. By contrast, genes which are expressed less well use more sub-optimal codons.

Where maximal expression of a chemically synthesized gene is desired it would be an easy matter to optimize codon choice in accordance with the observations of Ikemura (1981a,b). However, the correlation of cellular tRNA level with codon frequency may be more complex. A shortage of a charged tRNA during protein synthesis can lead to misincorporation of amino acids and frame-shifting. Thus the arrangement of the successive codons in the coding region of a natural mRNA as well as its corresponding overlapping triplets may have evolved in order to minimize such frame-shifts *in vivo*. Not surprisingly, a study by Grosjean and Fiers (1982) of the codon usage of several highly expressed mRNAs showed that most codons that are absent in the normal reading frame frequently are present in the two non-reading frames.

Analysis of the genetic code shows that triplets X_1X_2C and X_1X_2U always code for the same amino acid. Generally speaking, these two triplets are decoded by the same tRNA. Gouy and Gautier (1982) and Grosjean and Fiers (1982) have pointed out that there can be bias in the choice of pyrimidine at the third position of the codon and that the bias is correlated with gene expressivity. Thus, in highly expressed genes, if the first two bases of the codon are both A or U then C is the favoured third base, whereas if the first two bases are G or C, the preferred third base is U. The explanation for this preference for a particular pyrimidine is equalization of the codon–anticodon interaction energy. Selection of codons which have a very strong or very weak energy of interaction with their corresponding anti-codon leads to a decrease in the efficiency of translation.

The effect of plasmid copy number

A major rate-limiting step in protein synthesis is the binding of ribosomes to mRNA molecules. Since the number of ribosomes in a cell far exceeds any one class of messenger one way of increasing the expression of a cloned gene is to increase the number of the corresponding transcripts. Two factors affect the rate of transcription: first, the strength of the promoter as described above, and second, the number of gene copies. The easiest way of increasing the gene dosage is to clone the gene of interest on a high copy-number plasmid.

Most of the high copy-number vectors in common use, e.g. pBR322, are derivatives of ColE1. Two negatively acting components are known to be involved in the control of ColE1 replication (Fig. 7.10). One is a 108 nucleotide untranslated RNA

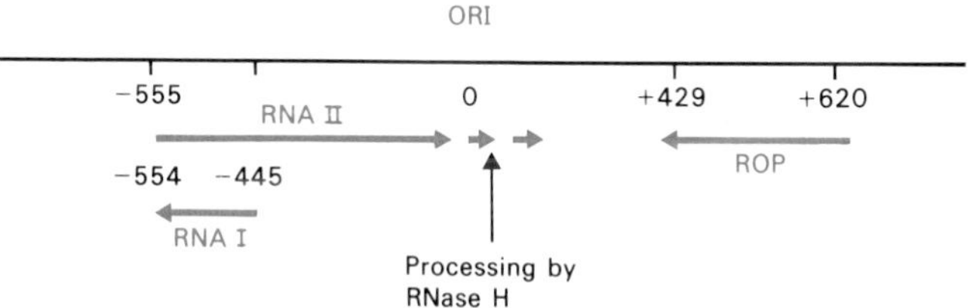

Fig. 7.10 The replication control region of plasmid ColE1. ROP controls the transcription of RNAII and RNAI inhibits the processing of RNAII by RNase H. The figures represent the base pair co-ordinates measured from the origin of replication.

molecule called RNAI (Tomizawa & Itoh 1981), and the other is a protein repressor called ROP (Cesarini *et al.* 1982). RNAII is a plasmid-encoded RNA molecule which is processed by RNase H to give a 555 nucleotide primer for the initiation of ColE1 replication (Tomizawa & Itoh 1982). RNAI inhibits DNA replication by base pairing with a complementary sequence on RNAII thereby preventing processing by RNase H. The ROP repressor is a 63 residue polypeptide which controls the initiation of transcription of RNAII. Deletion of the ROP gene, as in pAT153 (Twigg & Sheratt 1980), or mutations in RNAI Muesing *et al.* 1981) result in increased copy numbers. However, it should be noted that the host cell genetic background can also affect copy number (Nugent *et al.* 1983).

Mutant strains may also arise which result in a drop in copy number of plasmids which they harbour. These 'low cop' mutations are chromosomal in origin and may be counter-selected using increased antibiotic concentrations. The use of high antibiotic concentrations, whilst ensuring a population of 'high cop' plasmids, will not ensure that a high cop plasmid which no longer produces the recombinant protein does not arise.

Plasmid low cop mutants may arise which have a slower replication rate than normal. If such a mutation arose *in vivo* in one plasmid molecule out of a total cellular population of at least 50 (in the case of pBR322) then it is very unlikely that this mutant would ever take over in the population since the normal plasmids would replicate to compensate for the mutant and maintain the normal cellular copy number of pBR322. Under some circumstances, however, low cop plasmid mutants may be constructed *in vitro*.

Transcription termination

The presence of transcription terminators at the ends of cloned genes is important for a number of reasons. First, the synthesis of unnecessarily long transcripts will increase the energy drain on a cell which is expected to produce large amounts of non-essential protein. Second, undesirable secondary structures may form in the transcript which could reduce the efficiency of translation. Finally, promoter occlusion may occur, i.e. transcription from the promoter of the cloned gene may interfere with transcription of another essential or regulatory gene. Thus Stueber and Bujard (1982) found that certain strong promoters led to a decrease in plasmid copy number because of read-through into the ROP gene and interference with plasmid replication. Copy number control, and hence gene expression, was restored to normal by the inclusion of a transcriptional terminator at an appropriate point.

Plasmid stability

Having maximized the expression of a particular gene it is important to consider what effects this will have on the bacterium harbouring the recombinant plasmid. Increases in the levels of expression of recombinant genes lead to reductions in cell growth rates and may result in morphological changes such as filamentation and increased cell fragility. If a mutant arises which has either lost the recombinant plasmid, or has undergone structural rearrangement so that the recombinant gene is no longer expressed, or has a reduced plasmid copy number, then this will have a faster growth rate and may quickly take over and become predominant in the culture (Fig. 7.11).

Segregative instability

The loss of plasmids due to defective partitioning is called segregative instability. Naturally occurring plasmids are stably maintained because they contain a partitioning function, *par*, which ensures that they are accurately segregated at each cell division. Such *par* regions are essential for the stability of low copy-number plasmids. The higher copy-number plasmid ColE1 also contains a *par* region, but this region is deleted in pBR322 which is segregated randomly at cell division. Although the copy number of pBR322 is high and the probability of plasmid-free cells arising is very low, under certain conditions such as nutrient limitations or during rapid host cell growth, plasmid-free cells may arise (Jones *et al.* 1980, Nugent *et al.* 1983). This problem can be obviated by maintaining antibiotic selection. However, this may not be a desirable solution for large-scale culture because of cost and waste disposal considerations. The *par* region from a plasmid such as pSC101 may be cloned into pBR322-type vectors thus stabilizing the plasmid (Primrose *et al.* 1983, Skogman *et al.* 1983). The *par* region of pSC101 has been sequenced and does not appear to encode any proteins or to contain transcriptional or translational start signals. Regions of the *par* locus appear capable of forming regions of intra-strand secondary structure and these may play a role in partitioning (Miller *et al.* 1983).

An alternative strategy to counter segregative instability has

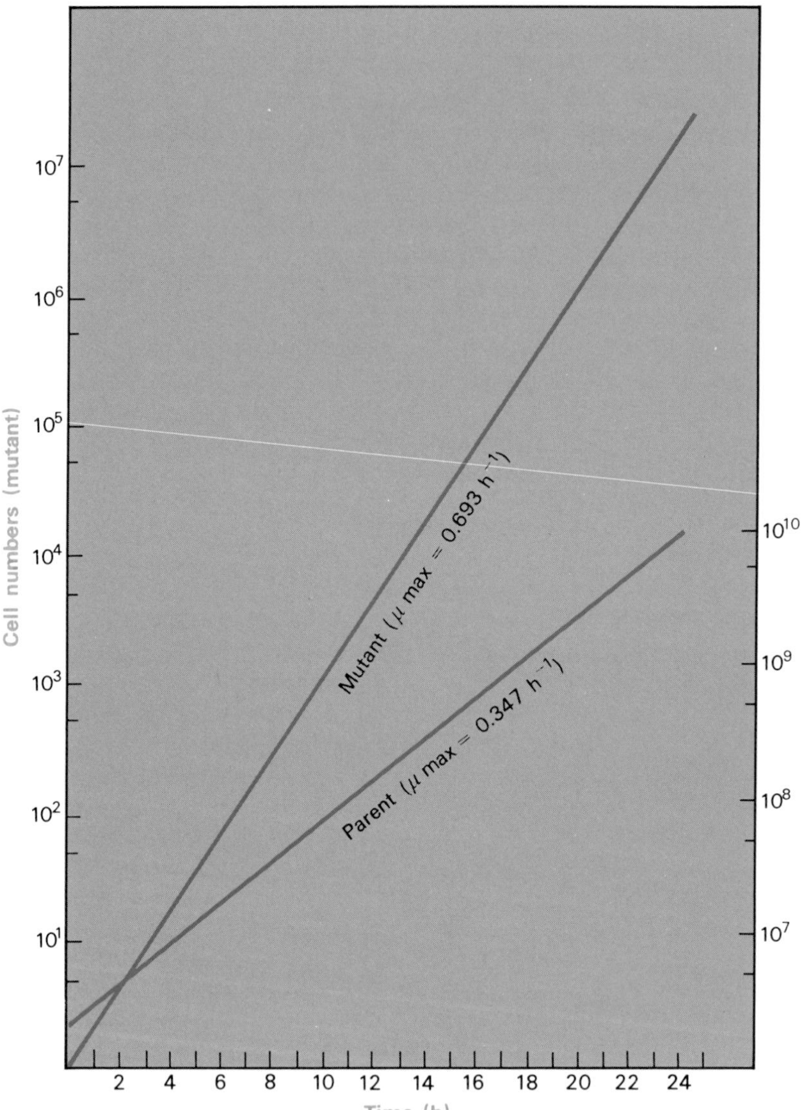

Fig. 7.11 Competition between a slow-growing production strain and a faster-growing mutant generated from it. At time zero there are 2×10^7 cells ml^{-1} of the parent strain and a single mutant cell per ml. Notice that when the parent cell density is 10^{10} cells ml^{-1} the mutant cell density is 10^7 cells ml^{-1}. On the next sub-culture the mutant cells will become predominant.

been described by Rosteck and Hershberger (1983). Unlike the use of a *par* region which *prevents* plasmid loss, their method *counter-selects* plasmid-free cells. In their method the plasmid carrying the gene of interest also carries the λ*c*I gene which encodes the λ repressor. Host cells carrying the chimaeric plasmids are then lysogenised with a repressor-defective mutant of bacteriophage λ. Loss of the plasmid from the lysogens causes concomitant loss of the λ repressor and hence cell death because the prophage is induced to enter the lytic growth cycle.

Plasmid instability may also arise due to the formation of multimeric forms of a plasmid. The mechanism which controls the copy number of a plasmid ensures a fixed number of plasmid origins per bacterium. Cells containing multimeric plasmids have the same number of plasmid origins but fewer plasmid molecules which leads to greater instability if these plasmids lack a partitioning function. These multimeric forms are not seen with ColE1 which has a natural method of resolving the multimers back to monomers. It contains a highly recombinogenic resolution site that resolves multimers in a way analogous to the resolution of transposon-induced co-integrates (Summers & Sherrat 1984). The resolution site of ColE1 has been cloned into pBR322-type plasmids and has eliminated problems due to multimerization.

Structural instability

Structural instability of plasmids may arise by deletion, insertion or rearrangements of DNA. Some of the earliest reports of deletions were in chimaeric plasmids which can replicate in both *E. coli* and *B. subtilis* (for reviews see Ehrlich *et al.* 1981, Kreft & Hughes 1981). Spontaneous deletions have now been observed in a wide range of plasmid, virus and chromosomal DNAs. A common feature of these deletions is the involvement of homologous recombination between short direct repeats (Jones *et al.* 1982). Artificial plasmids with multiple tandem promoters are particularly prone to deletion formation (Nugent *et al.* 1983) as are palindromes (Hagan & Warren 1983).

As well as homologous recombination between sites on a plasmid, structural instability may be mediated by insertion (IS) elements or transposons resident in the host chromosome or on plasmids. Both of these elements can cause spontaneous mutations by their insertion, adjacent deletion or inversion of DNA. There are many reports of plasmid instability due to insertion of IS elements from the chromosome; for example, transformation of *tyr*R strains of *E. coli* with multicopy plasmids carrying the tyrosine operon gave rise to modified plasmids with either insertions or deletions (Rood *et al.* 1980). These effects were due to IS1 insertion.

To prevent such problems of instability on scale-up it is desirable to minimize expression of the cloned gene until the organism is introduced into the final fermentation vessel. One strategy is to use controllable promoters, another is to use R1-derived runaway replication mutants (see Chapter 3). Better still, both techniques can be combined.

Host cell physiology can affect the level of expression

All of the factors affecting gene expression which we have discussed so far can be explained in terms of the sequence of bases in a nucleic acid and the way these sequences interact with a particular protein or protein complex. However, gene expression also is controlled by a less tangible factor—the physiology of the host cell. So far there has been no systematic study of the effect of different growth conditions on synthesis of foreign proteins in *E. coli*. Factors which will be important include the choice of nutrients and the way in which they are supplied to the culture and environmental parameters such as temperature and dissolved oxygen tension. Clearly there is still much to be learned.

SECTION 3

Cloning in Organisms other than *E. coli*

Chapter 8 Cloning in Bacteria other than *E. coli*

For many experiments it is convenient to use *E. coli* as a recipient for genes cloned from eukaryotes or other prokaryotes. In other experiments this would not be practicable because *E. coli* lacks some auxiliary biochemical pathways that are essential for the phenotypic expression of certain functions, e.g. degradation of aromatic compounds or plant pathogenicity. In such circumstances the genes have to be cloned back into species similar to those whence they were derived. Yet another reason for cloning in other organisms is that *E. coli* is not used in any of the traditional industrial fermentations. Rather the bacteria used tend to be Gram-positive species such as *Bacillus* (amylases, proteases), actinomycetes (antibiotics) or coryneforms (amino acids, steroid transformations). However, cloning in these organisms presents a different set of problems compared to cloning in *E. coli*.

An essential prerequisite for cloning is a method, preferably transformation, for introducing recombinant plasmids into the organism of choice. Normally, development of a transformation protocol is done using plasmids but failure to detect transformants need not be due to failure of the plasmid molecules to enter the cell; for example, the plasmid DNA might be taken up by the cell but fail to replicate. Alternatively, the plasmid-borne markers might not be expressed in their new environment. The solution to the latter problem is to use a plasmid carrying an easily selectable marker and which is indigenous to the chosen organism. However, in many instances the indigenous plasmids are cryptic, as with most *Bacillus* spp., and/or too large to handle easily. In the case of Gram-negative bacteria it is fortunate that plasmids belonging to three different incompatibility groups have a broad host range and can be subjugated as vectors. In the case of Gram-positive *Bacillus* spp. rapid progress in the development of vectors was facilitated by the discovery that antibiotic-resistance plasmids from *S. aureus* can replicate in them too and express their antibiotic resistance.

BROAD HOST RANGE VECTORS FOR CLONING IN GRAM-NEGATIVE BACTERIA

Plasmids belonging to any one of incompatibility groups P, Q or W have a broad host range among Gram-negative bacteria. Properties of representative plasmids from each group are shown in Table 8.1. Members of groups P and W are self-transmissible, i.e. they carry *tra* genes encoding the conjugative apparatus. Group Q plasmids are not self-transmissible and usually are introduced into the recipient by transformation. However, if the donor cell also carries a narrow host range conjugative plasmid they can be transferred to the recipient by conjugation. The ability of the broad host range plasmids to be mobilized into new hosts by means of conjugation is particularly useful, for many Gram-negative bacteria appear to be refractory to transformation. If genes have to be cloned in such organisms, the recombinant plasmids are first isolated in *E. coli* and then transferred by conjugation.

Table 8.1 Properties of representative broad host range plasmids.

Plasmid	Incompatibility group	Size (Mdal)	Copy No.	Self-transmissible	Markers
RP4 (RK2, RP1)	P	36	1–3	Yes	Ap^R Km^R Tc^R
RSF1010	Q	5.5	15–40	No	Sm^R Su^R
Sa	W	24	3–5	Yes	Km^R Sm^R Cm^R Su^R

Vectors derived from Q group plasmid RSF1010

Plasmid RSF1010 has been developed largely as a vector for use in *Pseudomonas* spp. The frequency of transformation of *Pseudomonas aeruginosa* and *P. putida* is 10^5 transformants/μg of RSF1010 DNA and *E. coli* is transformed at a ten times higher frequency. One drawback to the use of RSF1010 is that *P. putida* strains only can be transformed with RSF1010 DNA prepared in another *P. putida* strain (Bagdasarian *et al.* 1979). Apparently this is due to a powerful restriction system in *P. putida.*

Plasmid RSF1010 specifies resistance to two antimicrobial agents, sulphonamide and streptomycin. The plasmid contains unique cleavage sites for *Eco* RI, *Bst* EII, *Hpa* I, *Sst* I and *Pvu* II. There are two *Pst* I sites, about 750 base pairs apart, which flank the sulphonamide-resistance determinant (Fig. 8.1). There are no cleavage sites for restriction endonucleases *Bam* HI, *Bgl* II, *Hind* III, *Kpn* I, *Sma* I, *Xba* I, *Xho* I or *Xma* I. The usefulness of RSF1010 as a cloning vector is limited by the fact that it contains single cleavage sites for only a few endonucleases that ordinarily are used in cloning experiments. Nor are any of the unique endonuclease cleavage sites located within the antibiotic resistance determinants which could allow insertional inactivation. For this reason Bagdasarian *et al.* (1979, 1981) have constructed derivatives of RSF1010 that contain additional cleavage sites and antibiotic resistance determinants. Plasmids KT230 and KT231 encode Km^R and Sm^R and have unique sites for *Hind* III, *Xma* I, *Xho* I, *Eco* RI and *Sst* I which can be used for insertional inactivation. These two vectors have been used to clone in *P. putida* genes involved in catabolism of aromatic compounds (Franklin *et al.* 1981). More recently Bagdasarian *et al.* (1983) have shown that the hybrid *trp-lac* (*tac*) promoter functions in a controllable manner in *P. putida*.

Plasmid vectors that contain the *cos* site (cosmids, see p. 85) and which can be packaged *in vitro* into bacteriophage λ particles are of considerable utility for cloning large segments of DNA and for constructing gene banks. Bagdasarian *et al.* (1981) have produced a cosmid derivative of RSF1010 called pKT247.

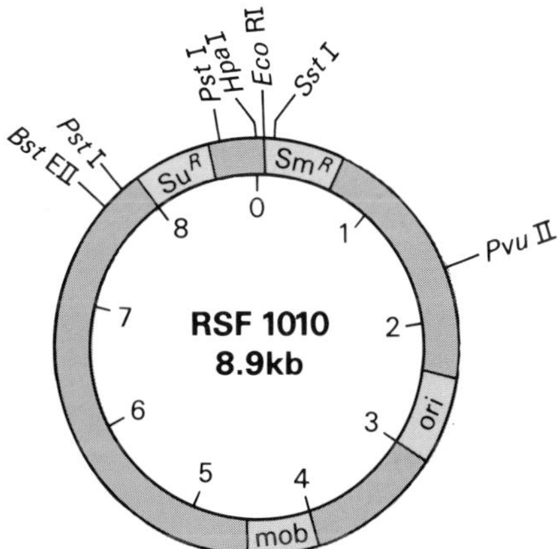

Fig. 8.1 The structure of plasmid RSF1010. The pale red tinted areas show the approximate positions of the Sm^R and Su^R genes. The region marked *ori* indicates the location of the origin of replication. The *mob* function is required for conjugal mobilization of a compatible self-transmissible plasmid.

The Sm^R structural gene lies to the right of the *Sst* I site and the promoter lies to the left of it (K. N. Timmis, pers. commun.).

Whereas normal RSF1010-derived vectors can be used to transform the recipient directly, *E. coli* has to serve as an intermediate host if cosmid cloning is employed. Cosmid pKT249 encodes Ap^R, Su^R and Sm^R and has unique sites for endonucleases *Eco* RI and *Sst* I. More recently, Frey *et al.* (1983) have described two improved cosmids. These new cosmids permit the selective cloning into their unique *Bam* HI site of 36 kb DNA fragments by a strategy that avoids the formation of polycosmids but does not require the cleaved vector to be dephosphorylated.

As indicated earlier, RSF1010-derived cloning vectors can be mobilized to other bacteria by certain conjugative plasmids. This characteristic is advantageous when the ultimate recipient is not transformable. However, for certain experiments regulatory authorities prefer vectors which have a low or non-existent frequency of conjugal transfer. For this reason Bagdasarian *et al.* (1981) constructed mobilization-defective derivatives of pKT230, pKT231 and pKT247, called pKT262, pKT263 and pKT264 respectively, by deleting the *mob* function (see Fig. 8.1).

Plasmid R300b is similar to, if not identical with, plasmid RSF1010. By cloning R300b in pBR322 a vector called pSS515 was constructed which encodes Ap^R and Tc^R while retaining the broad host range properties of R300b. This new vector has been used successfully to clone and express in the obligate methylotroph *Methylophilus methylotrophus* eukaryotic genes encoding dihydrofolate reductase (Hennan *et al.* 1982) and interferon α1 (de Mayer *et al.* 1982).

Broad host range vectors derived from P group plasmids

The best-studied P group plasmid is RP4 (also known as RP1 and RK2). It specifies resistance to Ap, Km and Tc and has single cleavage sites for *Eco* RI, *Bam* HI (in the Ap^R gene), *Bgl* II, *Hpa* I and *Hind* III (in the Km^R gene). Relatively little use has been made of RP4 as a cloning vehicle but Jacob *et al.* (1976) have used it for cloning DNA from *Rhizobium leguminosarum* and *Proteus mirabilis,* and Windass *et al.* (1980) for cloning the *E. coli* glutamate dehydrogenase gene in *M. methylotrophus.*

The basic problem with P group plasmids such as RP4 (RP1, RK2) is their size. Attempts to produce smaller derivatives have been hindered by the fact that non-continuous regions are required for replication (Fig. 8.2). Originally it was thought that three dispersed functions *ori*V, *trf*A and *trf*B were required. More recently it has been shown that a 0.7 kb seg-

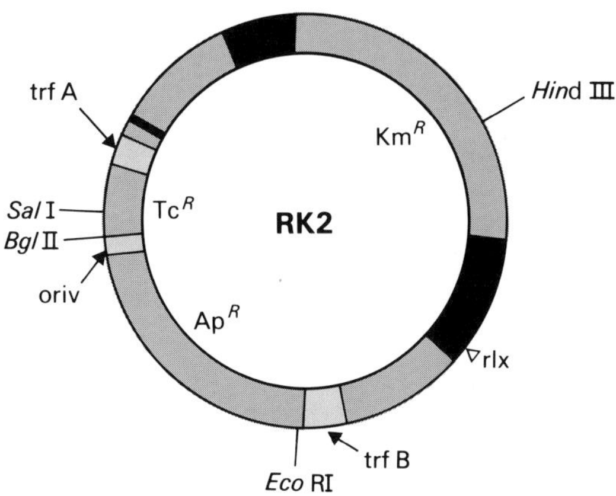

Fig. 8.2 The structure of P group plasmid RK2. The red tinted areas indicate those regions of RK2 originally believed to be essential for replication. The black areas indicate the location of genes encoding self-transmission of RK2. The site designated *rlx* must be present if conjugation-deficient derivatives of RK2 are to be mobilized by a compatible self-transmissible plasmid.

ment containing the replication origin and a 1.8 kb fragment designated *trf*A* are the minimal requirements for replication (Schmidhauser *et al.* 1983). Smaller derivatives of RK2 have been obtained by partial digestion with *Hae* II. One derivative, pRK248, is only 6.2 Mdal in size and specifies Tc^R (Thomas *et al.* 1979). A more useful derivative, pRK2501, has been constructed by inserting a *Hae* II fragment specifying Km^R into pRK248. Plasmid pRK2501 has unique cleavage sites for *Sal* I, *Hin*d III, *Xho* I, *Bgl* II and *Eco* RI, the first three being in either the Tc^R or Km^R genes. There are two problems associated with pRK2501. First, it is not stably maintained (Thomas *et al.* 1981). Second, it only can be transformed into recipients as the conjugative functions of RK2 have been deleted. It even cannot be mobilized from *E. coli* to other recipients by other conjugative plasmids because a region of RK2 called *rlx*, a cis-acting function necessary for conjugal transfer, also has been removed.

An improved RK2-derived vector system has been constructed by Ditta *et al.* (1980). In this system the transfer and replication functions are separated on different plasmids. Plasmid pRK290 contains a functional RK2 replicon and can be mobilized at high frequency by a helper plasmid because it retains the *rlx* locus. It encodes Tc^R and has single *Eco* RI and *Bgl* II sites for insertion of foreign DNA. The Km^R plasmid pRK203 consists of the RK2 transfer genes cloned onto a ColE1 replicon and its function is to mobilize RK290 into other hosts. Where the intended recipient of cloned genes is transformable, RK290 alone is used. If the recipient is not transformable, the vector RK290 containing the foreign DNA insert first is transformed into an *E. coli* strain carrying pRK2013 and then conjugated into the desired recipient.

Olsen *et al.* (1982) discovered a small (2×10^6 mol. wt) multicopy, broad host range plasmid which had arisen spontaneously from the P group plasmid RP1. Presumably this plasmid retains the *ori*V and *trf*A* functions which appear to be the minimum requirements for replication of RK2. From this plasmid two derivatives were constructed. The first has two *Pst* I sites and can be used for cloning DNA where there is direct selection for the acquired trait, e.g. acquisition of antibiotic resistance or reversal of auxotrophy. The second plasmid contains an entire pBR322 molecule and if genes are inserted at the unique *Hin*d III or *Bam* HI sites they can be detected by insertional inactivation of the Tc^R marker. Like pRK2501, these two vectors can be used only if the intended host is transformable.

Vectors derived from the broad host range group W plasmid Sa

Although a group W plasmid such as Sa (Fig. 8.3) can infect a wide range of Gram-negative bacteria, it has been developed

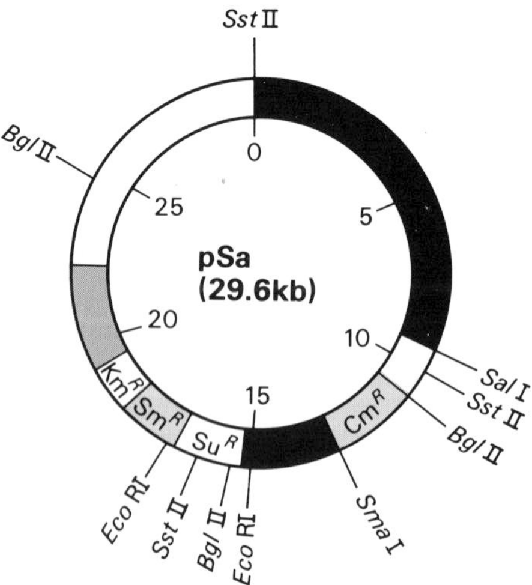

Fig. 8.3 The structure of plasmid pSa. The grey area encodes the functions essential for plasmid replication. The black areas represent the regions containing functions essential for self-transmission of pSa, the one between the *Sst* and *Sal* sites being responsible for suppression of tumour induction by *Agrobacterium tumefaciens*.

mainly as a vector for use with the oncogenic bacterium *Agrobacterium tumefaciens* (see p. 222). Two regions of the plasmid have been identified as involved in conjugal transfer of the plasmid and one of them has the unexpected property of suppressing oncogenicity by *A. tumefaciens* (Tait *et al.* 1982). Sufficient information for the replication of the plasmid in *E. coli* and *A. tumefaciens* is contained within a 4 kb DNA fragment. Leemans *et al.* (1982) have described four small (5.6–7.2 Mdal), multiply marked derivatives of plasmid Sa. The derivatives contain single target sites for a number of the common restriction endonucleases and at least one marker in each is subject to insertional inactivation. Although these Sa derivatives are non-conjugative they can be mobilized by other conjugative plasmids. Tait *et al.* (1983) also have constructed a set of broad host range vectors from plasmid Sa. The properties of their derivatives are similar to those of Leemans *et al.* (1982) but one of them also contains the bacteriophage λ *cos* sequence and hence is a cosmid.

A transposon as a broad host range vector

Transposons are mobile genetic elements which can insert at random into plasmids or the bacterial chromosome independently of the host cell recombination system. In addition to genes involved in transposition, transposons carry genes conferring new phenotypes on the host cell, e.g. antibiotic resistance. Grinter (1983) has devised a broad host range cloning vector based on transposon Tn7 which encodes Tp^R and Sm^R. Unlike the vectors just described, the transposon vector permits the cloned genes to be stably inserted into the chromosome. Such a vector could have useful industrial applications for it does not put an extra-genomic genetic load on the recipient cell.

Two compatible broad host range plasmids form the basis of the vector system. One of the plasmids is derived from RP1 but only encodes Tc^R. This plasmid is unstable and in the absence of selection for Tc^R is lost rapidly from host cells. A Tn7 derivative was inserted into this plasmid to generate pNJ5073. The Tn7 retains the original Tp^R and Sm^R markers but the DNA sequence which encodes transposition functions was replaced by a *Hin*d III fragment containing the *E. coli trp*E gene. The second part of the vector system is plasmid pNJ9279 which encodes Km^R and is derived from plasmid R300b (≡ RSF1010, see p. 156). It is used to provide the transposition functions missing from the Tn7 derivative.

If pNJ5073 alone is introduced into a bacterium then, as noted above, in the absence of antibiotic selection it will not be maintained stably. Not even the Tp^R and Sm^R markers will be retained, for the variant Tn7 is unable to transpose. If pNJ9297 also is present then it can provide in *trans* the necessary transposition functions: selection of Tp^R and Sm^R followed by screening for Tc^S will identify those cells in which the defective transposon has hopped into the chromosome. In this way Grinter (1983) was able to select strains of *M. methylotrophus* and *P. aeruginosa* which had incorporated the *E. coli trp*E gene into their chromosome. In theory any gene can be inserted into the chromosome by replacing the *trp*E gene with an appropriate *Hin*d III- generated DNA fragment.

As will be seen in Chapter 12 a formally analogous approach can be used to subjugate a eukaryotic (*Drosophila*) transposable element as a vector.

CLONING IN *B. SUBTILIS*

There are a number of reasons for cloning in *B. subtilis*. First, *Bacillus* spp. are Gram-positive and generally obligate aerobes compared with *E. coli* which is a Gram-negative facultative anaerobe. Thus the two groups of organisms may have quite different internal environments. Second, *Bacillus* spp. are able to sporulate and consequently are used as models for prokaryotic differentiation. The use of gene manipulation is facilitating these studies. Third, *Bacillus* spp. are widely used in the fermentation industry particularly for the production of exoenzymes. They can be tailored to secrete the products of cloned eukaryotic genes. Finally, from a biohazard point of view, *B. subtilis* is an extremely safe organism for it has no known pathogenic interactions with man or animals. Indeed it is consumed in large quantities in the East. Furthermore, in the literature between 1912 and 1983 there is only one authentic report of a human infection due to *B. subtilis* and that was in a severely compromised host—a drug addict.

Plasmid transformation in *B. subtilis*

An essential feature of any cloning experiment involving plasmids is transformation of a recipient cell with recombinant DNA. Although it is very easy to transform *B. subtilis* with fragments of chromosomal DNA there are problems associated with transformation by plasmid molecules. Ehrlich (1977) first reported that competent cultures of *B. subtilis* can be transformed with ccc plasmid DNA from *S. aureus* and that this plasmid DNA is capable of autonomous replication and expression in its new host. The development of competence for transformation by plasmid and chromosomal DNA follows a similar time course and in both cases transformation is first order with respect to DNA concentration suggesting that a single DNA molecule is sufficient for successful transformation (Contente & Dubnau 1979). However transformation of *B. subtilis* with plasmid DNA is very inefficient by comparison with chromosomal transformation for only one transformant is obtained per 10^3–10^4 plasmid molecules.

As will be seen later, much cloning in *B. subtilis* is done with bifunctional vectors that replicate in both *E. coli* and *B. subtilis*. Van Randen and Venema (1984) have shown that such bifunctional vectors can be transferred by replica-plating *E. coli* colonies containing them onto a lawn of competent *B. subtilis* cells. However plasmid transformation by replica plating differed in one respect from plasmid transformation in liquid. Whereas chromosomal integration of plasmid-borne chromosomal alleles with concomitant loss of plasmids occurred frequently during regular plasmid transformation of Rec^+ *B. subtilis*, this was a rare event during plasmid transfer by replica plating.

An explanation for the poor transformability of plasmid DNA molecules was provided by Canosi *et al.* (1978). They found that the specific activity of plasmid DNA in transformation of *B. subtilis* was dependent on the degree of oligomerization of the plasmid genome. Purified monomeric ccc forms of plasmids transform *B. subtilis* several orders of magnitude less efficiently than do unfractionated plasmid preparations or multimers. Furthermore, the low residual transforming activity of monomeric ccc DNA molecules can be attributed to low level contamination with multimers (Mottes *et al.* 1979). Using a recombinant plasmid capable of replication in both *E. coli* and *B. subtilis* (pHV14, see p. 170). Mottes *et al.* (1979) were able to show that plasmid transformation of *E. coli* occurs regardless of the degree of oligomerization, in contrast to the situation with *B. subtilis*. Oligomerization of linearized plasmid DNA by DNA ligase resulted in a subtantial increase of specific transforming activity when assayed with *B. subtilis* and caused a decrease when used to transform *E. coli*. An explanation of the molecular events in transformation which generate the requirement for oligomers has been presented by de Vos *et al.* (1981). Basically, the plasmids are cleaved into linear molecules upon contact with competent cells just as chromosomal DNA is cleaved during transformation of *Bacillus*. Once the linear plasmid enters the cell it is not reproduced unless it can circularize, hence, the need for multimers to provide regions of homology which can recombine. Michel *et al.* (1982) have shown that multimers, or even dimers, are not required provided part of the plasmid genome is duplicated. They constructed plasmids carrying direct internal repeats 260–2000bp long and found that circular or linear monomers of such plasmids were active in transformation.

Canosi *et al.* (1981) have shown that plasmid monomers will transform recombination proficient *B. subtilis* if they contain an insert of *B. subtilis* DNA. However the transformation efficiency of such monomers is still considerably less than that of oligomers. One consequence of the requirement for plasmid oligomers for efficient transformation of *B. subtilis* is that there have been very few successes in obtaining large numbers of clones in *B. subtilis* recipients (Keggins *et al.* 1978, Michel *et al.* 1980). The potential for generating multimers during ligation of vector and foreign DNA is limited. If the ratio of foreign to vector DNA is elevated in order to increase the proportion of recombinant molecules generated, the yield of transformants will decrease rapidly due to competition between vector–vector and vector–foreign DNA ligations.

Transformation by plasmid rescue

An alternative strategy for transforming *B. subtilis* has been suggested by Gryczan *et al.* (1980a). If plasmid DNA is linearized by restriction endonuclease cleavage, no transformation of *B. subtilis* results. However, if the recipient carries a homologous plasmid and if the restriction cut occurs within the homologous moiety then this same marker transforms efficiently. Since this rescue of donor plasmid markers by a homologous resident plasmid requires the *B. subtilis rec*E gene product it must be due to recombination between the linear donor DNA and the resident plasmid. Since DNA linearized by restriction endonuclease cleavage at a unique site is monomeric this rescue system (*plasmid rescue*) bypasses the requirement for a multimeric vector. The model presented by de Vos *et al.* (1981) to explain the requirement for oligomers (see previous section) can be adapted to account for transformation by monomers by means of plasmid rescue. In practice, foreign DNA is ligated to monomeric vector DNA and the in-vitro recombinants used to transform *B. subtilis* cells carrying a homologous plasmid. Using such a 'plasmid rescue' system, Gryczan *et al.* (1980a) were able to clone various genes from *B. licheniformis* in *B. subtilis*. One disadvantage of this system is that transformants contain both the recombinant molecule and the resident plasmid and they have to be streaked several times to allow plasmid segregation. Alternatively, the recombinant plasmids must be retransformed into plasmid-free cells.

Transformation of protoplasts

A third method for plasmid DNA transformation in *B. subtilis* involves polyethylene glycol (PEG) induction of DNA uptake in protoplasts and subsequent regeneration of the bacterial cell wall (Chang & Cohen 1979). The procedure is highly efficient and yields up to 80% transformants making the method suitable for the introduction of cryptic plasmids. In addition to its much higher yield of plasmid-containing transformants the protoplast transformation system differs in two respects from the 'traditional' system using physiologically competent cells. First, linear plasmid DNA and non-supercoiled circular plasmid DNA molecules constructed by ligation *in vitro* can be introduced at high efficiency into *B. subtilis* by the protoplast transformation system, albeit at a frequency 10^1–10^3 lower than the frequency observed for ccc plasmid DNA. Second, while competent cells can be transformed easily for genetic determinants located on the *B. subtilis* chromosome, no detectable transformation with chromosomal DNA is seen using the protoplast assay. One disadvantage of the protoplast system is that the regeneration medium is nutritionally complex necessitating a two-step selection procedure for auxotrophic markers.

The advantages and disadvantages of the three transformation systems are summarized in Table 8.2. It should be noted that although shotgun cloning of chromosomal DNA is difficult it is relatively easy to isolate recombinants containing fragments of bacteriophage or plasmid DNA. This is simply because of the small size of these genomes. Another point of interest is that in *B. subtilis* transforming DNA is taken up in a single-stranded form. A consequence of this is that it is not possible to clone recombinant molecules freshly constructed *in vitro* using homopolymer tailing. The reason is that recombinant molecules prepared in this way contain single-stranded regions because the homopolymer tails on the two components are generally of different lengths. With *E. coli* there is no problem because the molecules are taken up in a double-stranded form and the single-stranded gaps repaired *in vivo*.

Table 8.2 Comparison of the different methods of transforming *B. subtilis*.

System	Efficiency (transformants/μg DNA)		Advantages	Disadvantages
Competent cells	Unfractionated plasmid Linear ccc Monomer ccc Dimer ccc Multimer	2×10^4 0 4×10^1 8×10^3 2.6×10^5	Competent cells readily prepared. Transformants can be selected readily on any medium. Recipient can be Rec^-	Requires plasmid oligomers which makes shotgun experiments difficult unless high DNA concentrations and high vector/donor DNA ratios are used
Plasmid rescue	Unfractionated plasmid	2×10^6	Oligomers not required. Can transform with linear DNA. Transformants can be selected on any medium	Transformants contain resident plasmid and incoming plasmid and these have to be separated by segregation or retransformation. Recipient must be Rec^+
Protoplasts	Unfractionated plasmid Linear ccc Monomer ccc Dimer ccc Multimer	3.8×10^6 2×10^4 3×10^6 2×10^6 2×10^6	Most efficient system. Gives up to 80% transformants. Does not require competent cells. Can transform with linear DNA	Regeneration medium complex which limits selections which can be made. Efficiency lower with molecules which have been cut and religated

Plasmid vectors for cloning in *B. subtilis*

The development of *B. subtilis* cloning systems was hindered by the absence of suitable vector replicons; *E. coli* plasmids do not replicate efficiently in *B. subtilis* nor are their markers expressed. A large number of cryptic plasmids had been identified in *B. subtilis* (Le Hegarat & Anagnostopoulos 1977) but in the absence of a detectable phenotype they could not be used for cloning. Ehrlich (1977) showed that plasmids isolated from *S. aureus* that code for resistance to tetracycline (pT127) or chloramphenicol (pC194) can be transformed into *B. subtilis* where they replicate and express antibiotic resistance normally. Almost all the vectors used for cloning in *B. subtilis* are derived from plasmids like pT127 and pC194 which originated in *S. aureus*. Two antibiotic-resistance plasmids have been identified in *Bacillus* sp., pBC16 in *B. cereus* (Bernhard *et al.* 1978) and pAB124 in a thermophilic bacillus. Both confer tetracycline resistance in *B. subtilis*. Characterization of pAB124 has indicated that this plasmid could be used as a cloning vector (Bingham *et al.* 1980).

How do *S. aureus* plasmids rate as cloning vehicles? Well, they are small and are maintained in multiple copies like many *E. coli* plasmids, but are not chloramphenicol enrichable. The properties of these plasmids are shown in Table 8.3. In general, these plasmids are stable in *B. subtilis*. However, pE194 and pSAO501 are temperature sensitive and do not replicate above 35 °C. There have been reports that recombinants derived from plasmid pC194 can suffer deletions (Gryczan & Dubnau 1978) but this could be a consequence of the DNA fragments cloned. In *E. coli* the incidence of deletion formation in recombinant molecules depends on the origin and structure of the insert and the vector.

Table 8.3 *S. aureus* plasmids which have been transformed into *B. subtilis*.

Plasmid	Phenotype conferred on host cell	Molecular weight
pC194	Chloramphenicol-resistance	1.8×10^6
pE194	Erythromycin-resistance	2.3×10^6
pSA0501	Streptomycin-resistance	2.7×10^6
pUB110	Kanamycin-resistance	3.0×10^6
pT127	Tetracycline-resistance	2.9×10^6

Table 8.4 Some improved vectors for cloning in *B. subtilis*.

Plasmid	Origin	Size (Mdal.)	Markers	Single sites	Markers inactivated	Reference
pBD6	pSA0501 + pUB110	5.8	Km^R Sm^R	*Bam* HI *Tac* I *Hind* III *Bgl* II	none none Sm^R Km^R	Gryczan *et al.* (1980b)
pBD9	pE 194 + pUB110	5.4	Em^R Km^R	*Pst* I *Eco* RI *Bam* HI *Tac* I *Bgl* II *Hpa* I	none none none none Km^R Em^R	Gryczan *et al.* (1980b)
pBD12	pUB110 + pC194	4.5	Cm^R Km^R	*Hind* III *Eco* RI *Xba* I *Bam* HI *Tac* I *Bgl* II	none none none none none Km^R	Gryczan *et al.* (1980b)
pSL103	pUB110 + *trp* fragment	5.0	Km^R Trp^+	*Hind* III	Trp	Keggins *et al.* (1979)

Improved vectors for cloning in *B. subtilis*

As indicated in Chapter 3 the ideal vector confers more than one selectable phenotype on host cells, and in addition has unique restriction endonuclease cleavage sites in the corresponding genes. Since none of the *S. aureus* plasmids isolated so far carries more than one selectable marker, improved vectors have been constructed by gene manipulation *in vitro*. An example is the formation of pHV11 (Ehrlich 1978).

Plasmid pC194 has a single *Hin*d III site whereas pT127 has three sites. *Hin*d III cleavage destroyed the transforming activity of both plasmids and transforming activity could be restored by DNA ligase treatment of cleaved pC194 but not pT127. However, if the two cleaved DNAs were ligated together some Tc^R colonies could be recovered and these were Cm^R as well. The plasmid extracted from one of these Tc^RCm^R clones, designated pHV11, had a mol. wt of 3.3×10^6, equivalent to the sum of pC194 and the largest *Hin*d III fragment of pT127. In addition, *Hin*d III treatment of pHV11 DNA gave rise to two DNA segments which matched the *Hin*d III cleavage of pC194 and the slowest of the pT127 segments in electrophoretic mobility.

The properties of some improved vectors for cloning in *B. subtilis* are shown in Table 8.4. In each case the unique sites that do not permit insertional inactivation of an antibiotic-resistance determinant have been shown not to occur in essential plasmid genes since the sites can receive foreign DNA without interfering with replication of the chimaera. However, some of these sites may be in essential genes in the parental plasmids since the parental plasmids are not composite replicons.

Construction of *E. coli–B. subtilis* hybrid replicons

The work of Ehrlich (1978) described above indicated that pC194 can be used as a *Hin*d III cloning vector in *B. subtilis*. Earlier (p. 59) we described how pBR322 can be used as a *Hin*d III vector in *E. coli*. Thus by ligating pBR322 with pC194 it should be possible to construct a hybrid plasmid which could be used as a vector in both *E. coli* and *B. subtilis*. The beauty of such a vector is that with it it is possible to test the expression of a cloned gene in either host. Two such hybrid replicons, pHV14 and pHV15, have been constructed and these differ only in the orientation of pC194 with respect to pBR322.

To construct pHV14 and pHV15, pC194 and pBR322 were

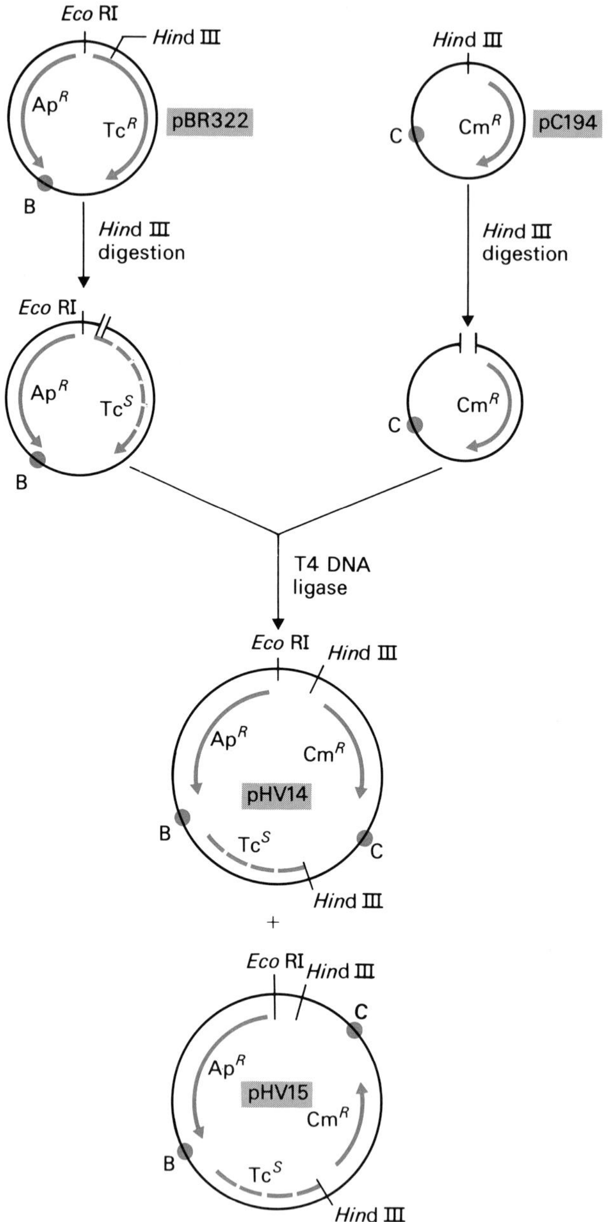

Fig. 8.4 Formation of PHV14 and pHV15 by the ligation of pBR322 and pC194. Note particularly the inactivation of the Tc^R gene and the orientations of the Cm^R gene in pHV14 and pHV15. O_B and O_C represent the origins of replication of pBR322 and pC194 respectively.

digested with *Hin*d III, mixed and ligated, and the mixture used to transform *E. coli* to chloramphenicol resistance (Fig. 8.4). Cm^R transformants were obtained and as expected were Tc^S, due to insertion of pC194, but still Ap^R. The plasmids isolated from these transformants were of a size equivalent to the sum of the parental plasmids and were cleaved by *Hin*d III into segments having mobilities that matched those of the cleaved parental DNAs.

In all cases the hybrid DNAs could transform *E. coli* to both ampicillin resistance and chloramphenicol resistance, confirming the observed phenotype of the colonies selected originally. The hybrid replicons also transformed *B. subtilis* to chloramphenicol resistance but the Ap^R gene of pBR322 was not expressed. The failure to get expression of ampicillin resistance was not due to fragmentation of the hybrid DNA for the *B. subtilis* transformants had acquired plasmids matching those from *E. coli* in both size and *Hin*d III restriction pattern. Nor was the lack of expression of the Ap^R gene related to its orientation relative to the pC194 part of the hybrid plasmid because neither pHV14 or pHV15 conferred ampicillin resistance in *B. subtilis.* The question concerning why the Ap^R gene from *E. coli* is not expressed in *B. subtilis* when the Cm^R gene from *B. subtilis* (originally *S. aureus*) is expressed in *E. coli* is particularly interesting and is discussed in more detail later in this chapter (see p. 178).

Plasmids pHV14 and pHV15 fail to express tetracycline resistance because the continuity of the promoter for the tetracycline-resistance gene is destroyed upon insertion of pC194 at the *Hin*d III site of pBR322 (Fig. 8.4). When *E. coli* cells carrying pHV14 are plated on tetracycline-containing media, rare Tc^R-derivatives can be isolated (Primrose & Ehrlich 1981). The plasmids isolated from these Tc^R cells confer resistance to Ap, Cm and Tc on *E. coli* recipients, i.e. tetracycline resistance is plasmid borne. These same plasmids still confer only Cm resistance on *B. subtilis* recipients. Two types of pHV14-derivative have been isolated in this way (Fig. 8.5): a derivative (pHV33) which appears to carry a point mutation which restores tetracycline resistance and a derivative (pHV32) which carries a small deletion.

The deletion derivative pHV32 is very convenient to use for genetic labelling of replication regions active in *B. subtilis* since:

1 the deletion destroys the replication functions of pC194 which are essential for replication of pHV14 in *B. subtilis;*

2 it can be prepared in large amounts free from any contaminating *B. subtilis* replicons in an appropriate *E. coli* strain;

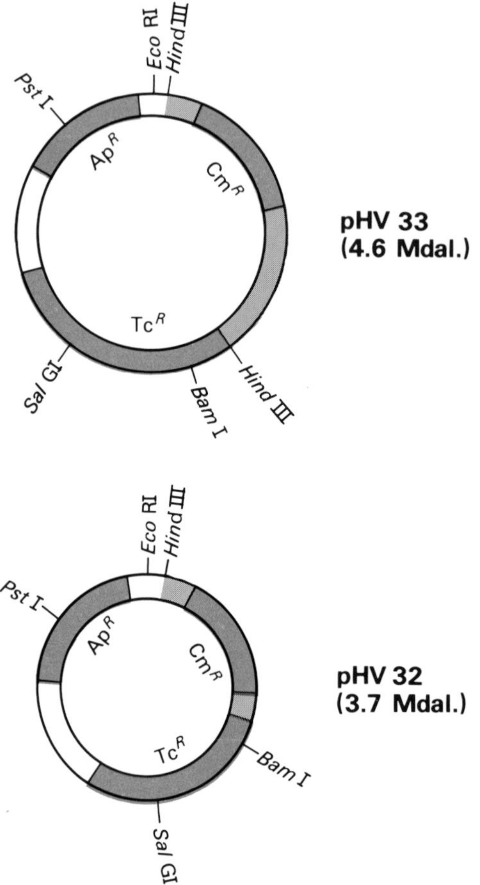

Fig. 8.5 The structure of pHV32 and pHV33. The grey areas represent DNA originally from pC194.

3 the chloramphenicol-resistance gene it contains can transform *B. subtilis* at a high efficiency.

Using this plasmid Niaudet and Ehrlich (1979) were able to label genetically *in vitro* the cryptic *B. subtilis* plasmid, pHV400.

A direct selection vector for *B. subtilis*

Earlier we described the development of direct selection vectors for use in *E. coli* (see p. 71). Gryczan and Dubnau (1982) have developed a similar system for *B. subtilis*. It is based on plasmid pBD214 which encodes Cm^R and carries a *thy*$^+$ gene, and on *B. subtilis* BD393, a highly competent thymine auxotroph. Thy^- strains are resistant to trimethoprim (Tp) and Thy^+ strains are sensitive. Inactivation of the pBD214 *thy*$^+$ gene by insertion of a foreign DNA fragment at unique *Eco* Ri, *Bcl* I, *Pvu* II or *Eco* RV sites permits selection of Cm^R Tp^R clones, all of which carry recombinant plasmids.

Shotgun cloning the *B. subtilis* genome in *E. coli*

Because of the problems associated with transformation of *B. subtilis,* discussed above, shotgun cloning in this organism is seldom very effective. An alternative approach is to use *E. coli* as an intermediate host and this kind of experiment requires bifunctional vectors which can replicate in both *E. coli* and *B. subtilis.* Rapoport *et al.* (1979) have used pHV33 in this way to construct a *B. subtilis* gene bank in *E. coli.* Detection of cloned recombinant DNA molecules was based on the insertional inactivation of the Tc^R-gene, transformants being selected by virtue of their ampicillin resistance. Recombinant plasmids were pooled in lots of 100 and used to transform auxotrophic mutants of *B. subtilis.* Complementation of these auxotrophic mutants was observed for several markers such as *thr, leu*A, *his*A, *gly*B and *pur*B. In several cases markers carried by the recombinant plasmids were lost from the plasmid and integrated into the chromosome unless a recombination-deficient (*rec* E) recipient was used.

The fate of cloned DNA in *B. subtilis*

When DNA from an organism which has little or no homology with *B. subtilis,* e.g. *B. licheniformis,* is cloned in *B. subtilis* the chimaeric plasmid appears to be stable. However, if the cloned insert has homology with *B. subtilis* the outcome is quite different. The homologous region may be excised and incorporated into the chromosome by the normal recombination process, i.e. substitution occurs via a double cross-over event. However, if only a single cross-over occurs the entire recombinant plasmid is integrated into the chromosome. It is possible to favour integration by using vectors which cannot replicate in *B. subtilis.*

B. subtilis phage ϕ3T contains a gene specifying the enzyme thymidylate synthetase and this gene has homology with the corresponding bacterial gene. The ϕ3T gene is carried on a particular *Eco* RI fragment which has been cloned in the *E. coli* vector pMB9 to generate the recombinant pCD1 (Fig. 8.6). When pCD1 is transformed into *E. coli* it complements *thy* auxotrophs and is maintained as a plasmid. When transformed into

B. subtilis pCD1 cannot replicate and the *thy* gene is excised from the plasmid and integrated into the chromosome (Duncan *et al.* 1978). In fact ccc pCD1 DNA transforms *B. subtilis* as well as linear pCD1 DNA. One spin-off from this is that it is possible to study the integration of only one gene under carefully controlled conditions.

The extent of the homology of pCD1 with the *B. subtilis* chromosome can be greatly reduced by excising a 1.5 Mdal. *Bgl* II fragment carrying the *thy* gene. This generates pCD2 (Fig. 8.6) in which the region homologous with the *B. subtilis* chromosome is only 0.5 Mdal. long. Like ϕ3T, phage β22 DNA also encodes thymidylate synthetase but the corresponding gene has no detectable homology with *B. subtilis*. The *thy* gene from β22 has been cloned in pCD2, using *Bgl* II, to generate pCD4 (Fig. 8.6). In *E. coli* pCD4 again complements *thy* auxotrophs and is maintained as a plasmid. The plasmid also transforms Thy$^-$ auxotrophs of *B. subtilis* to Thy$^+$ but in this case the entire plasmid is integrated into the genome, not just the *thy* gene. When the same *Bgl* II fragment from β22 was inserted into pMB9 rather than pCD2, Thy$^+$ transformants were obtained in *E. coli* but not *B. subtilis*, i.e. integration of the plasmid is dependent on the 0.5 Mdal. region of homology.

The *E. coli* chromosomal *thy* gene has been cloned in pMB9 and the resulting chimaera, pER2, has no detectable homology to the *B. subtilis* chromosome by Southern blotting. There must be some very low level of homology, however, for pER2 can transform thymine auxotrophs of *B. subtilis* to thymine inde-

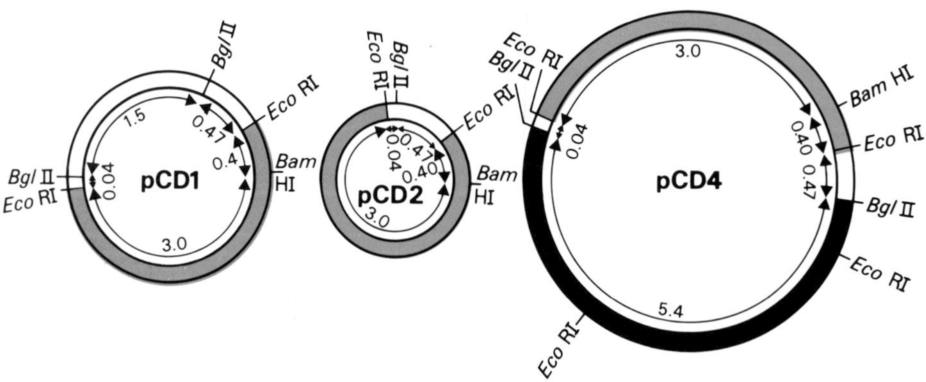

Fig. 8.6 The structures of pCD1, pCD2 and pCD4. The red areas represent material from pMB9, the unshaded areas material from ϕ3T and the black areas material from β22. In pCD1, the *thy* gene is carried on the 1.5 Mdal. fragment. In pCD4 in *thy* gene is carried on the 5.4 Mdal. fragment between the two *Eco* RI sites.

pendence. In all these Thy^+ transformants pER2 is integrated into the *B. subtilis* chromosome (Rubin *et al.* 1980).

Clearly, it is possible to integrate foreign genes into the chromosome of *B. subtilis* through the development of chimaeric plasmids with limited regions of homology. The addition of a foreign segment of DNA should provide an opportunity for the introduction of additional segments of foreign DNA through recombination in the new region of homology. This principle of 'scaffolding' is shown in Fig. 8.7. For example, the first segment of foreign DNA introduced could be pER2, as described above. Subsequently, another gene cloned in pBR322 could be used to transform the first recipient. In this way any gene previously cloned in pBR322 could be readily integrated into the chromosome of *B. subtilis*.

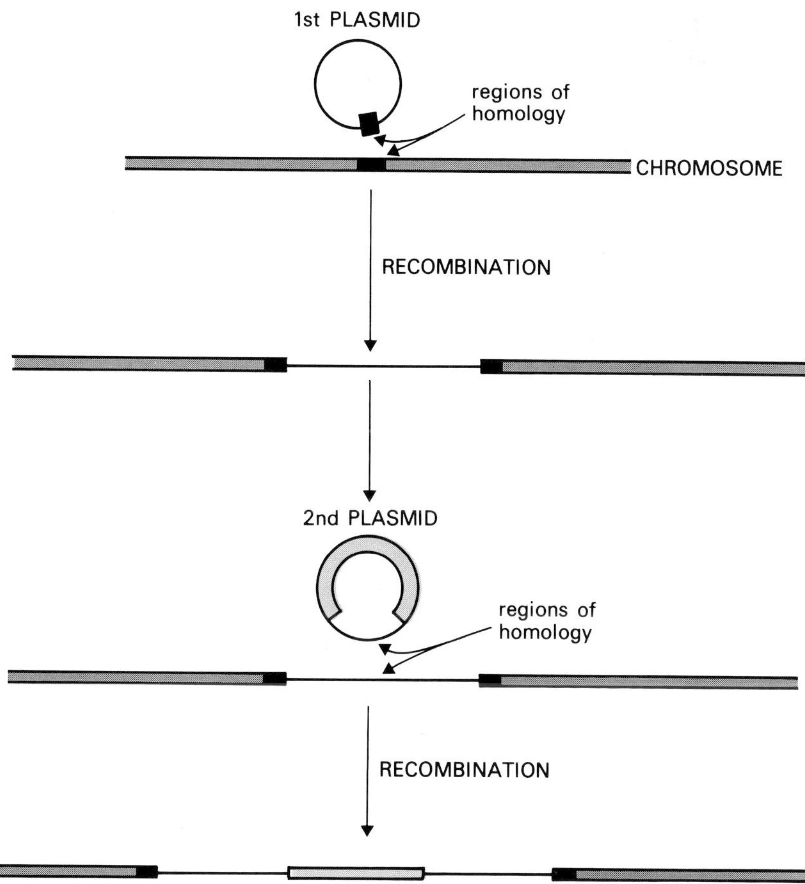

Fig. 8.7 The principle of 'scaffolding'.

Cloning sequences adjacent to the site of insertion

Once a recombinant plasmid has integrated into the chromosome it is relatively easy to clone adjacent sequences. Suppose, for example, that a pHV32 derivative carrying *B. subtilis* DNA in the *Bam* HI site (Fig. 8.8) has recombined into the chromosome. If the recombinant plasmid has no *Bgl* II sites it can be recovered by digesting the chromosomal DNA with *Bgl* II, ligating the resulting fragments and transforming *E. coli* to Ap^R. However, the plasmid which is isolated will be larger

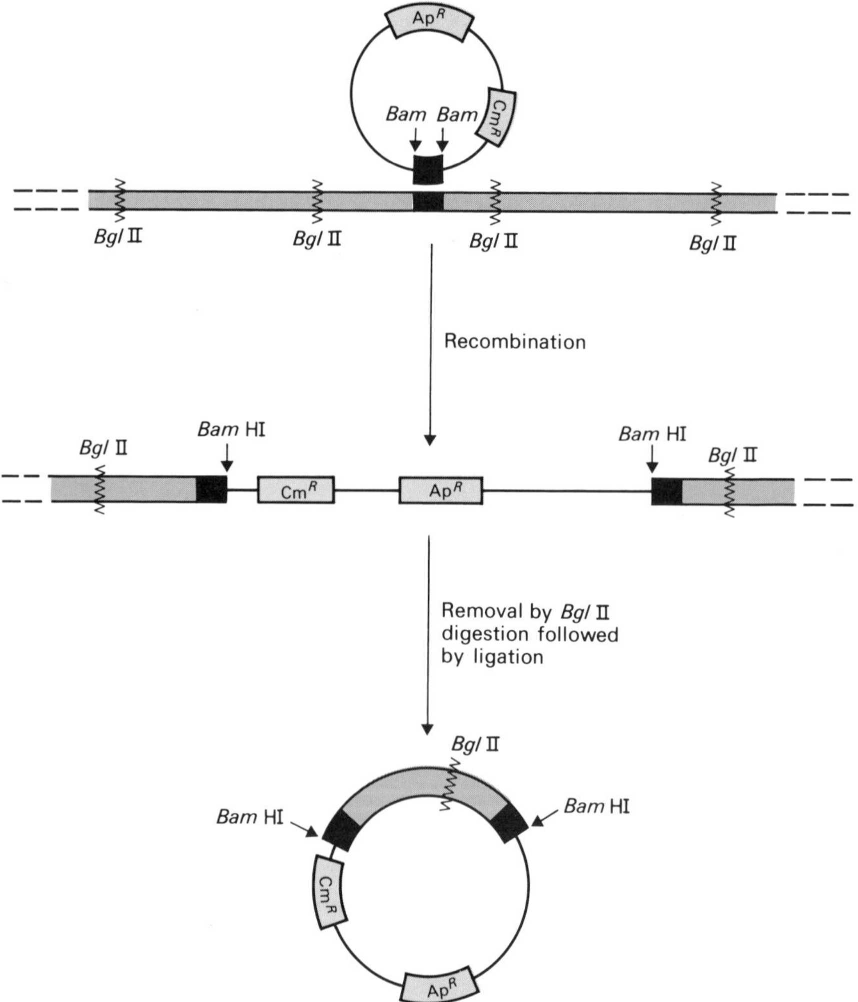

Fig. 8.8 Cloning DNA sequences flanking the site of insertion. The solid black bar on the plasmid represents a *Bam* HI fragment of *B. subtilis* chromosomal DNA. Note that the plasmid has no *Bgl* II sites. (See text for details.)

than the original one because DNA flanking the site of insertion will also have been cloned. In this way Niaudet *et al.* (1982) used a plasmid carrying a portion of the *B. subtilis ilv*A gene to clone the adjacent *thy*A gene.

Deleting regions of the chromosome

In the experiments described above the integration of non-replicating vectors into the chromosome of *B. subtilis* occurred via a single cross-over event (see Figs 8.7 and 8.8). However, bearing in mind that linear molecules may be formed during transformation of *B. subtilis* with plasmids, insertion of plasmid DNA into the chromosome can occur by a double cross-over. If

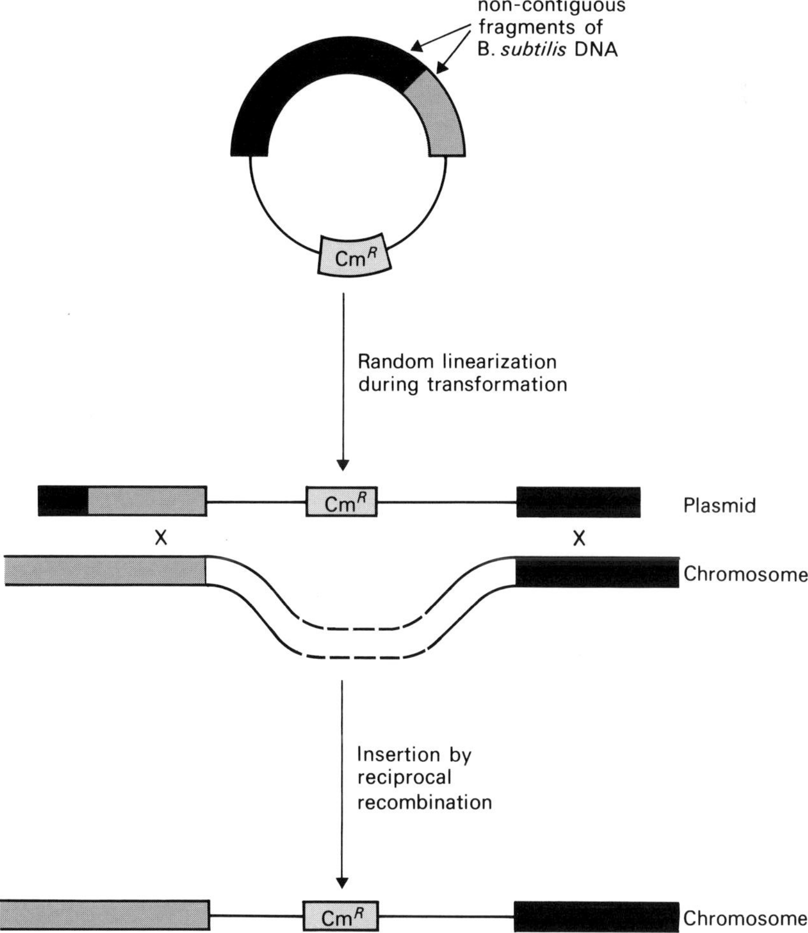

Fig. 8.9 Formation of deletions during insertion of a recombinant plasmid.

more than one fragment of *B. subtilis* DNA is cloned, and these fragments come from non-adjacent regions of the chromosome, then the intervening DNA may be deleted during integration of the plasmid (Niaudet *et al.* 1982). The basic principles of this technique, which are shown in Fig. 8.9, could be applied to many organisms.

Expression of cloned genes in *B. subtilis*

E. coli is promiscuous in its ability to recognize transcription and translation signals from a wide variety of organisms, e.g. the expression of drug-resistance genes from Gram-positive bacteria (Cohen *et al.* 1972; Ehrlich 1978) and amino acid biosynthetic genes from yeast (Struhl *et al.* 1976). By contrast, *B. subtilis* is restricted in its ability to express genes from other genera (Ehrlich 1978, S. D. Ehrlich *et al.* 1978). With one exception (Rubin *et al.* 1980), the only foreign genes expressed in *B. subtilis* are from other Gram-positive bacteria, e.g. drug resistance genes from *S. aureus* (Ehrlich 1977) and *Streptococcus* (Yagi *et al.* 1978). The lack of expression of *E. coli* genes in *B. subtilis* is not due to structural alterations, for these genes can be isolated from *B. subtilis* and are still functional when reintroduced into *E. coli*. What, then, are the barriers to the expression of *E. coli* genes in *B. subtilis*?

The first step in gene expression is the transcription of DNA by RNA polymerase. A comparison of the RNA polymerases from the two organisms has not revealed gross differences but the template specificity of the two polymerases differs in a very interesting way. The *E. coli* enzyme can transcribe equally well DNAs of *E. coli* bacteriophage T4 and of *B. subtilis* bacteriophage φe. By contrast, the *B. subtilis* polymerase is much less active with the *E. coli* than with the *B. subtilis* phage DNAs (Shorenstein & Losick 1973). This behaviour is consistent with the observation that *B. subtilis* genes can function in *E. coli* but the *E. coli* genes cannot function in *B. subtilis*. Moran *et al.* (1982) have determined the nucleotide sequence of two *B. subtilis* promoters and found no obvious differences with those of *E. coli*. They concluded that *B. subtilis* RNA polymerase may demand high fidelity to the canonical −35 and −10 promoter hexanucleotides and possibly other sequences as well. Two possibilities for these other sequences are an AT-rich region upstream from the −35 region and a PuTPuTG sequence at positions −18 to −14.

The translational apparatus of *E. coli* and various *Bacillus* spp. are considered to be essentially similar. Nevertheless, intriguing differences have been reported. *E. coli* ribosomes were found to support protein synthesis when directed by four out of five mRNAs from Gram-negative cells and six out of six mRNAs from Gram-positive cells. By contrast, *B. subtilis* ribosomes were not active with any of the mRNAs from Gram-negative bacteria nor three out of four mRNAs from Gram-positive bacteria (Stallcup *et al.* 1974). McLaughlin *et al.* (1981) and Moran *et al.* (1982) have analysed the mRNA sequences located downstream from a number of *B. subtilis* chromosomal and phage promoters. They have found that the sequence complementarity between the mRNA and the 3′ terminal region of the 16s ribosomal RNA is far greater than with *E. coli* Shine-Dalgarno sequences. Confirmation of this has been provided by Band and Henner (1984). In addition, they constructed a series of plasmids differing only in the sequence of the Shine-Dalgarno region preceding an interferon-alpha gene and used this series of plasmids to test the efficiency of interferon expression in both *B. subtilis* and *E. coli*. In *B. subtilis* interferon expression was much more sensitive to changes in the sequence of the Shine-Dalgarno region than in *E. coli*.

Expression vectors for *B. subtilis*

Because of the problems associated with heterologous gene expression in *B. subtilis* expression vectors developed for use in *E. coli* (see p. 128) cannot be used in this organism. Consequently specific *B. subtilis* expression vectors have been constructed. Williams *et al.* (1981a) generated a plasmid which can be used to clone promoters active in *B. subtilis*. This plasmid, pPL603, specifies Km^R and carries a *Bacillus* structural gene for chloramphenicol acetyl transferase (CAT) which lacks a promoter. There is a unique *Eco* RI site upstream of the CAT coding sequence (Fig. 8.10). Promoter-containing DNA is identified by cloning *Eco* RI-generated fragments of DNA from *B. subtilis* spp. into pPL603 with subsequent selection of Cm^R transformants. In such transformants the CAT is inducible by chloramphenicol regardless of the promoter fragment cloned because regulation seems to occur at the level of translation. A similar construction has been utilized by Goldfarb *et al.* (1981) but because their Cm^R gene was derived from *E. coli* the CAT is not inducible by Cm.

Plasmid pPL608 is a derivative of pPL603 containing a 0.2 Mdal. *Eco* RI* promoter fragment from *Bacillus* phage SPO2. This promoter fragment permits expression of the CAT gene enabling pPL608 to confer Cm^R on *B. subtilis*. Williams *et al.*

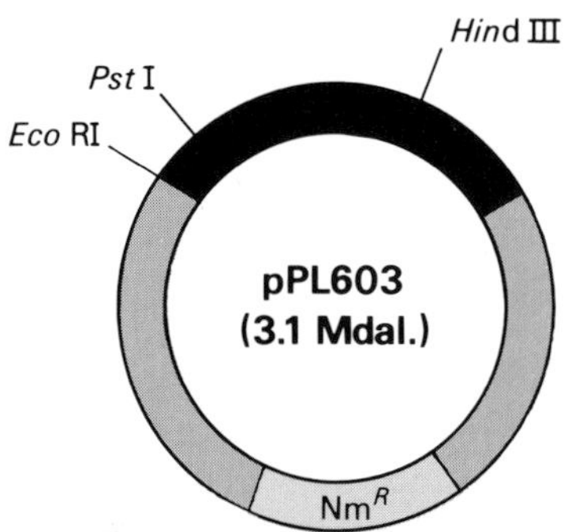

Fig. 8.10 The structure of plasmid pPL603. The black bar represents a DNA fragment from *B. pumilus* which includes a promoter-less gene specifying chloramphenicol acetyl transferase (CAT). The initiation codon for the CAT gene lies between the unique *Pst* I and *Hind* III sites.

(1981b) cloned the *E. coli trp*C gene in the *Hind* III site of pPL608 and obtained complementation of a *B. subtilis* tryptophan auxotroph. Expression of the *trp* gene was Cm inducible but the Cm^R phenotype was destroyed. Proof that the *trp* gene was under the control of the SPO2 promoter was provided by the observation that removal of the promoter fragment or reversing the orientation of the *E. coli* fragment did not yield Trp$^+$ transformants.

Williams *et al.* (1981b) also cloned a mouse dihydrofolate reductase (DHFR) gene in pPL608 but used the *Pst* I site instead of the *Hind* III site. Mammalian DHFR is resistant to trimethoprim (Tp) whereas the bacterial enzyme is Tp sensitive. Expression of the mouse DHFR in *B. subtilis* was detected by the Tp^R phenotype which is conferred. However, the transformants were Cm^R and Tp^R and the Tp^R was not Cm-inducible indicating that the *Pst* I site lies between the promoter fragment and the start of the CAT gene. Schoner *et al.* (1983) have identified a plasmid mutation lying between the *Eco* RI and *Pst* I sites which increases expression of the mouse DHFR tenfold. Nucleotide sequence analysis suggests that the SPO2 fragment contains sequences that correspond to a ribosome binding site.

Similar constructions to those of Williams *et al.* (1981b) have been made by Hardy *et al.* (1981). They fused a DNA fragment encoding all but the eight N-terminal amino acids of the foot and mouth disease virus (FMDV) VP1 antigen into the Em^R gene of a *B. subtilis* vector. *B. subtilis* cells transformed with the chimaeric plasmid synthesized a fusion protein consisting of the N-terminus of the Em^R gene product and VP1 protein minus the first eight amino acids. Just as expression of Em resistance is inducible by Em, so too was synthesis of the fusion protein. In addition, they placed the gene encoding the hepatitis B core antigen just downstream from the termination codon of the Em^R gene and in phase with it. Again, expression of the cloned gene was inducible by Em.

A different kind of repression system has been devised by Yansura and Henner (1984). They placed the *E. coli lac* operator on the 3′ side of the promoter of a *Bacillus* penicillinase gene thus creating a hybrid promoter controllable by the *E. coli lac* repressor. The *E. coli lac* repressor gene was placed under the control of a promoter and binding site functional in *B. subtilis*. When the penicillinase gene that contained the *lac* operator was expressed in *B. subtilis* on a plasmid that also produces the *lac* repressor, the expression of the penicillinase gene was modulated by IPTG, an inducer of the *lac* operon in the *E. coli*.

Secretion vectors for use in *B. subtilis*

As discussed in Chapter 7 most secretory proteins are synthesized initially as preproteins which have additional amino acids, the signal sequence, at their N-termini. Proteins secreted by *Bacillus* spp. are no different in that they have a signal sequence with a hydrophilic region at the N-terminus followed by a hydrophobic region. However *Bacillus* signal sequences are longer than those from Gram-negative bacteria and eukaryotes. This is reflected in an increase in size of the hydrophilic region and the presence of a peptide extension beyond the hydrophobic core (Kroyer & Chang 1981, Neugebauer *et al.* 1981, Palva *et al.* 1981). Despite these differences between Gram-positive and Gram-negative signal sequences they can be used in an identical way (see p. 132) to promote secretion of cloned-gene products. Thus Palva *et al.* (1982) fused an *E. coli*-derived β-lactamase gene to the promoter and signal sequence region of the α-amylase gene from *B. amyloliquefaciens*. The β-lactamase gene was expressed in *B. subtilis* and over 95% of the enzyme activity was secreted to the growth medium. Later Palva *et al.* (1983) showed that the amylase signal would also permit secretion of interferon from *B. subtilis*.

CLONING IN *STREPTOMYCES*

The only Gram-positive organisms other than *B. subtilis* in which there has been extensive in-vitro gene manipulation are

Streptomyces spp. Interest in these organisms stems largely from the fact that *Streptomyces* spp. produce most of the known antibiotics, including many of clinical utility. By using gene cloning it may be possible to improve the fermentation yields of these antibiotics and perhaps, if genes from different species are recombined, generate novel antibiotics. The techniques used for gene cloning in *Streptomyces* are similar to those used in other bacteria and consequently will not be discussed here (for reviews see Hopwood & Chater 1982, Hopwood *et al.* 1983). Three vectors are available:

1 a low copy-number, narrow host range plasmid,
2 a high copy-number, broad host range plasmid, and
3 a phage vector.

DNA of all three is taken up efficiently by *Streptomyces* protoplasts.

Little is known about the regulation of gene expression in *Streptomyces*. *Streptomyces* promoters do not appear to function in *E. coli*, whereas at least some *E. coli* and *B. subtilis* promoters permit transcription in *S. lividans*. However, the expression of totally heterologous genes in *Streptomyces* is probably of little applied interest. Much more important will be an understanding of those regulatory signals which switch on antibiotic synthesis after the organism has stopped growing.

Some industrial applications of cloning in *Streptomyces* spp. are discussed in Chapter 14.

Chapter 9 Cloning in Yeast and other Microbial Eukaryotes

The analysis of eukaryotic DNA sequences has been facilitated by the ease with which DNA from eukaryotes can be cloned in prokaryotes using the vectors described in previous chapters. Such cloned sequences can be obtained easily in large amounts and can be altered *in vivo* by bacterial genetic techniques and *in vitro* by specific enzyme modifications. To determine the effects of these experimentally induced changes on the function and expression of eukaryotic genes, the rearranged sequences must be taken out of the bacteria in which they were cloned and reintroduced into a eukaryotic organism, preferably the one from which they were obtained. Despite the overall unity of biochemistry there are many functions common to eukaryotic cells which are absent from prokaryotes, e.g. localization of ATP-generating systems to mitochondria, association of DNA with histones, mitosis and meiosis, and obligate differentiation of cells. The genetic control of such functions must be assessed in a eukaryotic environment. In this chapter we will discuss the potential for cloning in the yeast *Saccharomyces cerevisiae* and in later chapters we will discuss the possibilities for cloning in animal and plant cells. It should be borne in mind that as well as using yeast as a host for cloned genes from other eukaryotes it also can be used as a host for analysing cloned yeast genes, i.e. surrogate yeast genetics.

Transformation in yeast

Transformation of yeast was first achieved by Hinnen *et al.* (1978) who fused yeast spheroplasts (i.e. wall-less yeast cells) with polyethylene glycol in the presence of DNA and $CaCl_2$ and then allowed the spheroplasts to regenerate walls in a stabilizing medium containing 3% agar. The transforming DNA used was plasmid pYeLeu 10 which is a hybrid composed of the *E. coli* plasmid Col E1 and a segment of yeast DNA containing the Leu 2^+ gene. Spheroplasts from a stable Leu 2^-

auxotroph were transformed to prototrophy by this DNA at a frequency of 1×10^{-7}. Untreated spheroplasts reverted with a frequency of $<1 \times 10^{-10}$. When 42 Leu$^+$ transformants were checked by hybridization, 35 of them contained Col E1 DNA sequences. Genetic analysis of the remaining seven transformants indicated that there had been reciprocal recombination between the incoming Leu 2$^+$ and the recipient Leu 2$^-$ alleles.

Of the 35 transformants containing Col E1 DNA sequences, genetic analysis showed that in 30 of them the Leu 2$^+$ allele was closely linked to the original Leu 2$^-$ allele whereas in the remaining 5, the Leu 2$^+$ allele was located on another chromosome. These results can be confirmed by restriction endonuclease analysis since pYeLeu 10 contains no cleavage sites for *Hin*d III. When DNA from the Leu 2$^-$ parent was digested with endonuclease *Hin*d III and electrophoresed in agarose, multiple DNA fragments were observed but only one of these hybridized with DNA from pYeLeu 10. With the 30 transformants in which the Leu 2$^-$ and Leu 2$^+$ alleles were linked, only a single fragment of DNA hybridized to pYeLeu 10 but this had an increased size consistent with the insertion of a complete pYeLeu 10 molecule into the original fragment. This data is consistent with their being a tandem duplication of the Leu 2 region of the chromosome (Fig. 9.1). With the remaining

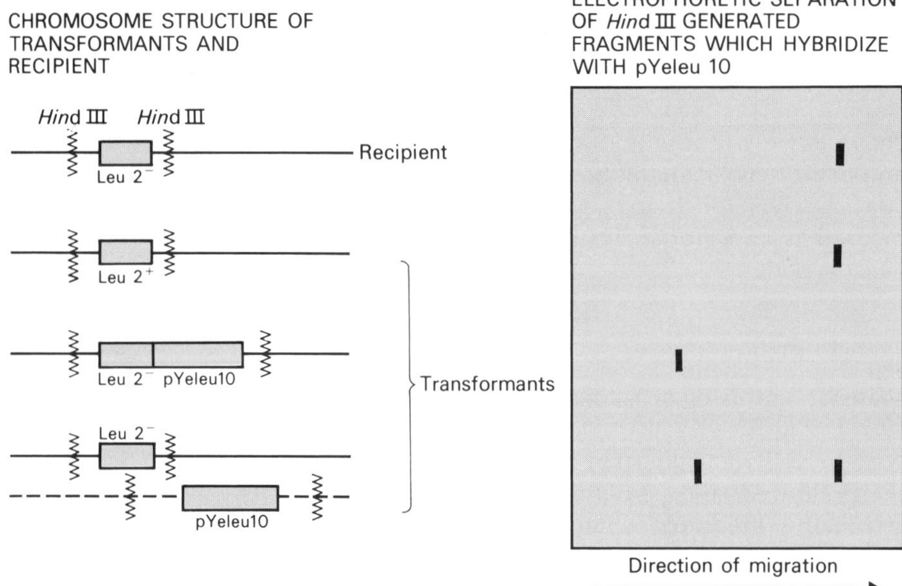

Fig. 9.1 Analysis of yeast transformants. See text for details.

five transformants, two DNA fragments which hybridized to pYeLeu 10 could be found on electrophoresis. One fragment corresponded to the fragment seen with DNA from the recipient cells, the other to the plasmid genome which had been inserted in another chromosome (Fig. 9.1). These results represented the first unambiguous demonstration that foreign DNA, in this case cloned Col E1 DNA, can integrate into the genome of a eukaryote. A plasmid such as pYeLeu 10 which can do this is known as a YIp-*y*east *i*ntegrating *p*lasmid.

The development of yeast vectors: 1. Common principles

Since the first demonstration of transformation in yeast a number of different kinds of yeast vector have been constructed. All of them have features in common. First, all of them can replicate in *E. coli*, often at high copy number. This is important because for many experiments it is necessary to amplify the vector DNA in *E. coli* before transformation of the ultimate yeast recipient. Second, all employ markers, e.g. Leu 2^+, His^+, Ura 3^+, Trp 1^+, that can be selected readily in yeast and which often will complement the corresponding mutations in *E. coli* as well. In addition to these selectable markers most of the vectors also carry antibiotic resistance markers for use in *E. coli*. Finally, all of them contain unique target sites for a number of restriction endonucleases.

The development of yeast vectors: 2. Yeast episomal plasmids (YEp)

Many strains of *S. cerevisiae* contain a plasmid, 2 μm long (6 kb), which has no known function. This plasmid replicates under nuclear control and there are approximately 50–100 copies per cell. It contains an inverted repeat sequence 600 base pairs long and is isolated as a mixture of two forms (Fig. 9.2) which differ from one another in the orientation of the two non-repeated segments of the molecule with respect to each other. The two forms are presumably due to recombination between the inverted repeat sequences (Hollenberg *et al.* 1976). Beggs (1978) constructed chimaeric plasmids containing this 2 μm yeast plasmid, fragments of yeast nuclear DNA and the *E. coli* vector pMB9. These chimaeras were able to replicate in both *E. coli* and yeast, transformed yeast with high frequency, and some were able to complement auxotrophic mutations in yeast.

The chimaeric plasmids were constructed in two stages.

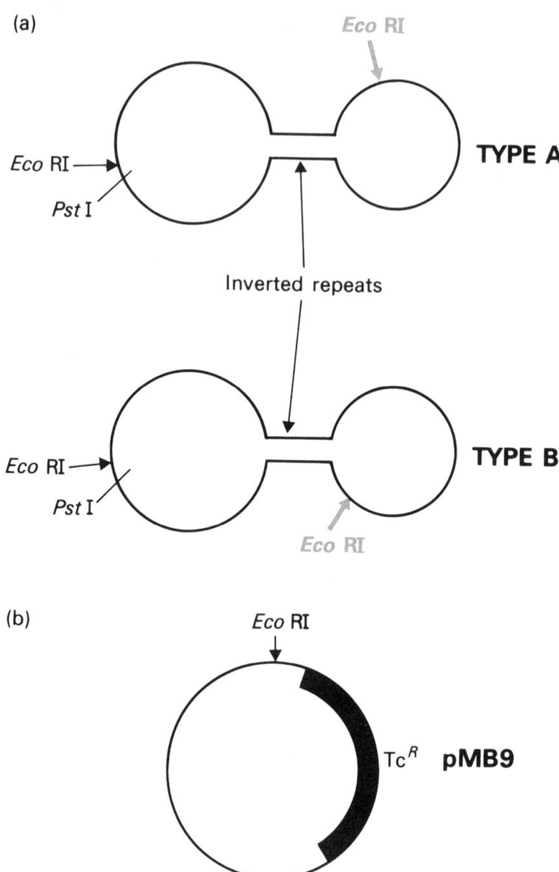

Fig. 9.2 Structures of (a) the yeast 2 μm plasmid, and (b) pMB9. Only the cleavage sites for *Eco* RI and *Pst* I are shown.

First, plasmid pMB9 (Fig. 9.2) and the 2 μm plasmid were joined by ligation of the DNA fragments produced by *Eco* RI endonuclease digestion. Tc^R clones were selected after transforming *E. coli* with the ligated DNAs and these were screened for a hybrid plasmid large enough to carry the complete yeast 2 μm plasmid sequence. Yeast–pMB9 hybrid plasmids theoretically have eight possible configurations. These are determined by the orientation of the insertion into pMB9, which of the *Eco* RI sites on the yeast plasmid is used for insertion and whether the yeast sequence is in the A or B configuration. Five of the eight possible configurations, which can be distinguished by restriction mapping, were found among seven complete hybrid plasmids examined. The exact configuration of the yeast 2 μm plasmid–pMB9 hybrids is probably not important.

For the second stage Beggs (1978) sheared nuclear DNA isolated from *S. cerevisiae* and linked it by means of poly (dA-dT) tails to a mixture of pMB9-yeast 2 μm plasmid hybrid plasmids which had been linearized by digestion with *Pst* I. Tc^R transformants were selected in *E. coli* and of the 21 000 obtained, two were found which complemented a Leu^- mutant. The plasmids from these clones, pJDB219 and pJDB248, transformed a *leu*B mutant of *E. coli* to Leu^+ or tetracycline resistance at the same frequency.

Both pJDB219 and pJDB248 transformed Leu 2 yeast mutants to Leu^+ with a frequency of 5×10^{-4} to 3×10^{-3} transformants per viable cell. This is several orders of magnitude greater than the frequency of transformation obtained by Hinnen *et al.* (1978). However, chromosomal integration of the transforming fragment was essential with the system of Hinnen *et al.* (1978). By contrast pJDB219 and pJDB248 could be recovered as plasmids from yeast cells and their inheritance in yeast was non-Mendelian.

Many other YEp vectors based on the 2 μm plasmid have been developed, e.g. Gerbaud *et al.* 1979, Storms *et al.* 1979, Struhl *et al.* 1979. They possess the same advantages as those of Beggs (1978), viz, they have a high copy number (25–100 copies $cell^{-1}$) in yeast and they transform yeast very well. Indeed, the transformation frequency can be increased 10- to 20-fold if single-stranded plasmids are used (Singh *et al.* 1982). The major problem associated with all YEp vectors is that novel rearranged recombinant plasmids are generated *in vivo* in *E. coli* and yeast.

The development of yeast vectors: 3. Yeast replicating plasmids (YRp)

Struhl *et al.* (1979) constructed a useful vector which consists of a 1.4 kb yeast DNA fragment containing the *trp*-1 gene inserted into the *Eco* RI site of pBR322. This vector transformed *trp*-1 yeast protoplasts to Trp^+ at high frequency and transforming sequences were always detected as ccc DNA molecules in yeast. No transformants were found in which the vector had integrated into the chromosomal DNA. Since pBR322 alone cannot replicate in yeast cells (Beggs 1978) a yeast chromosomal sequence must permit the vector to replicate autonomously and to express yeast structural genes in the absence of recombination with host chromosomal sequences. A similar vector based on pBR313 was developed by Kingsman *et al.* (1979). In both cases the yeast gene was linked to a centromere and initially it was thought that this was important. Since then it has been shown that a centromere is not essential; rather, the vector carries an *a*utonomously *r*eplicating *s*equence (*ars*) derived from the chromosome.

Although plasmids containing an *ars* transform yeast very efficiently the resulting transformants are exceedingly unstable, segregating at rates well above 1% per generation of growth in non-selective medium. Occasional stable transformants are found and these appear to be cases in which the entire YRp has integrated into a homologous region in a manner identical to that of YIps (Stinchcomb *et al.* 1979, Nasmyth & Reed 1980). The copy number of YRp vectors is much lower than that of YEp vectors.

The development of yeast vectors: 4. Yeast centromere plasmids (YCp)

Using a YRp vector Clarke and Carbon (1980) isolated a number of hybrid plasmids containing DNA segments from around the centromere-linked *leu*2, *cdc*10 and *pgk* loci on chromosome III of yeast. As expected for plasmids carrying an *ars* most of the recombinants were unstable in yeast. However, one of them was maintained stably through mitosis and meiosis. The stability segment was confined to a 1.6 kb region lying between the *leu*2 and *cdc*10 loci and its presence on plasmids carrying either of 2 *ars* tested resulted in those plasmids behaving like minichromosomes (Clarke & Carbon 1980, Hsiao & Carbon 1981). Genetic markers on the minichromosomes acted as linked markers segregating in the first meiotic division as centromere-linked genes and were unlinked to genes on other chromosomes. Stinchcomb *et al.* (1982) and Fitzgerald-Hayes *et al.* (1982) have isolated the centromeres from chromosomes IV and XI of yeast and found that they confer on plasmids similar properties to the centromere of chromosome III. The isolation of plasmids bearing centromeres functional in yeast is a particularly exciting development for it should permit a detailed molecular analysis of the events occurring at mitosis and meiosis. Already a detailed analysis has been made of the molecular architecture of the centromere (Bloom & Carbon 1982).

Surrogate genetics of yeast

For many biologists the primary purpose of cloning is to understand what particular genes do *in vivo*. Thus most of the applications of yeast vectors have been in the surrogate genetics of yeast. One advantage of cloned genes is that they can be analysed easily, particularly with the advent of DNA sequencing methods. Thus nucleotide sequencing analysis can reveal many of the elements which control expression of a gene as well as identifying the sequence of the gene product. In the case of the yeast actin gene (Gallwitz & Sures 1980, Ng & Abelson 1980) and some yeast tRNA genes (Abelson 1979, Olson 1981) this kind of analysis revealed the presence within these genes of non-coding sequences which are copied into primary transcripts. These *introns* subsequently are eliminated by a process known as *splicing* and this is discussed in more detail on page 200. Nucleotide sequence analysis also can reveal the direction of transcription of a gene although this can be determined *in vivo* by other methods; for example, if the yeast gene is expressed in *E. coli* using bacterial transcription signals, the direction of reading can be deduced by observing the orientation of a cloned fragment required to permit expression. Finally, if a single transcribed yeast gene is present on a vector the chimaera can be used as a probe for quantitative solution hybridization analysis of transcription of the gene.

The availability of different kinds of vectors with different properties (see Table 9.1) enables yeast geneticists to perform manipulations in yeast like those long available to *E. coli* geneticists with their sex factors and transducing phages. Thus cloned genes can be used in conventional genetic analysis by means of recombination using YIp vectors, or complementation using YEp, YRp or YCp vectors. These latter vectors also permit the formation of partial diploids and partial polyploids. Transposition of genes can be done using integrative transformation and the new position of the gene determined by conventional Mendelian analysis. Deletions, point mutations and frame shift mutations can be introduced *in vitro* into cloned genes and the altered genes returned to yeast and used to replace the wild-type allele. Excellent reviews of these techniques have been presented by Botstein and Davis (1982), Hicks *et al.* (1982) and Struhl (1983).

Targeted selection of recombinant clones

In an analogous fashion to that used for prokaryotic genes, many yeast genes have been identified by transforming a mutant yeast strain with a recombinant library followed by screening or selecting for clones that complement the mutation of interest. For this approach to work it is essential to have a mutation in the gene of interest and a genetic marker that

Table 9.1 Properties of the different yeast vectors.

Vector	Transformation frequency	Loss in non-selective medium	Disadvantages	Advantages
YIp	1–10 transformants per μg DNA	Much less than 1% per generation	(1) Low transformation frequency (2) Can only be recovered from yeast by cutting chromosomal DNA with restriction endonuclease which does not cleave original vector containing cloned gene	(1) Of all vectors this kind give most stable maintenance of cloned genes (2) An integrated YIp plasmid behaves as an ordinary genetic marker, e.g. a diploid heterozygous for an integrated plasmid segregates the plasmid in a Mendelian fashion. (3) Most useful for surrogate genetics of yeast, e.g. can be used to introduce deletions, inversions and transpositions (see Botstein & Davis 1982)
YEp	10^3–10^5 transformants per μg DNA	1% per generation	(1) Novel recombinants generated *in vivo* by recombination with endogenous 2 μm plasmid	(1) Readily recovered from yeast (2) High copy number (3) High transformation frequency (4) Very useful for complementation studies

YRp	10^2–10^3 transformants per μg DNA	Much greater than 1% per generation but can get chromosomal integration	(1) Instability of transformants	(1) Readily recovered from yeast (2) High copy number. Note that the copy number is usually less than that of YEp vectors but this may be useful if cloning gene whose product is deleterious to the cell if produced in excess (3) High transformation frequency (4) Very useful for complementation studies (5) Can integrate into the chromosome
YCp	10^2–10^3 transformants per μg DNA	Less than 1% per generation	(1) Low copy number makes recovery from yeast more difficult than that of YEp or YRp vectors	(1) Low copy number is useful if product of cloned gene is deleterious to cell (2) High transformation frequency (3) Very useful for complementation studies (4) At meiosis generally shows Mendelian segregation

can be used to select the desired transformants. However, despite the fact that *S. cerevisiae* has been intensively studied genetically only a small fraction of the genes encoding enzymes has been identified by mutation. Furthermore, there are several examples of two or more yeast genes encoding either the same product or functionally homologous products or multiple coding sequences for the same product, e.g. alcohol dehydrogenase. Such duplication makes it difficult to obtain suitable mutants. Rine *et al.* (1983) have presented an alternative approach to complementation for isolating genes. Their method is based on the observation that it is possible to overcome the effects of an inhibitor by increasing the dosage of the gene encoding the target protein. In practice, the concentration of the inhibitor which is just sufficient to prevent growth of wild-type yeast cells is determined and the transformants are selected which are resistant to this concentration of inhibitor. In this way Rine *et al.* (1983) were able to clone genes conferring resistance to tunicamycin (an inhibitor of UDP-N-acetyl glucosamine-1-P transferase), compactin (an inhibitor of 3-hydroxy-3-methyl-glutamyl-CoA reductase), and ethionine (an inhibitor of S-adenosyl methionine synthetase).

Linear vectors for cloning yeast telomeres

All three autonomous plasmid vectors are maintained in yeast as circular DNA molecules—even the YCp vectors which possess yeast centromeres. Thus none of these vectors can model the eukaryotic chromosome which is a linear structure. A model system is of particular interest as the biochemical mechanism by which the ends of linear DNA molecules are replicated is unknown (see Watson 1972, for detailed discussion). Szostak and Blackburn (1982) have developed such a model system by using a linear vector to clone yeast telomeres.

The linear yeast vector was constructed from two components. The first was a YRp, pSZ213 (Fig. 9.3) which has no *Bam* HI sites and a single *Bgl* II site. The second component was fragments of an unusual, linear, rRNA-encoding plasmid found in the protozoan *Tetrahymena*. This plasmid DNA was cleaved with *Bam* HI and the end fragments were ligated to *Bgl* II-cut pSZ213. Since *Bam* HI-cut and *Bgl* II-cut ends of DNA are complementary they can be joined by ligase but the product of ligation is not a substrate for either enzyme. Consequently both *Bam* HI and *Bgl* II endonucleases were present in the ligation mixture to cut circularized or dimerized vector molecules and

(a)

(b)

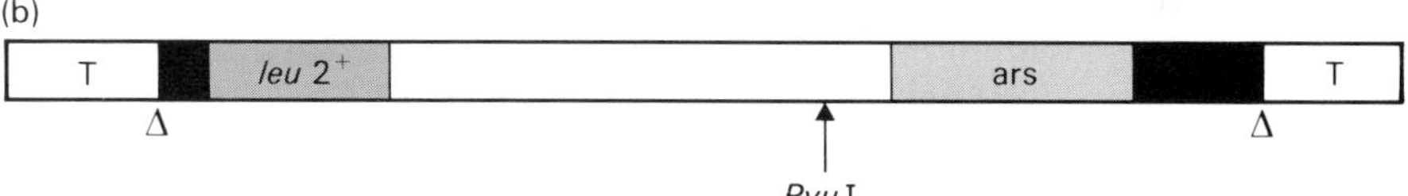

Fig. 9.3 (a) Simplified map of plasmid pSZ213 used for cloning yeast telomeres. The plasmid has no *Bam* HI sites and insertion of telomeric DNA at the unique *Bgl* II site inactivates the *his*3$^+$ gene. (b) Structure of a linear plasmid constructed from pSZ213 showing the asymmetric location of the unique *Pvu* I site. The open triangles indicate the location of *Bam*/*Bgl* joints created when the telomeric DNA (indicated with a T) was added to *Bgl* II-digested pSZ213.

dimerized end fragments. In this way the desired linear vector carrying *Tetrahymena* telomeres accumulated in the reaction mixture and it was purified by agarose gel electrophoresis.

The linear plasmid containing the *Tetrahymena* telomeres retains the *Leu*2$^+$ marker and this was used for selection of transformants in yeast. Since YRp vectors are capable of integration, those transformants in which the linear plasmid was replicating autonomously were detected by their mitotic instability. Restriction mapping showed that the plasmid in the unstable transformants is a linear molecule identical in structure to the linear molecule constructed *in vitro* and used in the transformation. Furthermore, the ends of *Tetrahymena* rDNA have three unusual structural features: a variable number of short, 5′-CCCCAA-3′ repeat units, specific single-strand interruptions within the repeated sequences and a cross-linked terminus. All three structural features were maintained when the *Tetrahymena* sequences were cloned in yeast.

The linear plasmid carrying the *Tetrahymena* telomeres has a single, asymmetrically placed target site for endonuclease *Pvu* I. This enzyme cuts yeast DNA into approximately 2000 fragments and since yeast has 17 chromosomes 34 of these fragments should contain telomeres. After ligating *Pvu* I-digested chromosomal and plasmid DNA unstable Leu^+ transformants were selected. Many of the plasmids generated in this way would have multiple yeast *Pvu* I fragments ligated onto the vector followed by either a yeast end or an end derived from the vector. These plasmids would be as large as or larger than the original vector. In contrast, the ligation of a single *Pvu* I telomere fragment from yeast, smaller than the *Tetrahymena*-containing fragment being replaced, would yield a linear plasmid smaller than at the start. After size screening of the plasmids from the transformants three plasmids carrying a yeast telomere at one end were identified. Analysis of these plasmids shows that yeast telomeres have at least some of the structural features of *Tetrahymena* telomeres. Furthermore, their use as hybridization probes suggests that all yeast telomeres are structurally similar.

Stability of yeast cloning vectors

As noted earlier, YRp vectors are not stably maintained by yeast cells and, in the absence of selection, are quickly lost from the population. By contrast, the segregational stability of YCp vectors which carry a yeast centromere is considerably greater than that of YRp vectors. Murray and Szostak (1983a) have found that YRp vectors have a strong bias to segregate to the mother cell at mitosis. This segregation bias explains how the fraction of plasmid-bearing cells can be small despite the high average copy number of YRp vectors. The presence of a centromere eliminates segregation bias thus accounting for the increased stability of YCp vectors relative to YRp vectors. YEp vectors are stably maintained in yeast cells provided the strain contains endogenous intact 2 μm circles. In the absence of endogenous 2 μm circles YEp vectors show maternal segregation bias.

Despite the fact that YCp vectors are relatively stable, they are still 1000 times less stable than *bona fide* yeast chromosomes. However, YCp vectors and YRp vectors are circular molecules whereas chromosomes are linear molecules. A linear plasmid vector carrying a centromere would be much more representative of a yeast chromosome. Dani and Zakian (1983) constructed linear yeast plasmids but found that stability was reduced relative to the circular plasmid. However, Murray and Szostak (1983b) found that stability of linear yeast plasmids was related to size. Thus artificial yeast chromosomes which were 55 kb long and contained cloned genes, *ars*, centromeres and telomeres had many of the properties of natural yeast chromosomes. When the artificial chromosomes were less than 20 kb in size, centromere function was impaired.

Whereas the stability of linear YCp vectors is dependent on size this is not true of YRp vectors. Murray and Szostak (1983a) constructed a linear YRp vector and found that it did not exhibit maternal segregational bias. The model used by them to explain these results is as follows. During DNA replication there is an association between *ars* elements and fixed nuclear sites. Replicated molecules remain attached to this site which is destined to segregate to the mother cell. The replicated plasmids will exist initially as catenated dimers which subsequently are resolved. If the dimers are attached to the putative segregation site by only one of their constituent monomers, their resolution by topoisomerase activity would release one monomer from each dimer and this monomer would be free to segregate at random to either the mother or the daughter cell. With linear plasmids, replication will produce two linear molecules which are not interconnected. Thus prior to mitosis there will be at least one freely segregating molecule and this explains the increased stability of linear YRp vectors.

Expression of cloned genes in yeast

As might be expected, most cloned yeast genes are expressed when reintroduced into yeast. More surprising, some bacterial genes are also expressed in yeast (Cohen *et al.* 1980, Jimenez & Davies 1980) and in one instance expression was dependent on a bacterial promoter (Breunig *et al.* 1982). Since Struhl and Davies (1980) showed that a yeast promoter is functional in *E. coli* it might be thought that transcription signals such as promoters can be active in prokaryotes and eukaryotes. However, this clearly is not the case for a number of workers failed to get expression of foreign genes in yeast. Thus Rose *et al.* (1981) obtained expression of β-galactosidase in *E. coli* when it was under the control of either the *E. coli* Tc^R promoter or the yeast *ura* 3 promoter but achieved expression in yeast only with the latter promoter. When Beggs *et al.* (1980) introduced the rabbit β-globin gene into yeast, β-globin-specific transcripts were obtained but transcription started at a position downstream from the usual initiation site. Finally, even though a *Drosophila* gene corresponding to the yeast *ade*8 locus has been identified by complementation, *Drosophila* genes complementing mutants at other yeast loci have not been obtained (Henikoff *et al.* 1981).

Use of yeast promoters

Because of difficulties in obtaining heterologous gene expression in yeast a number of groups have turned to the use of yeast promoters and translation initiation signals. Thus expression of the *E. coli* β-galactosidase gene was obtained by fusing it to the N-terminus of the *ura*3, *cyc*1 and *arg*3 genes (Guarente & Ptashne 1981), Rose *et al.* 1981, Crabeel *et al.* 1983). Expression in yeast of an interferon-alpha gene was obtained by fusing it to either the *pgk* or *trp*1 genes (Tuite *et al.* 1982, Dobson *et al.* 1983). In many instances, expression of a mature protein rather than a fusion protein is required. To achieve this Hitzeman *et al.* (1981) started with a plasmid carrying the promoter and part of the coding sequence of the *adh*1 gene (Fig. 9.4). This plasmid was cut with endonuclease *Xho* I for which there is a unique cleavage site downstream from the initiating ATG codon. The linearized vector was digested with nuclease *Bal* 31 to remove 30–70bp of DNA from each end of the molecule. The DNA then was incubated with the Klenow fragment of DNA polymerase and deoxynucleoside triphosphates to fill in the ends and synthetic *Eco* RI linkers added. After cleavage with endonucleases *Eco* RI and *Bam* HI the assorted fragments containing variously deleted *adh*1 promoter sequences were isolated by preparative electrophoresis. In this way six different promoter fragments were isolated, joined to an interferon-alpha gene and used to direct the synthesis of mature interferon in yeast. In a similar fashion synthesis of mature hepatitis B virus surface antigen was achieved from the *adh*1 promoter (Valenzuela *et al.* 1982) and mature interferon-gamma from the *pgk* promoter (Derynck *et al.* 1983).

Factors affecting expression

In all three instances where a mature foreign protein was synthesized in yeast the levels of expression were considerably less than that of yeast alcohol dehydrogenase I (*adh*I) or 3-phosphoglycerate kinase (*pgk*). There are a number of explanations; for example, Derynck *et al.* (1983) found that transcription was greatly reduced. Since the distance between

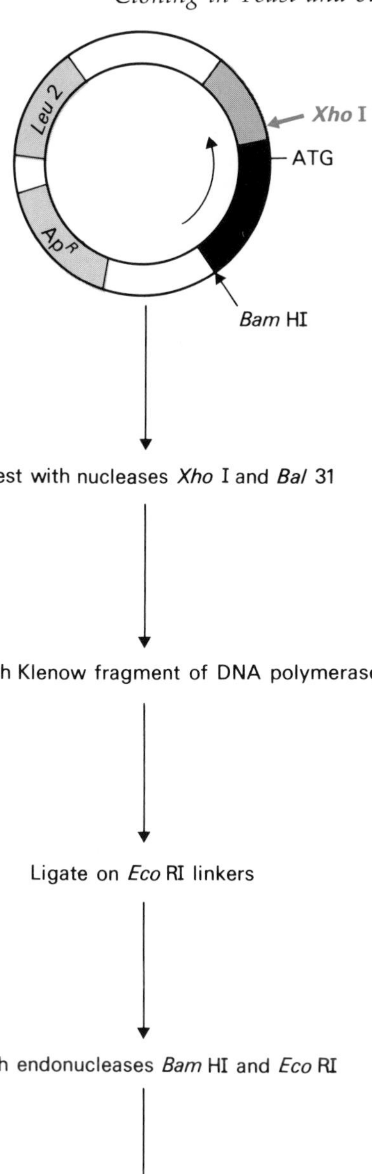

Fig. 9.4 The procedure used to generate *adh* promoter fragments. The *adh* promoter region in the starting plasmid is indicated by the black area. The grey area represents another gene fused to the N-terminus of the *adh* gene. The distance between the unique *Xho* I site and the initiating ATG codon is 17 bp. The arrow inside the plasmid circle (top) and above the *adh* promoter fragment (bottom) indicates the direction of transcription.

the promoter and initiating ATG codon is less than usual and the base sequences in the vicinity of the ATG codon are different, translation initiation might be reduced (c.f. *E. coli*, p. 141). Also, translation might be inefficient due to strong differences in codon usage between human and yeast genes, particularly since there is an extreme codon bias in highly expressed yeast genes (Bennetzen & Hall 1982). Stepien *et al.* (1983) were unable to obtain detectable expression of proinsulin from a synthetic gene under the control of the yeast *adh* promoter, in marked contrast to the results obtained with interferon and hepatitis B surface antigen. Since proinsulin-specific mRNA was abundant, poor translation due to the absence of preferred codons could be an explanation. A more likely explanation is proteolysis of the proinsulin because detectable synthesis of proinsulin was obtained when the proinsulin gene was fused in phase with the yeast galactokinase gene. This result is analogous to that obtained in *E. coli* with β-galactosidase-somatostatin gene fusions (see p. 64).

In an attempt to identify those structural features which control expression of yeast genes, Dobson *et al.* (1982) compared the nucleotide sequence of the 5′ flanking regions of 17 yeast genes. One sequence thought to be important for transcription initiation in eukaryotes was found in most of the yeast genes. This is the TATA box which is an AT-rich region with the canonical sequence TATAT/AAT/A usually located 25–32bp upstream from the transcription initiation site. Functionally it is equivalent to the Pribnow box of prokaryotes (see p. 139). Other sequences thought to be important for transcription initiation in eukaryotes were missing. With regard to translation initiation all the yeast genes had an adenine residue at position −3 and all except one had a pyrimidine, usually thymine, at position +6. The poor expression of mature hepatitis antigen and interferons alpha and gamma could be explained by poor translation initiation since in each case one of these residues was not conserved.

Ideally, cloned genes should be placed downstream from a controllable promoter and this is what Kramer *et al.* (1984) have done. They constructed a suitable vector using the promoter/regulator region of the repressible acid phosphatase gene of yeast. An interferon gene was inserted into this vector. Yeast cells transformed with the resulting plasmid chimaera produced significant amounts of interferon only when grown in medium lacking inorganic phosphate. Furthermore, mutants in two acid phosphatase regulator genes (coding for a defective repressor and a temperature-sensitive positive regulator) were used to develop a yeast strain that grew well at 35 °C but produced interferon only at 23 °C, independent of the phosphate concentration.

Stetler and Thorner (1984) have developed a method for identifying yeast genes whose transcription is differentially regulated. The method is based on incorporation of the analogue 4-thiouridine into nascent RNAs which are purified by affinity chromatography on phenylmercury agarose. The purified RNAs are used to prepare cDNA copies for screening of genomic DNA libraries by hybridization. Using this procedure several cloned yeast DNA segments were found whose transcription *in vivo* was modulated by exposure to mating hormone. These hormone-responsive genes fell into 3 major classes: first, genes expressed in vegetatively growing cells that were no longer transcribed after administration of mating factor; second, genes whose expression was increased 10 to 20-fold after exposure to the mating hormone; finally, genes which were expressed only after mating factor treatment.

It must be realized that control of gene expression in yeast is much more complicated than is apparent from the above discussion. In-vitro mutagenesis of cloned genes followed by their reintroduction into yeast has resulted in the identification of *cis*-acting elements that modulate transcription despite the fact that they are located hundreds of nucleotides upstream of the site of transcription initiation. For a review of this topic the reader should consult Guarente (1984).

Secretion of proteins by yeast

Secretion of proteins synthesized in yeast can be achieved by the addition of a signal sequence. Hitzeman *et al.* (1983) constructed a series of plasmids in which either mature interferon or pre-interferon genes were placed downstream from a *pgk* promoter such that native, as opposed to fused, proteins were produced. Yeast cells carrying such plasmids synthesized interferon but only in those encoding pre-interferon was any interferon activity found in the medium. Sequencing of interferons purified from the growth medium showed that they had the same amino termini as the natural mature interferons. This result shows that yeast cells can recognize and correctly process a signal sequence from a higher eukaryote.

Targeting of proteins to the nucleus

The yeast nucleus contains a discrete set of proteins which are synthesized in the cytoplasm. In order to elucidate the mechanism governing nuclear protein localization Hall *et al.* (1984) constructed a set of hybrid genes by fusing the yeast MAT alpha 2 gene, encoding a presumptive nuclear protein, and the *E. coli lac*Z gene. A segment of the MAT alpha 2 gene product which was 13 amino acids long was sufficient to localize beta-galactosidase activity in the nucleus. The nuclear location of the beta-galactosidase was confirmed by immunofluorescence.

Excision of introns by yeast

Many eukaryotic genes contain non-coding regions called introns (see p. 189). Introns have two common structure features: their sequences begin with the dinucleotide 5′-GT-3′ and end with the dinucleotide 5′-AG-3′. Besides these invariant nucleotides there is limited structural similarity at and around the intron–exon junction and consensus sequences have been derived from comparison of more than 100 junctions. Because of the similarity of these junction sequences in widely different species, e.g. yeast and man, it was thought that the mechanism of RNA processing to remove introns might be universal. Support for this idea came from the observation that monkey cells can splice out introns from mouse and rat genes, and mouse cells correctly splice transcripts of rabbit and chicken genes. This raises the question whether yeast cells can remove introns from genes of higher eukaryotes. If so this would be of great practical value, for the presence of introns prevents shotgun cloning of functional eukaryotic genes in *E. coli*.

To test the ability of yeast cells to excise introns from foreign genes Beggs *et al.* (1980) transformed *S. cerevisiae* with a hybrid plasmid containing a cloned rabbit chromosomal DNA segment including a complete β-globin gene with two intervening sequences and extended flanking regions. Yeast cells transformed with this chimaera produced β-globin specific mRNA. However, these globin transcripts were about 20–40 nucleotides shorter at the 5′ end than normal globin mRNA, contained one intron and extended only as far as the first half of the second intron. This result could be taken to indicate that the splicing mechanisms in yeast and rabbit differ but it could be argued that a complete transcript is a prerequisite for RNA splicing and that the prematurely terminated globin RNA was not a substrate for the yeast splicing enzyme(s). Consequently, Langford *et al.* (1983) inserted into the intron-containing yeast actin gene an intron-containing fragment from either *Acanthamoeba* or duck. In both instances yeast cells removed the natural yeast intron but not the foreign intron from the chimaeric transcript.

Why are introns in foreign genes not removed by yeast cells? The sequence of the coding regions surrounding the splice site cannot be important, for Teem and Rosbash (1983) have inserted a yeast intron into the *E. coli* β-galactosidase gene and found that it is removed correctly in yeast cells. The most likely explanation is that sequences within the intron are required for correct splicing in yeast. In this context Langford and Gallwitz (1983) have found an octanucleotide (5′-TACT AACA-3′) which occurs 20 to 55 nucleotides upstream from the 3′ splice site in all split protein-coding genes of *S. cerevisiae* analysed and which is absent from most introns of higher eukaryotes. Single A→C transversions in the fourth or eighth position of this sequence prevent splicing from occurring (Langford *et al.* 1984).

CLONING IN OTHER MICROBIAL EUKARYOTES

Compared with yeast there has been relatively little development of gene cloning systems in filamentous fungi. In an analogous experiment to that of Hinnen *et al.* (1978), Case *et al.* (1979) were able to transform *Neurospora crassa* with a cloned *qa*-2^+ gene, encoding catabolic dehydroquinase. As with yeast, three types of transformation event were distinguished: replacement of the *qa*-2^- gene with the *qa*-2^+ gene, linked insertion of the *qa*-2^+ gene, and unlinked insertion. Shuttle vectors have been developed by Stohl and Lambowitz (1983) and Hughes *et al.* (1983) for the transfer of genes between *E. coli* and *N. crassa*. Southern blots show that these shuttle vectors are present in nuclear and cytosolic fractions of *Neurospora* transformants. As might be expected the frequencies of transformation of *Neurospora* with these shuttle vectors is 5- to 10-fold higher than for plasmids that transform mainly by integration.

By a modification of the procedure of Hinnen *et al.* (1978) Ballance *et al.* (1983) were able to transform *Aspergillus nidulans* with a pBR322-borne *pyr*-4 gene of *N. crassa*. The available evidence suggests that the *pyr*-4 gene and at least some of the pBR322 had integrated into the chromosome of the *A. nidulans* transformants. In a more definitive study Yelton *et al.* (1984) have shown that a cloned *A. nidulans trp*C gene becomes integrated into the *A. nidulans* chromosome following transformation.

Transformation of *Podospora* has been achieved with a hybrid plasmid consisting of *E. coli* plasmid pBR325 and defective mtDNA (Stahl *et al.* 1982). This hybrid plasmid replicated autonomously in *Podospora* where it expressed the *Podospora* senescence trait which is the successful competition of defective mtDNA with wild-type mtDNA.

Plasmids that replicate autonomously in the unicellular green alga *Chlamydomonas reinhardii* have been constructed by inserting random DNA fragments from this alga into a plasmid containing the yeast *arg* 4 locus. These plasmids transformed an *arg* auxotroph of *C. reinhardii* to prototrophy (Rochaix *et al.* 1984). The presence of free plasmids in the transformants was demonstrated by hybridisation with a specific plasmid probe and by recovering the plasmids in *E. coli*. Analysis of the plasmids showed that they contained chloroplast DNA sequences functionally equivalent to the *arg* sequence of yeast.

Chapter 10 Cloning in Plant Cells

Cloning of exogenous DNA in prokaryotic and yeast cells is so simple that now it is routinely performed in hundreds of laboratories around the world. Cloning in higher eukaryotes, particularly plants, is at a much more primitive stage because until recently there was a lack of suitable vectors. In plants, just as with *E. coli*, these vectors are being developed from DNA viruses (the Caulimoviruses) and plasmids (the Ti plasmids of *Agrobacterium tumefaciens*). In the last two years there have been extremely rapid developments in this area and these will continue for at least the next two to five years. Consequently the aim of this chapter is to provide readers with sufficient background information to enable them to follow developments as they appear in the scientific literature.

DNA plant viruses

There are two groups of plant viruses which contain DNA—the Caulimoviruses, which have double-stranded DNA, and the Geminiviruses, which have single-stranded DNA. Until recently little was known about the Geminiviruses and this, coupled with the fact that most of the technology of gene manipulation involves double-stranded DNA, has meant that most work on virus vectors has centred on the Caulimoviruses. Detailed reviews of the literature on Caulimoviruses and Geminiviruses have been provided by Shepherd (1979), Goodman (1981) and Hull and Davies (1983).

Geminiviruses

These viruses are so called because they have double or 'geminate' particles. The DNA genome is small, of mol. wt about 0.75×10^6, and consists mainly of covalently closed circles. Restriction enzyme analysis suggested that there are two non-identical molecules, i.e. a bipartite genome, and this has been confirmed following sequencing of the complete genome of Cassava Latent Virus (Stanley & Gay 1983). It has been shown that the individual DNA species of Cassava Latent Virus, separated by cloning, are not infectious on their own but only when combined. Each geminate particle carries only one component molecule of the genome. A bipartite genome could have some advantages in deriving a vector; for example, if only one component carries genes involved in replication, insertions into the other molecule might not interfere with replication. Little progress will be made in developing vectors from the Geminiviruses until more is known about their replication, transcription and translation. However, because Geminiviruses infect a number of economically important crops including two cereals, maize and wheat, they undoubtedly will receive a lot of attention in the near future. Disadvantages of the Geminiviruses as vectors are that they are largely confined to the vascular tissue of infected plants, some are not mechanically transmissible, e.g. maize streak virus, and they have spherical particles which may limit the amount of DNA that can be inserted.

Biological properties of Caulimoviruses

Only a small number of Caulimoviruses are known and these are listed in Table 10.1. They all have a similar particle size and in-vivo behaviour and several are serologically related. Caulimoviruses have restricted host ranges and are confined to a few closely related plants in nature. There appears to be little, if any, overlap between the host ranges of the individual viruses within the group, in spite of some of the close serological affinities. Cauliflower mosaic virus (CaMV), for example, infects only members of the Cruciferae in nature although it is experimentally transmissible to a few plants outside this family. The Caulimoviruses are widely distributed throughout the temperate regions of the world and are responsible for a number of economically important diseases of cultivated crops.

Most of the isolates of Caulimoviruses that have been tested are transmitted by aphids in a non-persistent or stylet-borne manner. Successful transfer of CaMV by aphids requires the presence of a transmission factor in infected cells. This factor is not part of the virus particle but is encoded in the CaMV genome (Woolston *et al.* 1983). Two non-transmissible isolates of CaMV have been identified (Lung & Pirone 1974) and turnip leaves infected with them fail to synthesize a polypeptide (P18) of molecular weight 18 000 (Woolston *et al.* 1983).

The Caulimoviruses induce in infected cells the formation

Table 10.1 Host range of some Caulimoviruses. Plant families shown in parentheses have only been infected in the laboratory.

Virus	Host range	Comments
Cauliflower Mosaic Virus (CaMV)	Cruciferae (Solanaceae)	Serologically related to CERV and DaMV
Carnation Etched Ring Virus (CERV)	Caryophyllaceae	Serologically related to CaMV. Has same nucleic acid structure as CaMV
Dahlia Mosaic Virus (DaMV)	Compositae (Amaranthaceae Solanaceae Chenopodiaceae)	Serologically related to CaMV and CERV. Has the same nucleic acid structure as CaMV
Figwort Mosaic Virus (FMV)	Scrophulariaceae (Solanaceae)	DNA has same structure as that from CaMV
Mirabilis Mosaic Virus (MMV)	Nyctaginacéae	Serologically unrelated to CaMV or DaMV. DNA has same structure as that from CaMV
Strawberry Vein Banding Virus (CVBV)	Roseaceae	
Thistle Mottle Virus (ThMV)	Compositae	DNA has same structure as that from CaMV
Possible Caulimoviruses:		
Blueberry Red Ring Spot Virus	Ericaceae	
Cassava Vein Mosaic Virus	Euphorbiaceae	
Cestrum Virus	Solanaceae	
Petunia Vein Clearing Virus	Solanaceae	
Plantago Virus 4	Plantaginaceae	

of refractile, round inclusions which consist of many virus particles embedded in a protein matrix. The function of these inclusions is not known but the matrix protein is virus coded and can account for up to 5% of the protein in infected cells.

The remainder of our discussion on Caulimoviruses will be restricted to the best-studied example, CaMV. However, it is reasonable to assume that CaMV does not differ significantly from the other Caulimoviruses; for example the unusual structural features of CaMV DNA (see p. 207) are found in many other members of the Caulimovirus group (Hull & Donson 1982, Donson & Hull 1983, Richins & Shepherd 1983).

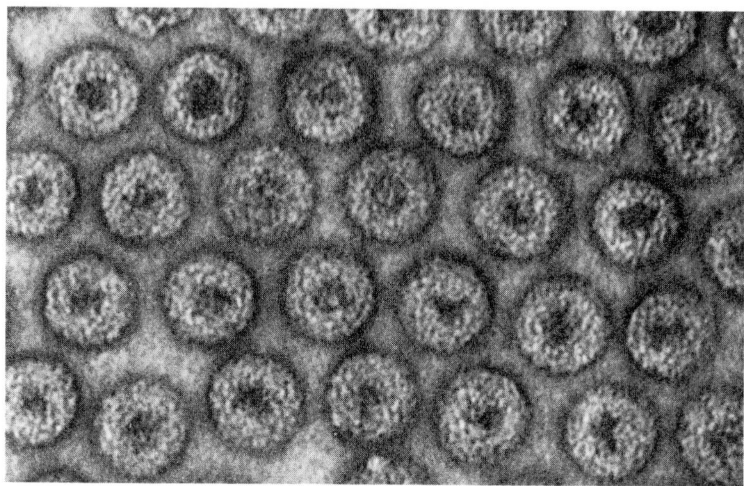

Fig. 10.1 Semi-crystalline array of cauliflower mosaic virus purified from turnip (approx. magnification ×200 000). The dark spots in the centres of the particles are typical and are the result of the outer protein shell being sucked into the hollow core during preparation for electron microscopy. (Photograph courtesy of M. Webb, National Vegetable Research Station.)

The structure of CaMV

CaMV particles are isometric and about 50 nm in diameter (Fig. 10.1). The literature concerning the exact molecule architecture of CaMV virions is rather confused. Estimates of the number of structural polypeptides vary from 2 to >7 and the corresponding mol. wt from 30 000 to 85 000. Now it is believed that in reality the structure is very simple and that there is only one major virion protein, a polypeptide of mol. wt 42 000 (Al Ani *et al.* 1979). This polypeptide, called p42, is processed from a precursor polypeptide of mol. wt 58 000 (Hahn & Shepherd 1982).

Properties of CaMV DNA

The genome of CaMV consists of double-stranded circular DNA. Three isolates of CaMV have been completely sequenced. The DNA of the Cabb S isolate is 8024 bp long (Franck *et al.* 1980), that of isolate CM1841 is 8031 bp long (Gardner *et al.* 1981) and isolate D/H is 8016 bp long (Balazs *et al.* 1982). The sequence variation between these isolates is about 5% and the differences can be accounted for by substitutions, small deletions and small insertions. Other isolates have been partially sequenced and show a similar degree of variation. Of particular interest is isolate CM4-184 which is non-aphid transmissible and has a 421 bp deletion in the gene encoding polypeptide p18 (Howarth *et al.* 1981).

An unusual feature of CaMV DNA is the presence of single-stranded discontinuities, often referred to as 'gaps' (Hull & Howell 1978, Volvovitch *et al.* 1978). These can be demonstrated by gel electrophoresis of DNA which either has been denatured or treated with the single-strand specific nuclease S1. The DNA from most isolates of CaMV has three discontinuities, two in one strand and one in the other (Fig. 10.2). The positions of the discontinuities are fixed with respect to restriction endonuclease cleavage sites. Originally the discontinuities were thought to be either nicks or gaps but detailed sequencing has shown that they are, in fact, regions of sequence overlap (Frank *et al.* 1980, Richards *et al.* 1981). A short sequence at one terminus of each discontinuity is displaced from the double helix by an identical sequence at the other boundary of the

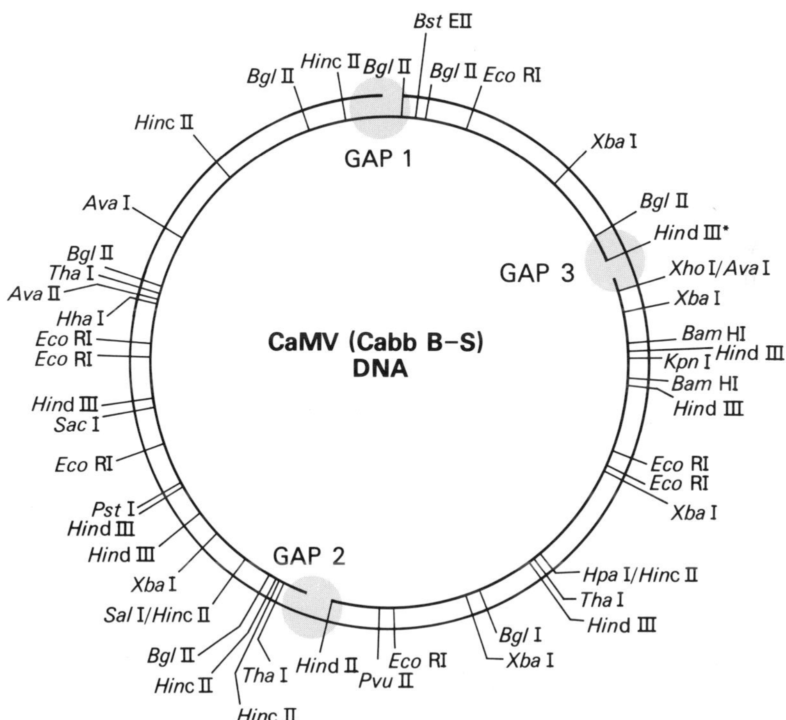

Fig. 10.2 Restriction endonuclease map of the Cabb B-S strain of CaMV.

discontinuity. The location of the 5′ extremity of each discontinuity is fixed but there is considerable variation in the position of the termini at the 3′ extremities (Fig. 10.3). The sequences around these genome discontinuities probably play a vital role in replication (for reviews see Hull & Covey 1983a,b).

Another unusual feature of CaMV DNA is a number of short sequences of ribonucleotides in a small proportion of the population of molecules (Hull & Shepherd 1977). The function of these ribonucleotides, which comprise less than 1% of the total nucleotides, is unknown. Recently it has been shown that the 5′ nucleotide at the discontinuity is sometimes a ribonucleo-

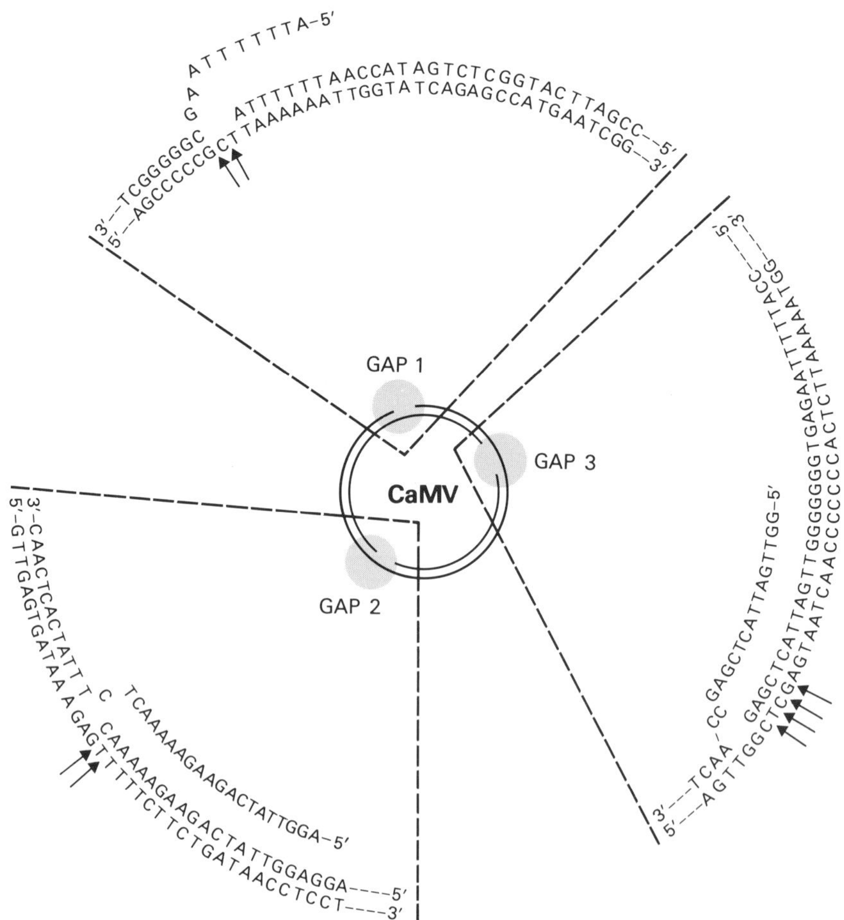

Fig. 10.3 DNA sequence in the vicinity of the three single-stranded discontinuities ('gaps') in CaMV DNA. The arrows indicate other possible positions for the 3′ termini.

tide (Guilley *et al.* 1983). The above and other observations have led Hull and Covey (1983a,b) and Pfeiffer and Hohn (1983) to suggest that CaMV replication involves reverse transcription.

Restriction endonuclease cleavage maps of CaMV DNA

The genomes of over 30 strains of CaMV have been mapped using restriction endonucleases (Meagher *et al.* 1977a, Hull & Howell 1978, Lebeurier *et al.* 1978, Volvovitch *et al.* 1978, Hohn *et al.* 1980, Hull 1980, Delseny & Hull 1983). All these isolates differ from one another in the number and location of the cleavage sites for a number of different restriction enzymes: a map is shown in Fig. 10.2 for only one isolate, Cabb S, the first one to be sequenced. Comparison of the maps of different isolates indicates that some regions of the genome are variable and others are non-variable (Hull 1980). Some of the isolates have DNA molecules which appear to differ in size from those of the others. In most instances these apparent size differences are in fact due to failure to detect minor restriction fragments although one isolate, CM4-184, does have a genuine deletion.

As well as variation in restriction endonuclease patterns between isolates there is evidence of heterogeneity within an isolate, restriction fragments being produced in less than stoichiometric amounts. There are two sources of these intra-strain variations. One is the formation of linear molecules by ring breakage at or near the gap in the α-strand (Hull & Howell 1978). The other is the fact that fragments carrying the variable-length single-strand discontinuities give rise to diffuse weak bands difficult to recognize in gels (Hohn *et al.* 1980).

There appears to be no correlation between restriction endonuclease maps and biological characteristics such as aphid transmissibility, types of symptoms produced or geographical distribution. It is possible that variants of CaMV arise rather frequently and are not suppressed by the parental strains. Support for this idea comes from the observation that the different isolates can be related to one another as losses or gains of single cleavage sites (Hull 1980).

One potential source of confusion in the literature is the designation of coordinates for restriction maps of CaMV DNA. Meagher *et al.* (1977a) set a precedent by assigning a particular *Eco* RI site as position zero. This was a fortunate coincidence for it is found in the DNAs of all the isolates studied. According to Hull (1980) there is no alternative cleavage site which could be designated zero. By contrast, Hohn *et al.* (1980) designated gap 1, the single break in the transcribed α-strand, as the zero point of their restriction map. Recently, most CaMV workers have agreed to accept the designation of Hohn *et al.* (1980) since it appears that all Caulimoviruses have an α-strand with a single break.

Transcription and translation of CaMV DNA

Transcription of CaMV DNA is asymmetric. Virus-specific RNA from infected turnip leaves or protoplasts hybridizes exclusively with the α-strand of CaMV DNA (Howell & Hull 1978, Hull *et al.* 1978, Hull *et al.* 1979). The conclusion that CaMV transcription is asymmetric has been substantiated by DNA sequencing. Analysis of the distribution of TGA, TAG and TAA termination codons in the three possible reading frames of the β-strand (i.e. homologous with the α-strand RNA transcript) indicates six open reading frames (Franck *et al.* 1980). Figure 10.4 reveals that, apart from a region of about 1000 nucleotides in the vicinity of gap 1, virtually the entire sequence is free of nonsense codons for considerable distances in one or another of the three reading frames. By way of contrast, the sequence complementary to the β-strand, a sequence which is not transcribed, contains no uninterrupted triplet reading frame of more than 370 nucleotides. McKnight and Meagher (1981) have cloned small *Eco* RI* fragments of CaMV DNA which act as promoters in *E. coli* for the tetracycline resistance gene of pBR322 (see p. 59). Each of the promoter active fragments was located at the 5′-terminal portion of one of the six open-reading regions predicted from the DNA sequence.

There are two major RNA transcripts in CaMV infected cells. Both are polyadenylated and are 8.2 kb and 1.9 kb in size (Covey & Hull 1981, Odell *et al.* 1981). The 1.9 kb transcript is the mRNA for the protein encoded by open reading region VI, the larger RNA species is a transcript of the entire genome but has a terminal repeat of 180 nucleotides (Fig. 10.4). A eukaryotic in-vitro transcription system derived from Hela cell extracts recognizes two promoters in CaMV DNA which coincide with the transcription start sites of the two polyadenylated RNA messengers (Guilley *et al.* 1982).

The polypeptides encoded by three of the open-reading regions have been identified. Hybrid arrest translation and hybrid select translation experiments have shown that open-reading region VI encodes a 62 000 mol. wt polypeptide which accumulates in CaMV-infected plants (Al Ani *et al.* 1980, Odell & Howell 1980, Covey & Hull 1981). The CaMV coat protein (p42) is unusual because of its high lysine content (18% of the

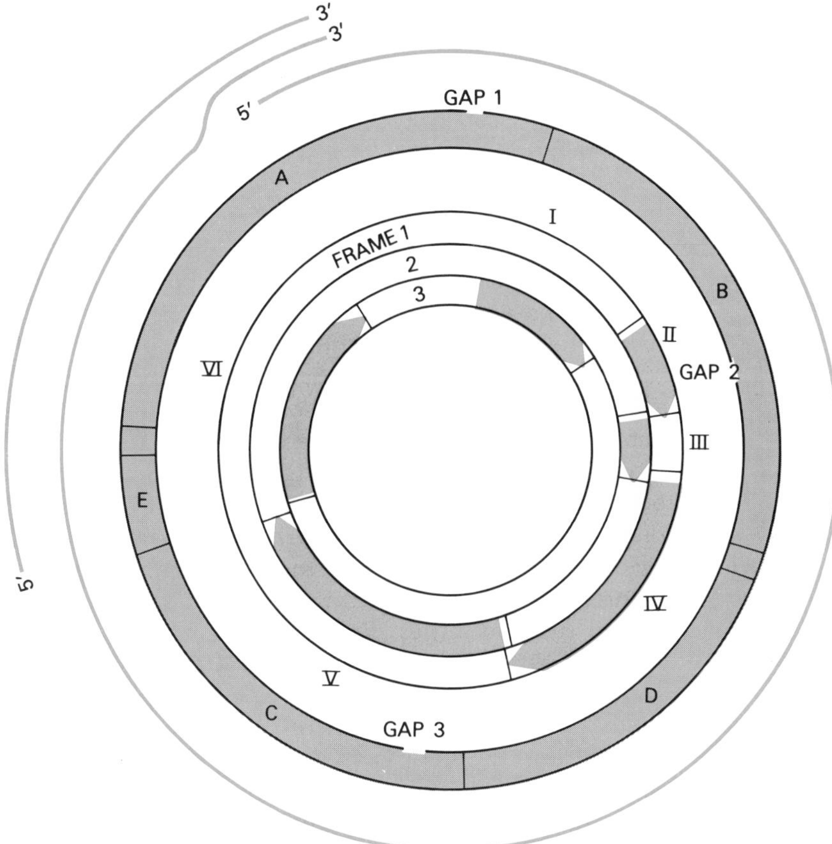

Fig. 10.4 Transcription and translation of CaMV DNA. The outer grey circle represents the viral DNA and gives the positions of *Eco* RI fragments A–E and the three 'gaps'. The inner circles give the positions of the six long open-reading frames and the two intergenic regions. The outer bold red lines represent the two major polyadenylated transcripts found in CaMV-infected cells.

total virion protein on a molar basis). Examination of the DNA sequences of each of the open-reading regions indicates that only region IV has the potential to code for a protein approaching this degree of richness in lysine. When various segments of the CaMV genome were cloned in *E. coli*, open-reading region IV alone was sufficient for synthesis of material cross-reacting with CaMV antisera (Daubert *et al.* 1982). Open-reading region IV encodes a protein which would be larger than p42 but Hahn and Shepherd (1982) have shown that the protein is processed from a larger precursor molecule. Open-reading region II probably encodes a polypeptide involved in aphid transmission since non-aphid transmissible isolates have a deletion of part of this region and fail to synthesize a polypeptide of mol. wt 18 000 (Woolston *et al.* 1983).

Infectivity of CaMV DNA

If CaMV is to be used as a vector then it is essential that its DNA is still transformable/transfectable after manipulation *in vitro*. In this context an early report by Szeto *et al.* (1977) was disappointing. These workers found that whereas purified CaMV DNA was infectious, viral DNA that had been cleaved at a unique *Sal* I site and religated was not. The reason for this is not clear. Later results were much more promising. Howell *et al.* (1980) found that CaMV DNA digested with *Sal* I was infectious even in the absence of religation. When CaMV DNA was cloned in a plasmid the chimaera was not infectious for plants. However, on restriction at the cloning site, the DNA became infectious again despite the fact that there were no single-strand discontinuities. Progeny virus from such infections contains native CaMV DNA, i.e. with the three gaps. Thus the peculiar structural features of CaMV DNA are not essential for infectivity but result from passage through the plant. In addition, the viral DNA, whether derived from the plasmid or directly from the virion does not need to be circular to infect plants. The plant host is capable of recircularizing the DNA as long as it has cohesive ends. Removal of the cohesive ends with S1 nuclease destroys infectivity.

The results of Howell *et al.* (1980b) have been confirmed by Lebeurier *et al.* (1980) and Delseny and Hull (1983). Lebeurier *et al.* (1980) further showed that the DNA is infectious regardless of the restriction endonuclease site at which it was cloned or cleaved. Plant cells are also able to religate CaMV DNA which has been cut into two separate fragments, whereas the individual fragments are not infectious (Lebeurier *et al.* 1982, Walden & Howell 1982). Pairs of non-infectious recombinant CaMV genomes can also recombine to create full-length parental DNA molecules following simultaneous inoculation of a sensitive host. This could give rise to difficulty in using CaMV DNA as a plant cloning vector in conjunction with helper viral DNA. Whereas vector and helper virus combinations have been used successfully in other systems, e.g. SV40 (see p. 241) with CaMV the foreign DNA may be expelled by recombination between the vector and the helper.

Cloning of foreign DNA in CaMV

There are two problems associated with cloning foreign DNA in CaMV. First, because the virus has a spherical capsid there will be a physical limitation to the size of DNA which can be inserted unless non-essential DNA can be deleted. However, as a general rule, small viruses are parsimonious in their use of DNA and contain little, if any, non-essential parts in their genome. Second, the viral genes often overlap or, as with CaMV the promoters of one gene lie within the coding region of the preceding gene. Thus most of the attempts to perturb the structure of CaMV DNA by either insertion or deletion have caused loss of infectivity (Howell *et al.* 1981, Delseny & Hull 1983).

Two sites have been found in the CaMV genome where foreign DNA can be inserted without destroying infectivity. Howell *et al.* (1981) inserted an 8 bp *Eco* RI linker into intergenic region 1 just clockwise of gap 1. The viral DNA was still infectious and the linker was retained. Gronenborn *et al.* (1981) made use of the fact that open-reading frame II encodes a protein for aphid transmission which is not essential for viral replication. They inserted bacterial DNA fragments into the *Xho* I site which lies within this reading frame. Initially they cloned a 65 bp fragment derived from the *lac* promoter/operator region. Two different constructs, differing in the orientation of the *lac* insert, were obtained and both were infectious when rubbed on the leaves of turnip plants and symptoms typical of systemic infection developed. Four individual local lesions induced by DNA containing both types of inserts were transferred to several plants and after systemic symptoms developed viral DNA from each of the individual transfers was analysed. After three successive transfers of virus derived from systemically infected tissues, all independent insert lines still contained the *lac* operator fragment. However, after five successive transfers and extended growth of the plants the *lac* insert was lost regardless of its orientation. In addition, insertions up to 30 bp can be made in the amino-terminal portion of open-reading frame VI without destroying infectivity, although the severity of the symptoms is reduced (Daubert *et al.* 1983). Longer DNA segments when inserted into open-reading frame VI destroyed infectivity.

To determine the maximum insertion capacity at the *Xho* I site Gronenborn *et al.* (1981) inserted fragments of 256, 531 and 1200 bp respectively. Infectivity assays using CaMV DNA containing these inserts were performed as before. After development of systemic symptoms viral DNA was prepared from infected leaves and analysed. This analysis showed that only the 250 bp insertion was propagated stably. CaMV DNA with an insertion of 531 bp gave rise to deletion mutants, presumably by recomination, whereas insertion of 1200 bp did not give rise to infectious CaMV DNA molecules.

Other workers have carried out similar experiments and these are documented by Hull and Davies (1983).

The potential of CaMV DNA as a cloning vector

So far, no foreign genes have been cloned in CaMV. Given the limitation of the size of nucleic acid which can be packaged, it may not be possible to clone entire genes without making the virus defective. Normally this problem can be circumvented by the use of a helper virus but, as indicated above, this may not be possible unless recombination between the vector and the helper virus can be suppressed. Whatever viral functions are destroyed by insertion of exogenous DNA those regions involved in replication must be kept intact if the chimaera is to be maintained in a cytoplasmic state. As well as replication of the foreign gene, transcription and translation are required. As yet there are no clear indications as to what promoters, plant or viral, might be used nor whether they will be constitutive or inducible. What little we know about transcription and translation of CaMV in infected cells indicates that more detailed molecular analysis is required before CaMV can be subjugated as a vector. Finally, all good cloning vectors carry at least one selectable marker. In the case of wild-type CaMV the only method which can be used to detect transfection with viral DNA is the appearance of symptoms. Other, better markers are needed.

Having manipulated a gene into a CaMV-derived vector it is necessary to introduce the chimaera into suitable plant recipients. The DNAs of Caulimoviruses are infectious to plants and the viruses which form spread systemically. This might not happen with a recombinant CaMV genome. An alternative approach would be to infect protoplasts of susceptible plants; for example, turnip protoplasts can be infected with CaMV DNA, albeit with some difficulty. However, there are many problems in regenerating plants from protoplasts and it is not known if a virus such as CaMV would be retained during the process.

Given the problems which need to be solved before CaMV

can be turned into a useful vector the reader might wonder if the effort is justified. The answer is easy. If a CaMV-derived vector can be used to improve a single major crop, then time and money will have been well spent. Even if CaMV is not subjugated as a vector it may provide a source of strong promoters for use in plant cells.

AGROBACTERIUM AND GENETIC ENGINEERING IN PLANTS

Crown gall disease

Crown gall is a plant tumour, a lump of undifferentiated tissue, which can be induced in a wide variety of gymnosperms and dicotyledonous angiosperms by inoculation of wound sites with the Gram-negative soil bacterium *Agrobacterium tumefaciens* (Fig. 10.5). The disease was first described long ago, and the involvement of bacteria was recognized as early as 1907 (Smith & Townsend 1907). It was subsequently shown that the

Fig. 10.5 Crown gall on blackberry cane. (Photograph courtesy of Dr C. M. E. Garrett, East Malling Research Station.)

crown gall tissue represents true oncogenic transformation; callus tissue can be cultivated *in vitro* in the absence of the bacterium and yet retain its tumorous properties. These properties include the ability to form an overgrowth when grafted onto a healthy plant, the capacity for unlimited growth as a callus in tissue culture in media devoid of the plant hormones necessary for in-vitro growth of normal plant cells, and the synthesis of *opines*, which are unusual amino acid derivatives not found in normal tissue (Fig. 10.6). The most common of these opines are octopine and nopaline. In addition, agropine or the agrocinopine family of opines may be present. Crown gall tumour cells continue to synthesize opines in tissue culture, and shoots or whole plants regenerated from tumour cell lines may also continue to synthesize opine (Braun & Wood 1976, Schell & Van Montagu 1977, Wullems *et al.* 1981a, Wullems *et al.* 1981b).

The metabolism of opines is a central feature of crown gall disease. Opine synthesis is a property conferred upon the plant cell when it is transformed by *A. tumefaciens*. The type of opine produced is determined not by the host plant but by the bacterial strain. In general, the bacterium induces the synthesis of an opine which it can catabolize and use as its sole energy, carbon and/or nitrogen source. Thus, bacteria that utilize

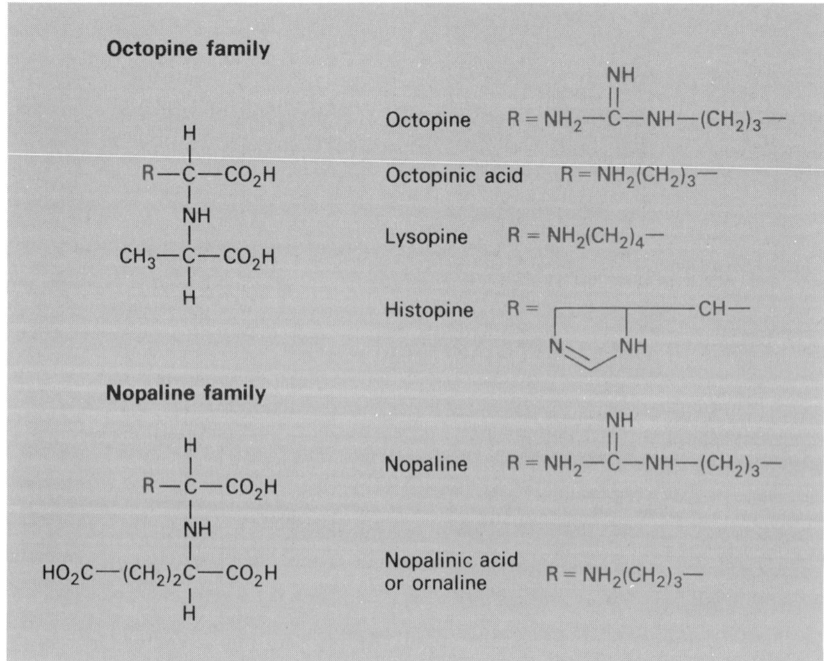

Fig. 10.6 Structures of some known opines. The structures of the agrocinopine family opines have not been determined.

octopine induce tumours that synthesize octopine, and those that utilize nopaline induce tumours that synthesize nopaline (Bomhoff *et al.* 1976, Montaya *et al.* 1977). Bacterial mutants have been isolated from a number of either octopine- or nopaline-utilizing strains which were defective in the genes specifying either octopine or nopaline oxidase, the enzymes primarily responsible for opine catabolism by the bacteria. These mutants still retained the bacterial permease for the opine, were still virulent, i.e. caused tumours, and the tumours which they induced still synthesized the opine. Clearly, the bacterial determinants of octopine or nopaline biosynthesis by tumour cells are distinct from the genes that direct opine catabolism in bacteria.

Investigation of the molecular biology of crown gall disease has revealed that *A. tumefaciens* has evolved a natural system for genetically engineering plant cells so as to subvert them for its own ends. Recent research, mainly using tobacco, tomato and petunia as the experimental plants, has enabled gene manipulators to exploit this natural system. This will be of great value in analysing gene expression in plants, and there is the ultimate goal of generating novel, agriculturally useful, varieties of a range of dicotyledonous crop plants.

The tumour-inducing principle and the Ti-plasmid

As it was clear that the continued presence of bacteria is not required for transformation of the plant cells, attention was focused on the nature of the 'tumour-inducing principle', the name given to the putative genetic element that must be transferred from the bacterium to the plant at the wound site. For a long time it was believed, correctly, that DNA is transferred from the bacterium to the plant cell. Attempts to detect such bacterial DNA in the tumour cells failed, simply because the techniques used were not sufficiently sensitive, until plasmids were detected in *Agrobacterium*. Zaenen *et al.* (1974) first noted that virulent strains of *A. tumefaciens* harbour large plasmids (140–235 kb), and it is now clear that the virulence trait is plasmid borne. Virulence is lost when the bacteria are cured of the plasmid and with at least one strain this can be achieved by growing the cells at 37 °C instead of 28 °C. Cured strains also lose the capacity to utilize octopine or nopaline (Van Larebeke *et al.* 1974, Watson *et al.* 1975). Virulence is acquired by avirulent strains when a virulence plasmid is reintroduced by conjugation (Bomhoff *et al.* 1976, Gordon *et al.* 1979). If the plasmid from a nopaline strain is transferred to an avirulent derivative of a previously octopine strain, the avirulent strain then acquires the ability to induce nopaline tumours and catabolize nopaline. The virulence plasmid can also be transferred to the legume symbiont *Rhizobium trifolii* which becomes oncogenic and acquires the ability to utilize either octopine or nopaline, depending on the donor.

From the above information it is clear that plasmids are essential for virulence and for this reason they are referred to as Ti (tumour-inducing) plasmids. Furthermore, the genetic information specifying bacterial utilization of opines, and their synthesis by plants, must also be plasmid-borne. It should be remembered, however, that the presence of a plasmid in *A. tumefaciens* does not mean that the strain is tumorigenic. Many strains contain very large cryptic plasmids that do not confer virulence, and in some natural isolates a cryptic plasmid is present together with a Ti-plasmid.

Plasmids in the octopine group have been shown to be closely related to each other while those in the nopaline group are considerably more diverse (Currier & Nester 1976, Sciaky *et al.* 1978). Between these two groups there is little DNA homology, except for four limited regions, one of which includes the genes directly responsible for crown gall formation (Drummond & Chilton 1978, Engler *et al.* 1981).

A related organism, *A. rhizogenes*, incites a disease, hairy root disease, in dicotyledonous plants in a manner very similar to *A. tumefaciens*. A large plasmid, the Ri (root inducing) plasmid, is responsible for pathogenicity and the induction of opine synthesis (White & Nester 1980, Chilton *et al.* 1982). The Ri plasmids share little homology with Ti plasmids. Research into them has not proceeded as far as with the Ti plasmids but they are of interest because tissue transformed by *A. rhizogenes* readily regenerates into plantlets which continue to synthesize opine (see p. 227).

Incorporation of T-DNA into the nuclear DNA of plant cells

Complete Ti-plasmid DNA is not found in plant tumour cells but a small, specific segment of the plasmid, about 20 kb in size, is found integrated in the plant nuclear DNA (Chilton *et al.* 1977). This DNA segment is called T-DNA (transferred DNA; Fig. 10.7). The structure and organization of the integrated T-DNA in tumour cells has been studied in detail by means of Southern hybridization to restriction enzyme digests of nuclear DNA from transformed cells. The main conclusions (Thomashow *et al.* 1980, Zambryski *et al.* 1980) of these studies are listed as follows.

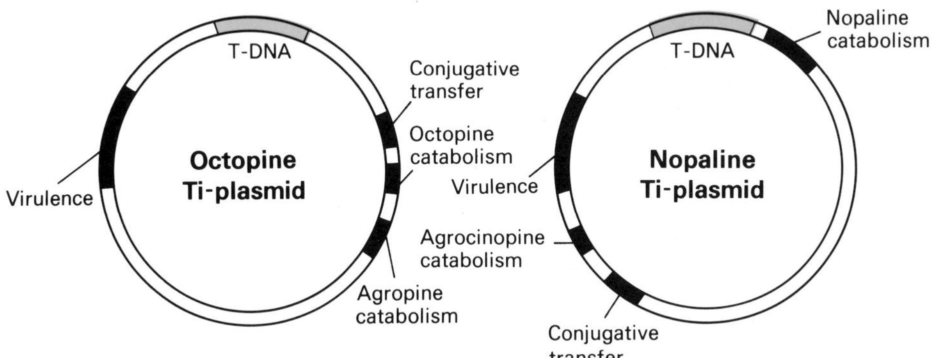

Fig. 10.7 Ti-plasmid gene maps (after Nester & Kosuge 1981).

1 Integration of the T-DNA can occur at many different, possibly nearly random, sites in the plant nuclear DNA.
2 The organization of the integrated octopine T-DNA is more complicated than that of integrated nopaline T-DNA.
3 In some octopine cell-lines the T-DNA exists as two segments (Fig. 10.7). The left end of the T-DNA (T_L) includes gene affecting tumour morphology and usually includes the octopine synthase gene (*ocs*). The right end of the T-DNA (T_R) is not necessary for tumour maintenance since it is absent in some cell lines. Usually only one copy of T_L is present, per diploid cell equivalent, but as many as ten copies have been found. T_R may be present in high copy numbers. The significance of these complications in integrated octopine T-DNA structure is not clear. It is likely that rearrangements, amplifications and deletions of the T-DNA may follow the initial transformation event.
4 Integrated nopaline T-DNA occurs as a single segment. Regions including the junctions of nopaline T-DNA with plant DNA have been cloned in phage λ vectors and analysed at the nucleotide sequence level. The righthand junction with the plant DNA appears to be precise to the nucleotide, whereas the lefthand junction can vary by about 100 nucleotides (Yadav *et al.* 1982, Zambryski *et al.* 1982). A 25 nucleotide direct repeat has been found at the junctions of nopaline DNA with host DNA. A related sequence has been found at a junction of octopine T-DNA, strongly suggesting some functional significance of these sequences in the integration mechanism (Simpson *et al.* 1982, Yadav *et al.* 1982). Whether a site-specific recombination mechanism exists for T-DNA remains to be seen and the mechanism by which only the T-DNA segment of the Ti-

plasmid becomes integrated remains unknown. Two observations are important here. First, deletion of the righthand copy of the repeat in a nopaline plasmid does not completely eliminate tumour induction (Holsters *et al.* 1980, Lemmers *et al.* 1980). Second, exhaustive mutagenesis of T-DNA by deletion and transposon mutagenesis has failed to reveal a function in T-DNA essential for integration in the plant genome. Any plasmid-borne recombination system must be encoded outside the T-DNA region, possibly in the virulence region.

Gene maps and expression of T-DNA

Genetic maps of Ti-plasmids and T-DNA have been obtained by the study of spontaneous deletions and by transposon mutagenesis (Koekman *et al.* 1979, Garfinkel *et al.* 1980, Holsters *et al.* 1980, Ooms *et al.* 1980, De Greve *et al.* 1981, Ooms *et al.* 1981). Such studies have revealed a large region mapping outside the T-DNA which is necessary for virulence. This region is probably required for DNA transfer and/or integration. As expected, regions of the T-DNA itself were found to affect virulence and tumour morphology (Fig. 10.7). The loci affecting tumour morphology are designated 'large' (*tml*), 'shooty' (*tms*) and 'rooty' (*tmr*) to indicate the phenotype of the callus obtained when the loci are inactivated. Ooms *et al.* (1981) have proposed that certain mutations in T-DNA appear to affect plant hormone concentrations in the resulting tumours so as to produce rooty and shooty types. The basis for this proposal was the observation that although the tumour grows in media lacking added auxins and cytokinins, the callus actually contains high concentrations of these hormones. Indeed, uncloned primary tumours contain some normal, untransformed cells growing as undifferentiated callus owing to the presence of these endogenous hormones (Gordon 1980). Normal tobacco tissue requires a particular ratio of auxins to cytokinins for growth as undifferentiated callus. Increasing the ratio in favour of auxin causes root proliferation, decreasing the ratio promotes shoot development. T-DNA may encode enzymes that directly synthesize phytohormones or, alternatively, T-DNA genes may disturb normal host mechanisms for hormone synthesis or breakdown.

Transcripts of the T-DNA in tumour cells have been examined (Bevan & Chilton 1982, Gelvin *et al.* 1982, Willmitzer *et al.* 1982). Sensitive techniques, such as Northern blot analysis, are required for a comprehensive analysis since the total abundance of the T-DNA transcripts in octopine tumours is only about 0.001% of tumour polyadenylated RNA (Willmitzer *et al.* 1982). Seven distinct T-DNA transcripts are present in octopine tumours (Fig. 10.8). The transcript encoding octopine synthase (now renamed lysopine dehydrogenase) has been identified (transcript 3 in Fig. 10.8) and its gene located at the right end of T_L. DNA sequencing has shown that this gene has promoter elements that are eukaryotic in character, that it lacks introns, and that it contains the eukaryotic polyadenylation signal near its 3′ end (De Greve *et al.* 1982). Essentially similar results have been obtained for the nopaline synthase gene (Depicker *et al.* 1982, Bevan *et al.* 1983). This suggests that genes of the T-DNA are eukaryotic in origin and have been 'captured' by the Ti-plasmid during its evolution.

Five of the seven identified transcripts are transcribed from sequences common to octopine and nopaline T-DNA. Two of the transcripts (1 and 2) map to genes (*tms*) identified as possibly producing auxin-like substances. Another transcript (4) maps to a gene (*tmr*) possibly producing a cytokinin-like substance. The common T-DNA encodes two other transcripts

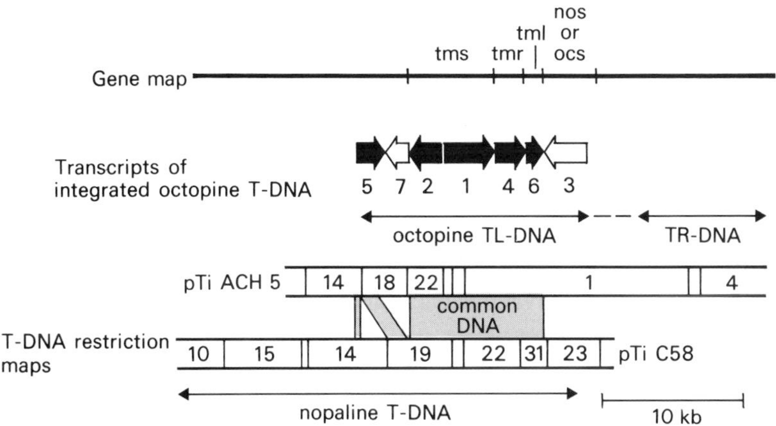

The T-regions of octopine and nopaline Ti-plasmids have been aligned; red areas indicate the DNA sequences that are common to both T-regions. Numbers refer to fragments produced following digestion with *Hin*d III. Arrows above the figure represent different transcripts of T-DNA in *octopine crown gall tumours.* Black arrows represent transcripts derived from common DNA sequences

Genetic loci as defined by deletion and transposon mutagenesis

ocs: octopine synthase
nos: nopaline synthase
tmr: rooty tumour
tms: shooty tumour
tml: large tumour

Fig. 10.8 Structure and transcription of T-DNA.

(5 and 6) which appear to suppress differentiation only in cells in which they are expressed, i.e. in a non-hormonal manner.

Ti-plasmid derivatives as plant vectors

We have seen that the Ti-plasmid is a natural vector for genetically engineering plant cells because it can transfer its T-DNA from the bacterium to the plant genome. The overall strategy of a genetic engineering programme might, therefore, involve insertion of a foreign DNA into T-DNA. This T-DNA would then carry the foreign DNA into the plant genome where one would often want to obtain expression of the foreign gene(s). It would then be desirable to regenerate whole plants from transformed cells, and, finally, it would be most useful if the foreign DNA were transmitted normally by the sexual process through the flowers of the plant to subsequent generations. Let us consider these stages in turn.

Insertion of foreign DNA into T-DNA

Wild-type Ti-plasmids are not suitable as experimental gene vectors because their large size means that it is not possible to find adequate unique restriction sites in the T-region. Their size also makes other procedures cumbersome. Intermediate vectors (abbreviated iv) have therefore been developed in which T-DNA has been subcloned into conventional small plasmid vectors of *E. coli* (Matzke & Chilton 1981). Standard procedures can then be used to insert any desired DNA into the T-region of such an iv. This iv may have the additional property of being a shuttle vector, capable of stable replication in *E. coli* and *A. tumefaciens*. Plasmid pBR322, and its relatives, cannot replicate in *A. tumefaciens* but this property can be conferred upon it by insertion of an origin of replication derived from the broad host-range plasmid pSa (see p. 160) of the W incompatibility group (Leemans *et al.* 1982).

The iv, containing foreign DNA in the T-region, can be transferred to *A. tumefaciens* by conjugation. Since the ivs are conjugation-deficient, conjugation must be mediated by the presence in the donor *E. coli* of a helper, conjugation-proficient plasmid which can mobilize the iv. Suitable helper plasmids are pRK2013 (Ditta *et al.* 1980) which consists of the transfer genes from the naturally occurring plasmid pRK2 (see p. 158) cloned onto a Col E1 replicon, or pRN3 (Shaw *et al.* 1983). Neither of these helper plasmids is capable of replicating in *Agrobacterium*. These transfers are conveniently brought about by 'triparental'

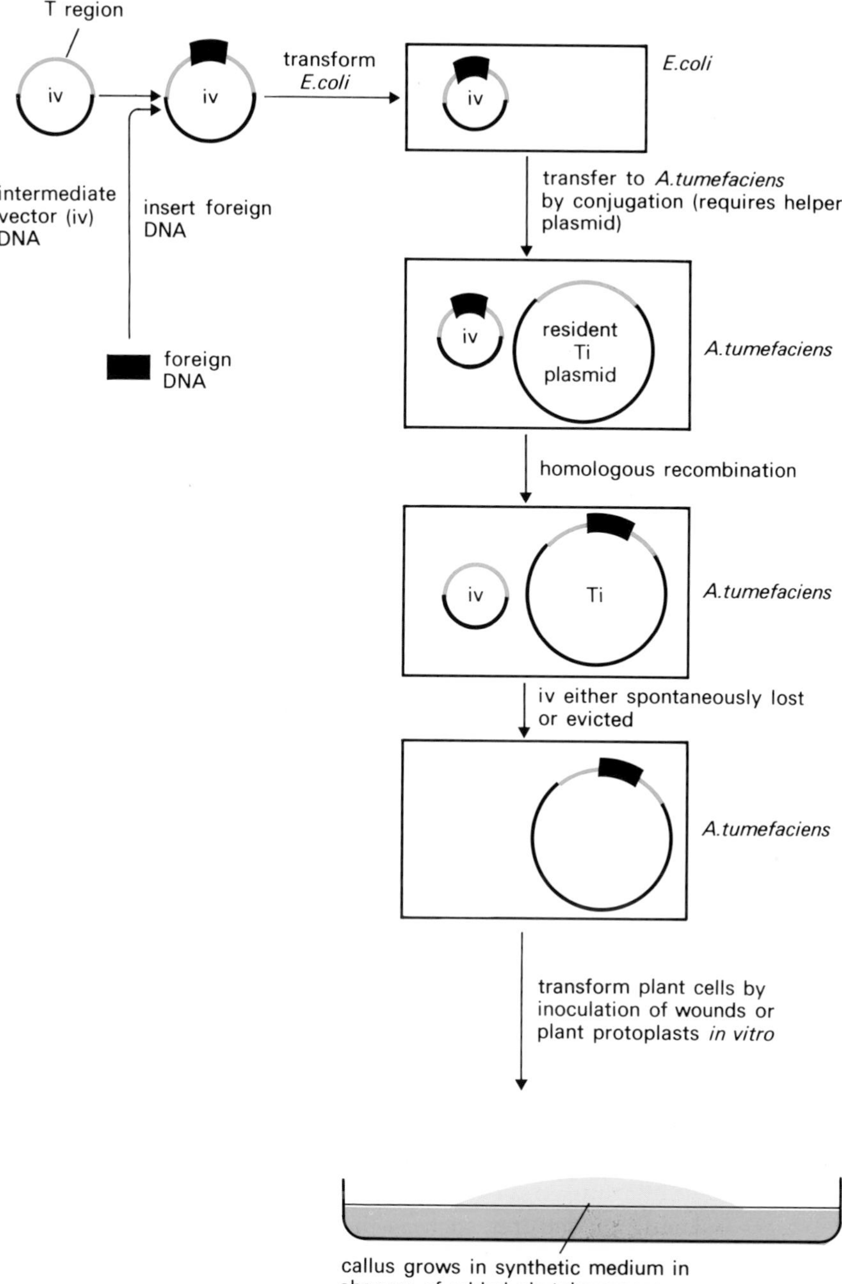

Fig. 10.9 Insertion of foreign DNA into an intact Ti-plasmid. See text for details.

matings. In these matings three bacterial strains are mixed together. These are (i) the *E. coli* carrying the conjugation-proficient helper plasmid, (ii) the *E. coli* strain carrying the recombinant iv, and (iii) the recipient *Agrobacterium*. During the course of the incubation the helper plasmid transfers to the strain carrying the recombinant iv which is then mobilized and transferred to *Agrobacterium*. Once the iv has been introduced into *Agrobacterium*, there are two alternative strategies available.

The recipient *A. tumefaciens* may already contain a resident, non-recombinant Ti-plasmid. If the incoming iv is not a shuttle vector, it will not be maintained in *Agrobacterium*. However, rare double homologous recombination events can occur which substitute the T-DNA region of the iv, including the foreign DNA, for the unmanipulated T-DNA of the resident Ti-plasmid (Fig. 10.9). In order to facilitate the recovery of such rare recombinants it is usual to include a drug resistance marker with the foreign DNA in the manipulated T-DNA of the iv (Herrera-Estrella *et al.* 1983a). If the iv is a shuttle vector, it will be maintained in the recipient *Agrobacterium*, hence allowing greater opportunity for the homologous recombination events. The shuttle iv must then be 'evicted' by the conjugative introduction of another plasmid with which it is incompatible (Matzke & Chilton 1981, Ruvkun & Ausubel 1981, Murai *et al.* 1983).

The aim of the procedure outlined above, and in Fig. 10.9, is to reconstruct an entire recombinant Ti-plasmid. It is now realized that this is unnecessary. A T-DNA region carrying foreign DNA can be constructed, as described above, in a small plasmid that replicates in *E. coli*. This small plasmid, designated mini-Ti, contains full-length nopaline T-DNA including the repeated border sequences. When such a plasmid is transferred conjugatively to *A. tumefaciens* it can induce nopaline tumours if the *Agrobacterium* simultaneously contains a plasmid carrying the virulence genes but no T-DNA of its own. Thus the *vir* functions can be supplied by a separate plasmid to give a fully tumorigenic phenotype (de Framond *et al.* 1983). The mini-Ti approach considerably simplifies the transfer of foreign genes to plant cells.

Using procedures such as these, insertion of foreign DNA within the T-region leads to its co-transfer and integration in the plant genome. To date, foreign DNA inserts of up to 50 kb have successfully been transferred. An upper size limit has not yet been determined (Herrera-Estrella *et al.* 1983a).

Expression of foreign genes in plant cells

Just as with foreign gene expression in bacterial hosts there may be barriers to foreign gene expression in plants and these barriers may occur at any stage of the gene-to-phenotype pathway. Several bacterial, animal and unrelated plant genes have been inserted into various sites of the T-DNA and successfully transferred into the plant genome. However, as examples in Table 10.2 show, foreign genes successfully transferred have not all been expressed.

In order to overcome barriers to transcription, intermediate expression vectors have been designed in which the foreign gene is positioned next to a suitable promoter; for example, chimaeric genes have been constructed consisting of promoter sequences derived from the nopaline synthase (*nos*) gene linked to coding sequences of foreign bacterial genes such as the neomycin phosphotransferase of Tn5 or Tn601 (which confer resistance to the aminoglycoside antibiotics, kanamycin, neomycin and G418 when expressed in plant cells), methotrexate-resistant dihydrofolate reductase of plasmid pR67, and chloramphenicol acetyltransferase of pBR325 (Bevan *et al.* 1983, Fraley *et al.* 1983, Herrera-Estrella *et al.* 1983a,b, Horsch *et al.* 1984). In each of these cases expression of the foreign gene was detected in callus tissue. In the cases of neomycin phosphotransferase and dihydrofolate reductase, the plant cells became resistant to lethal concentrations of aminoglycoside antibiotic or methotrexate. These constructions now provide very useful selectable markers for future experimentation (see p. 228). In a related example, coding sequences of the bean seed-protein (phaseolin) were inserted into the coding region of the octopine synthase (*ocs*) gene and were, therefore, put under transcriptional control of the *ocs* promoter. When transferred to plant cells this construct directed the expression of immunoreactive phaseolin polypeptides. It appears that the intact phaseolin fusion protein was degraded in this tissue (Murai *et al.* 1983).

The *nos* and *ocs* promoters are constitutively expressed. For future genetic engineering experiments where regeneration of whole plants is involved, appropriate differentiated expression of a gene may be required. Evidence from animal systems suggests that in many instances signals for differentiated expression lie associated with, and upstream from, the promoter region and that these signals together with tissue-specific enhancer elements, confer tissue-specificity of transcription (see Chapters 11 and 12). It is likely that the same will

Table 10.2 Expression of foreign genes transferred to plant cells in T-DNA.

Foreign gene	Expression in plant cells	Promoter controlling expression of foreign gene	Regeneration of plants	Transmission through gametes	Reference
Bacterial chloramphenicol transacetylase	+	nopaline synthase (*nos*) promoter	–	–	Herrera-Estrella *et al.* (1983a)
Bacterial neomycin phosphotrans-ferase	+	*nos* promoter	–	–	Bevan *et al.* (1983) Herrera-Estrella *et al.* (1983a) Fraley *et al.* (1983)
	+		+	+	Horsch *et al.* (1984)
Bacterial dihydrofolate reductase	+	*nos* promoter	–	–	Herrera-Estrella *et al.* (1983b)
Yeast alcohol dehydrogenase	–	ADH promoter	+	+	Barton *et al.* (1983)
Bean phaseolin	+	octopine synthase promoter	–	–	Murai *et al.* (1983)
Bean phaseolin	+ (low)	phaseolin promoter	–	–	Murai *et al.* (1983)
Leghaemoglobin	–	leghaemoglobin promoter	–	–	Quoted in Herrera-Estrella *et al.* (1983a)
Interferon	–	interferon promoter	–	–	Quoted in Marx (1982)

be the case in plants. A good candidate for experimentation is the nuclear gene for the small subunit of ribulose bisphosphate carboxylase, whose expression is under regulation by light. By using Ti-derived vectors it has been shown that light-regulated transcription of this gene in transformed callus is dependent upon upstream DNA sequences (Broglie *et al.* 1984, Herrera-Estrella *et al.* 1984).

Regeneration of plants and transmission of T-DNA to progeny

One of the problems with the natural T-region of Ti-plasmids is that when the region successfully carries the foreign DNA to plant cells there is the accompanying transformation of these cells to tumorous growth. In general, such tumour cells have proven recalcitrant to attempts to induce regeneration either into normal plantlets or into normal tissue that can be grafted onto healthy plants. However, tobacco callus transformed with wild-type Ti-plasmid does occasionally spontaneously regenerate shoots that can be grafted onto healthy plants. When such shoots were examined, it was found that they synthesized opine and failed to form roots. Some grafted shoots were fertile and produced seed that developed into apparently normal plants; however, these plants lacked opine and all or most of the T-DNA had been deleted from them (Braun and Wood 1976, Turgeon *et al.* 1976, Lemmers *et al.* 1980, Yang *et al.* 1980, Buins *et al.* 1981, Wullems *et al.* 1981a,b). A single case has been reported of opine-positive complete plants regenerated from a tumour. This tumour had originally been induced by a shooty mutant of octopine T-DNA (Leemans *et al.* 1982). In breeding experiments these plants transmitted the octopine trait in a simple Mendelian fashion, giving apparently healthy progeny. Analysis of the T-DNA revealed that little T-DNA was present except for the *ocs* gene (De Greve *et al.* 1982).

The above results, together with the genetic data on the functions of T-DNA, indicate that it is necessary to disarm the T-DNA by inactivating or deleting one or more of the tumour-inducing genes. Barton *et al.* (1983) have shown that inactivation of the rooty locus of a nopaline T-DNA effectively disarms it, at least with respect to tobacco plants. In their experiments T-DNA was engineered by insertion of a model eukaryotic gene, the yeast alcohol dehydrogenase (ADH) gene, at the rooty locus. From the resulting transformed cell lines, healthy regenerated plantlets with roots and shoots were obtained. The mature plants were fertile and transmitted the ADH gene to progeny obtained by self-fertilization. The progeny contained intact multiple copies of the engineered T-DNA but expression of the ADH gene was not detected.

An extreme method of disarming the T-DNA would consist of deleting all the oncogenic genes. Leemans *et al.* (1981) have constructed such a deletion derivative which, although avirulent, nevertheless inserted its T-DNA into plant cells as was evident by opine synthesis. This strain presumably could produce tissue capable of regenerating into healthy plants but since it is avirulent no tumour is formed and there is no selection for transformed cells: individual clones must be screened for opine production. Such a deletion, however, could be combined with the aminoglycoside or methotrexate-resistance markers mentioned previously, to give rise to a useful, selectable, disarmed vector. Using an approach similar to this Horsch *et al.* (1984) have demonstrated sexual transmission of a functional neomycin phosphotransferase gene.

Transformation of protoplasts

Experimental transformation of plant cells by *Agrobacterium* occurs when bacteria are painted onto wounds or abrasions made on the plant leaves or stem. The major obstacle to DNA uptake in plant cells is the cell wall, and *Agrobacterium* clearly has a natural mechanism, largely unknown at present, for overcoming this obstacle and delivering DNA to the plant cell nucleus. The plant cell wall can be removed experimentally by digesting the wall with enzymes so as to produce protoplasts. An in-vitro system for transforming plant protoplasts with *A. tumefaciens* has been devised (Marton *et al.* 1979). Transformed cells are subsequently recovered free of bacteria by treatment with the antibiotic carbenicillin, to which *A. tumefaciens* is sensitive. Transformants are selected by their ability to grow in synthetic culture media without the addition of hormones. This procedure has application because it is efficient, with as many as 10% of the plant cells being transformed (Fraley *et al.* 1983).

In-vitro transformation of tobacco protoplasts with pure Ti-plasmid DNA has also been reported (Krens *et al.* 1982). The procedure involves incubating protoplasts and DNA in the presence of polyethylene glycol, followed by incubation with a Ca^{2+} solution. Clearly, *A. tumefaciens* itself is not a prerequisite for obtaining tumour cells. Furthermore, this procedure may be adapted for introducing other DNAs into plant cells of diverse species, and possibly into monocotyledonous plants (see p. 229).

Future prospects

Investigation of the Ti-plasmid has proceeded rapidly. A great deal has been learned, but many interesting questions remain unanswered. However, as Schell has stated, it is already true that 'Any DNA sequence can be transferred without further ado.' The practical applications of this research will, no doubt, be wide ranging and are currently difficult to predict. Much of the uncertainty will be removed with the rapid advance of plant molecular biology. In the short term the applications may consist of expressing enzymes of bacterial origin in plants. Soil bacteria contain genes which are responsible for detoxifying various herbicides. It may prove possible to include such genes in crop plants so as to create varieties specifically protected from certain herbicides. Other applications may include the improvement of the quality of seed storage proteins which are used for animal or human consumption. This would entail altering the amino acid composition of these proteins which are often deficient in methionine or lysine (see Chapter 13). Further applications may depend upon the fact that agriculture produces biomass cheaply and conventionally. Ultimately the production of commercially valuable and/or therapeutic proteins may be undertaken not in expensive fermentors with microorganisms such as yeast or bacteria as the host, but in fields of genetically engineered plants.

The Ti-plasmid may have an unexpected role in plant molecular biology as a consequence of its suspected ability to integrate T-DNA at many sites in the plant genome. Thus it may be possible to exploit T-DNA as a mutagen in a procedure analogous to transposon mutagenesis. If T-DNA can integrate in and inactivate any plant gene and particular, non-essential, genes are thus identified as being of interest, then the integrated T-DNA can act as a molecular marker. Isolation of the gene sequence would be possible using radioactive T-DNA to probe a genomic library prepared from the mutant genome.

A major limitation of the Ti-plasmid is its inapplicability to cereals: monocotyledonous plants are insensitive to crown gall disease.* However, this may simply be due to the failure of *Agrobacterium* to attach effectively to the cell walls of these

*At the time of going to press it has been reported that T-DNA enters the cells of at least some monocot species but is not oncogenic (Hooykaas-Van Slogteren *et al.* (1984) *Nature* **311**, 763–4). Ti plasmids may therefore be very well suited to genetic engineering of these plants for which they appear to be naturally disarmed.

plants. Research is being directed towards introducing the Ti-plasmid to these cells by artificial means such as microinjection or protoplast fusion. The selectable markers now available in T-DNA should make these approaches experimentally amenable. It is also conceivable that plant cells may be induced to take up and integrate foreign DNA without the involvement of any vector. This would be analogous to the transfection/transformation technique widely practised with mammalian cells. The ability to transform tobacco protoplasts with Ti-plasmid DNA shows that DNA uptake at least, can be achieved. The availability of good selectable markers should be of value here also.

Finally, genetic engineering of cereals may be approached using transposable elements such as the Ac-Ds family of elements described in maize by McClintock (1951). (For an excellent review of maize transposable elements see Federoff 1984.) Both elements can transpose around the maize nuclear genome. The Ds (*'Dissociation'*) element cannot transpose unless the *trans*-acting Ac (*'Activator'*) element is present anywhere in the nuclear DNA. In the absence of Ac, the Ds element is stable. Ds elements have been cloned in bacteria (Doring *et al.* 1984) and their possible exploitation as vectors is being actively investigated. Analogous elements may be found in other cereals.

Chapter 11 Introducing Genes into Mammalian Cells in Culture

Analysis of the regulation of gene expression in animal cells is one of the central themes of current molecular biology. For this and other reasons procedures for introducing manipulated genes into animal cells, where their regulation can be assayed, have been the subject of intensive research. There has been a great diversity of approaches. Direct microinjection of DNA into the nucleus of amphibian oocytes allows foreign eukaryotic genes to be faithfully expressed. By direct microinjection of DNA into fertilized eggs of *Xenopus*, or the mouse, or into early *Drosophila* embryos, it has been possible to incorporate foreign DNA into the genomes of the resulting adult animals (see Chapter 12). Here we shall consider methods for introducing foreign DNA into animal cells in culture. Almost all of this research has made use of established mammalian cell lines.

Integration of DNA into the genome of mammalian cells

The ability of mammalian cells to take up exogenously added DNA, and to express genes included in that DNA, has been known for many years. Szybalska and Szybalski (1962) were the first to report DNA-mediated transfer. They transfected* $HGPRT^-$ mutant human cells to $HGPRT^+$ using total, uncloned, human nuclear DNA as the source of the wild-type gene. The rare $HGPRT^+$ transformants were selected by means of the HAT selection system which they had devised (Fig. 11.1). Much later, it was appreciated that successful DNA transfer in these

* The term transformation commonly has two different meanings in the context of this chapter. (1) An inherited change in genotype due to the uptake of foreign DNA, analogous to bacterial transformation. (2) A change in properties of an animal cell possessing normal growth characteristics to one with many of the characteristics of a cancer cell, i.e. growth transformation. In order to avoid confusion, in this chapter the term *transfection* will be used for the first meaning although originally transfection applied to the uptake of viral DNA. The term transformation will be used only with meaning (2).

PURINE BIOSYNTHESIS

Endogenous pathway

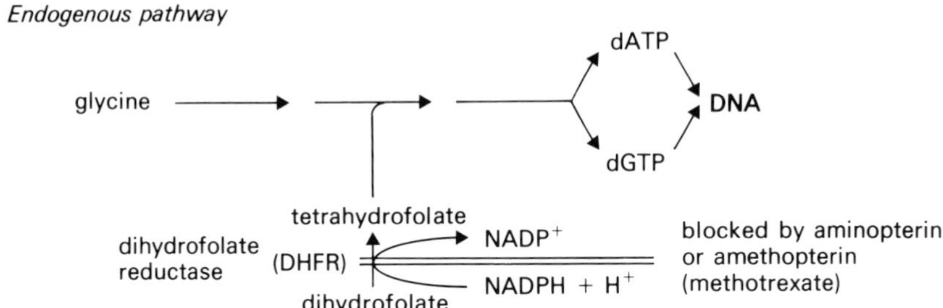

Salvage pathway

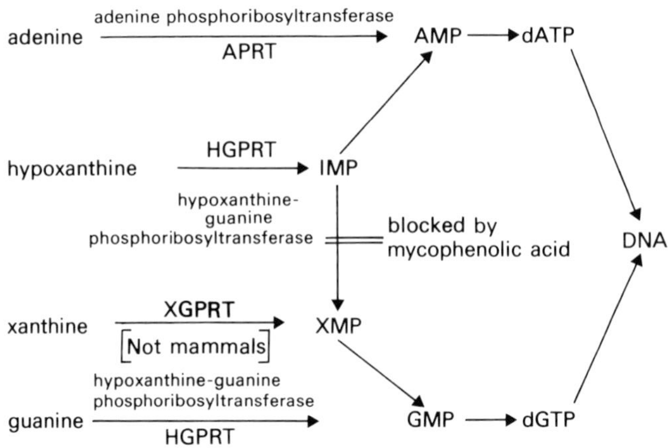

PYRIMIDINE BIOSYNTHESIS

Endogenous pathway

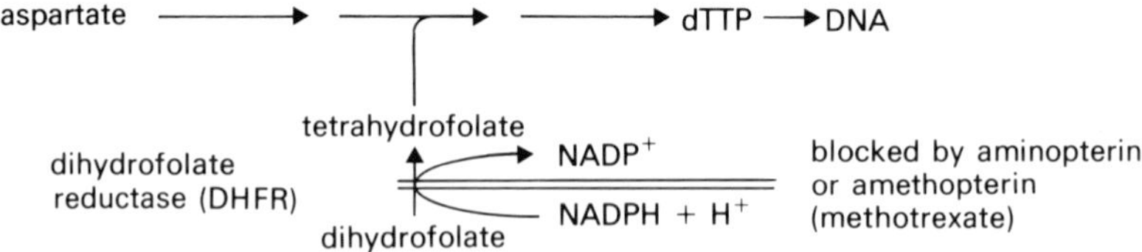

Salvage pathway

thymidine $\xrightarrow[\text{TK}]{\text{thymidine kinase}}$ dTMP → dTTP → DNA

Fig. 11.1 Commonly used mutants and inhibitors in cell culture. dATP, dGTP and dTTP have two synthetic pathways. A loss of either pathway for any one nucleotide, therefore, is not lethal, so mutants can be isolated which lack one of these pathways. $APRT^-$ cells can be isolated because they are *resistant* to toxic base analogues (2,6-diamino-purine, 2-fluoroadenine) but are *killed* in medium containing experiments was dependent upon the formation of a co-precipitate of the DNA with calcium phosphate, which is insoluble, and must be formed freshly in the presence of the DNA when the transfection mixture is assembled. Apparently the calcium phosphate granules are phagocytosed by the cells and in a small proportion of the recipients some of the DNA becomes stably integrated into the nuclear genome. The technique became generally accepted after its application by Graham and Van der Eb (1973) to the analysis of infectivity of viral DNA.

The calcium phosphate co-precipitate provides a general method for introducing any DNA into mammalian cells. It can be applied to relatively large numbers of cells in a culture dish but, as originally described, it is limited by the variable and usually rather low (1–2%) proportion of cells that take up exogenous DNA. Only a sub-fraction of these cells will be stably transfected. Procedures have been designed to increase to about 20% the proportion of cells that take up the DNA (Chu & Sharp 1981). A related procedure involves fusion of cultured cells with bacterial protoplasts containing the exogenous DNA (Schaffner 1980, Rassoulzadegan *et al.* 1982). Here, the proportion of cells taking up exogenous DNA can approach 100%. An alternative approach is to microinject DNA into the nucleus of cultured cells (see, for example, Kondoh *et al.* 1983), a procedure which has the advantage that 'hits' are almost certain but which cannot be applied to large numbers of cells.

Direct application of calcium phosphate-mediated transfection: transient assays without eukaryotic vector sequences

An illustrative example of the direct application of calcium phosphate-mediated transfection is the work of Rutter's group on DNA sequences controlling cell-specific expression of insulin and chymotrypsin genes (Walker *et al.* 1983). The genes

---

adenine and hypoxanthine plus azaserine, whereas wild-type cells survive. Azaserine inhibits several reactions in the endogenous pathway of purine biosynthesis. HGPRT$^-$ cells are *resistant* to toxic base analogues (thioguanine, azaguanine), but are *killed* in medium containing aminopterin and hypoxanthine in which wild-type cells survive. TK$^-$ cells are *resistant* to the thymidine analogue BUdR, but are *killed* in medium containing aminopterin and thymidine. HAT medium contains hypoxanthine, aminopterin and thymidine and is commonly used to select against TK$^-$ and HGPRT$^-$ cells.

XGPRT is an enzyme that is found in *E. coli* but not in mammalian cells.

coding for human and rat insulin, and the rat chymotrypsin gene had been cloned and characterized. These genes are expressed at a high level only in the pancreas. Each is expressed in clearly distinct cell types: insulin is synthesized in endocrine β-cells and chymotrypsin in exocrine cells. DNA sequences containing the promoter and 5′-flanking sequences of these genes were linked to the coding sequence of bacterial chloramphenicol acetyltransferase (CAT). This enzyme activity can be assayed very sensitively, and was used here as a 'reporter enzyme' whose activity was taken to be a measure of the transcription of the insulin or chymotrypsin genes. Recombinant plasmids containing such genes were constructed and grown in *E. coli*. These plasmids were not designed to be replicating eukaryotic vectors. Although it is possible that some exogenous DNA replication may occur when they are introduced into mammalian cells (this was not tested), they were expected to persist just long enough in the cell for their *transient* expression to be assayed. Therefore, plasmid DNA was introduced into either pancreatic endocrine or pancreatic exocrine cell lines in culture and after a subsequent 44-hour incubation cell extracts were assayed for CAT activity. It was found that the constructs retained their preferential expression in the appropriate cell type. The insulin 5′-flanking DNA conferred a high level of CAT expression in the endocrine, but not the exocrine cell line, with the converse being the case for the chymotrypsin 5′-flanking DNA.

The analysis was extended by creating deletions in the 5′-flanking sequences and testing their effects on expression. From such experiments it could be concluded that there are sequences located upstream of the promoter, between 150 and 300 bp of the transcription start site, which are essential for appropriate cell-specific gene transcription.

Co-transfection (= co-transformation)

Following the general acceptance of the calcium phosphate method, subsequent experiments showed that the thymidine kinase gene of Herpes simplex virus was effective in transfecting Tk^- mammalian cells (Fig. 11.1) to a stable Tk^+ phenotype which can be selected in HAT medium (Wigler *et al.* 1977). However, the isolation of cells transfected by other genes which do not encode selectable markers remained problematic. A breakthrough was made when it was discovered that cells can be simultaneously co-transfected by a mixture of two *physically unlinked* DNAs in the calcium phosphate precipitate (Wigler *et al.* 1979). To obtain co-transfectants, cultured cells were exposed to the thymidine kinase gene in the presence of a vast excess of a well-defined DNA, such as pBR322 or bacteriophage ϕX174 DNA, for which hybridization probes could be readily prepared. In order to achieve a suitably high DNA concentration for the formation of an effective co-precipitate, 'carrier' DNA was also included. This often consisted of total cellular DNA isolated from salmon sperm. Tk^+ transfectants were selected and scored by molecular hybridization for the co-transfer of the *unselected*, pBR322 or ϕX174, DNA.

Wigler *et al.* (1979) demonstrated the co-transfection of mouse Tk^- cells with pBR322, bacteriophage ϕX174 DNA and rabbit β-globin gene sequences and we shall use their ϕX experiments for illustrative purposes. ϕX replicative form DNA was cleaved with *Pst* I which recognizes a single site in the circular genome. 500 pg of the purified thymidine kinase gene were mixed with 10 μg of *Pst* I-cleaved ϕX replicative form DNA. This DNA mixture was added to Tk^- mouse cells and 25 Tk^+ transfectants were observed per 10^6 cells after 2 weeks in HAT medium. To determine if these Tk^+ transfectants also contained ϕX DNA sequences, high mol. wt DNA from the transfectants was cleaved with *Eco* RI which recognizes no sites in the ϕX genome. The cleaved DNA was fractionated by agarose gel electrophoresis, transferred to nitrocellulose filters by 'Southern blotting' and annealed with labelled ϕX DNA. These annealing experiments demonstrated that 14 out of 16 Tk^+ transfectants had acquired one or more ϕX sequences. Subsequent studies of the integrated DNA tell us something about the molecular events taking place during co-transfection. Although the added DNAs in the calcium phosphate co-precipitate are not physically linked, Southern blot analysis of the DNA integrated into the host genome reveals large concatemeric structures, up to 2000 kb long. These concatemers include copies of the selectable marker, the co-transfecting DNA and fragments of carrier DNA (Perucho *et al.* 1980). Therefore at some stage during the co-transfection process the DNAs must be physically ligated. The foreign DNA can probably integrate virtually anywhere in the genome (Robbins *et al.* 1981, Scangos *et al.* 1981).

The co-transfection phenomenon allows the stable introduction into cultured mammalian cells of any cloned gene. There is no requirement for a vector capable of replication in the host cell. Originally, co-transfection was discovered using two unlinked DNAs. However, this is an unimportant feature; analogous results can be obtained if the selected and unselected genes are ligated in prior manipulations. Originally, Herpes simplex virus DNA provided pure tk DNA. Subsequently this tk gene has been cloned in *E. coli* plasmids, hence providing a very convenient source. Initially the requirement for a tk^- recipient cell line was a serious limitation. This has been overcome by the development of so-called *dominant* selectable markers that can be used with *non-mutant* cell lines.

Selectable markers other than tk

These markers incude:
1 dihydrofolate reductase and associated methotrexate resistance;
2 rodent CAD;
3 bacterial XGPRT, and
4 bacterial neomycin phosphotransferase.

1 *Dihydrofolate reductase*

Dihydrofolate reductase (Fig. 11.1) is sensitive to the inhibitor methotrexate (Mtx). Cultured wild-type cells are sensitive to concentrations of the drug at about 0.1 μg/ml. Mtx-resistant cell lines have been selected and have been found to fall into three categories:
1 cells with decreased cellular uptake of Mtx;
2 cells overproducing DHFR;
3 cells having structural alterations in DHFR, lowering its affinity for Mtx.

It has been found that cells overproducing DHFR contain increased copy numbers of the gene (Schmike *et al.* 1978). The Chinese hamster ovary 'A29' cell line is notable because it is extremely resistant to Mtx and has been shown to synthesize *increased amounts* of an *altered* DHFR (Flintoff *et al.* 1976). Wigler *et al.* (1980) have used genomic DNA of the A29 cell line as a donor of the DHFR gene in co-transfection experiments with Mtx-sensitive cells. This system has the advantage that the high gene copy number in donor DNA gives efficient transfection. Additionally, the selection system is powerful and can be applied to *non-mutant* Mtx-sensitive cell lines.

There is one further advantage of the Mtx system: highly resistant variants of the transfected cell line can be selected so as to give concomitant amplification of the unselected DNA. In order to explore this possibility, Wigler *et al.* (1980) first showed that *Sal* I digestion of A29 DNA did not destroy its ability to transfect cells to Mtx-resistance. The *Sal* I-cleaved A29 DNA was then ligated to *Sal* I-linearized pBR322 and used to transfect a Mtx-sensitive mouse cell line. Resistant colonies were picked, grown and exposed to increasing concentrations of the drug so as to select highly resistant variants. DNAs of certain variants resistant to 40 μg/ml were analysed by Southern blot hybridization with pBR322 DNA as the probe. This analysis showed that the pBR322 sequence had undergone a substantial amplification of at least 50-fold.

Another variation on the use of DHFR depends upon the fact that bacterial DHFR is intrinsically resistant to Mtx (although sensitive to trimethoprim; see Chapter 6). O'Hare *et al.* (1981) have constructed plasmids in which the bacterial gene is transcribed from an SV40 promoter. Such plasmids transfected mouse cells to a Mtx-resistant phenotype by integration into the mouse genome.

2 *CAD*

The CAD protein is a multifunctional enzyme catalysing the first three steps of de-novo uridine biosynthesis: carbamyl phosphate synthetase; aspartate transcarbamylase; and dihydroorotase (Swyryd *et al.* 1974). One of these activities, aspartate transcarbamylase, is inhibited by N-phosphonacetyl-L-aspartate (PALA). PALA-resistant mammalian cells overproduce CAD from highly amplified copies of the CAD gene, in a manner analogous to Mtx-resistance (Wahl *et al.* 1984). The CAD gene of the Syrian hamster has been cloned on cosmid vectors in *E. coli* and has been shown to provide a dominant, amplifiable genetic marker that can be selected in non-mutant cells on the basis of resistance to high concentrations of PALA (De Saint Vincent *et al.* 1981, Wahl *et al.* 1984).

3 *XGPRT*

The *E. coli* enzyme xanthine-guanine phosphoribosyltransferase, XGPRT, is a bacterial analogue of mammalian HGPRT. However, by contrast with HGPRT, it has the additional ability to convert xanthine to XMP and hence ultimately to GMP (Fig. 11.1). Only hypoxanthine and guanine are substrates of HGPRT. Mulligan and Berg (1980, 1981a,b) have cloned the XGPRT gene and incorporated it into a variety of vectors in which transcription of the bacterial gene is directed by a SV40 promoter (see p. 251). Such constructs are capable of transfecting $HGPRT^-$ cells to $HGPRT^+$, but, more importantly, provide a selectable marker for non-mutant cells in medium containing adenine, mycophenolic acid and xanthine (Fig. 11.1). This selection can be made more effective by adding aminopterin, which blocks endogenous purine biosynthesis.

4 *Neomycin phosphotransferase*

Bacterial transposons Tn5 and Tn601 encode distinct neomycin phosphotransferases, whose expression confers resistance to aminoglycoside antibiotics (kanamycin, neomycin and G418), which are active in bacterial or eukaryotic cells. Berg (1981) has incorporated the neomycin phosphotransferase gene into constructs analogous to those containing XGPRT. Other constructs have linked the neomycin phosphotransferase gene to the Herpes simplex virus tk promoter in an *E. coli* plasmid (Colbère-Gorapin *et al.* 1981). Transfectants of non-mutant mammalian cells which contain such constructs can be selected by antibiotic resistance. Colbère-Gorapin *et al.* (1981) demonstrated the application of their construct to the co-transfection of a variety of cell lines from different mammalian species. Grosveld *et al.* (1982) have also constructed cosmid cloning vectors which include selective markers for growth in the host bacterium (β-lactamase) and animal cells. The markers for animal cells were neomycin phosphotransferase, HSV thymidine kinase or XGPRT. Such cosmids can be used to construct libraries of eukaryotic genes from which a particular recombinant can be isolated. The recombinant cosmid DNA can then be transfected into animal cells at high efficiency where transfectants in which the DNA has integrated into the nuclear genome can be readily selected. The power of aminoglycoside antibiotic resistance as a selective system in eukaryotes is now very evident. It has application in yeast (Jimenez & Davis 1980; see Chapter 9) and plants (Chapter 10). It has recently been applied to the development of an efficient transfection system in the amoebae of *Dictyostelium discoideum* (Barclay & Mellow 1983).

VIRAL VECTORS

Although many animal viruses now have been subjugated as vectors, those based upon SV40 have been the most widely used. Development of SV40 derivatives preceded others because knowledge of its molecular biology was the most advanced. The genome is a small covalently closed circular duplex DNA of about 5.2 kb (Fig. 11.2). It was the first animal virus genome to be completely sequenced, and there was a wealth of information concerning its replication and transcription. Sufficient information will be presented here for the reader to appreciate its use as a vector. For other aspects of SV40 biology, excellent reviews are available (Tooze 1980, Martin 1981).

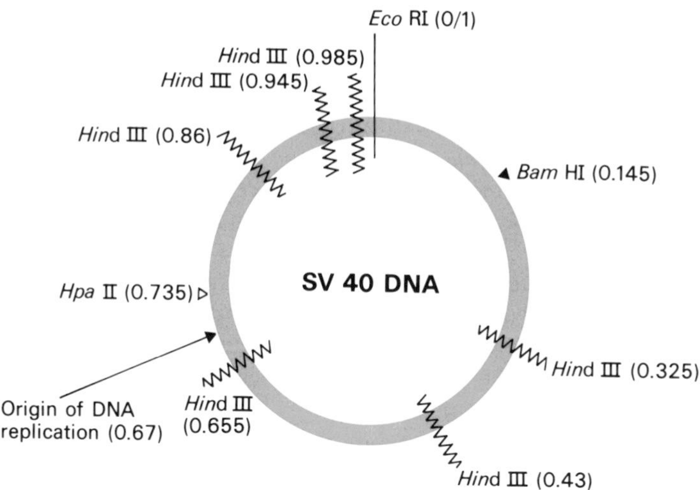

Fig. 11.2 Restriction endonuclease sites on SV40 DNA. The map coordinates of each site are shown in parentheses.

Basic properties of SV40

The virus particle contains the circular DNA genome associated with the four histones H4, H2a, H2b and H3 in a mini-chromosome. H1 is absent. The capsid is composed of 420 subunits of the 47 K polypeptide VP1. Two minor polypeptides VP2 and VP3, which consist largely of identical amino acid sequences, are also present.

SV40 can enter two types of life cycle depending upon the host cell. In *permissive* cells, which are usually permanent cell lines derived from the African green monkey, virus replication occurs in a normal infection. In *non-permissive* cells, usually mouse or hamster cell lines, there is no lytic infection because the virus is unable to complete DNA replication. However, growth transformation of non-permissive cells can occur. In such cells SV40 DNA sequences are integrated into the host genome. The SV40 sequences are often amplified and rearranged in such transformed cell DNA.

The lytic infection of monkey cells by SV40 can be divided into three distinct phases. During the first eight hours the virus particles are uncoated and the DNA moves to the host cell nucleus. In the following four hours, the *early* phase, synthesis of early mRNA and early protein occur and there is a virus-induced stimulation of host cell DNA synthesis. The *late* phase occupies the next thirty-six hours and during this period there is synthesis of viral DNA, late mRNA and late protein and culminates in virus assembly and cellular disintegration.

A functional map of the SV40 genome is given in Fig. 11.3. The region of about 400 bp around the origin of DNA replication is extremely interesting. Closely associated with the origin are control signals regulating the initiation of early and late primary transcripts. The early transcription is stimulated by two tandem, 72 bp, enhancer sequences located within this region.

An important feature of SV40 gene expression is the complex pattern of RNA splicing (Fig. 11.3). Alternative splicing pathways process the early primary transcript into two different early mRNAs. The late primary transcript can be processed into three different late mRNAs.

The two early mRNAs encode the large T and small t proteins (*Tumour* proteins or antigens). The late mRNAs, designated according to their sedimentation coefficients as the 16*s*, 18*s* and 19*s* mRNAs, encode VP1, VP3 and VP2 respectively.

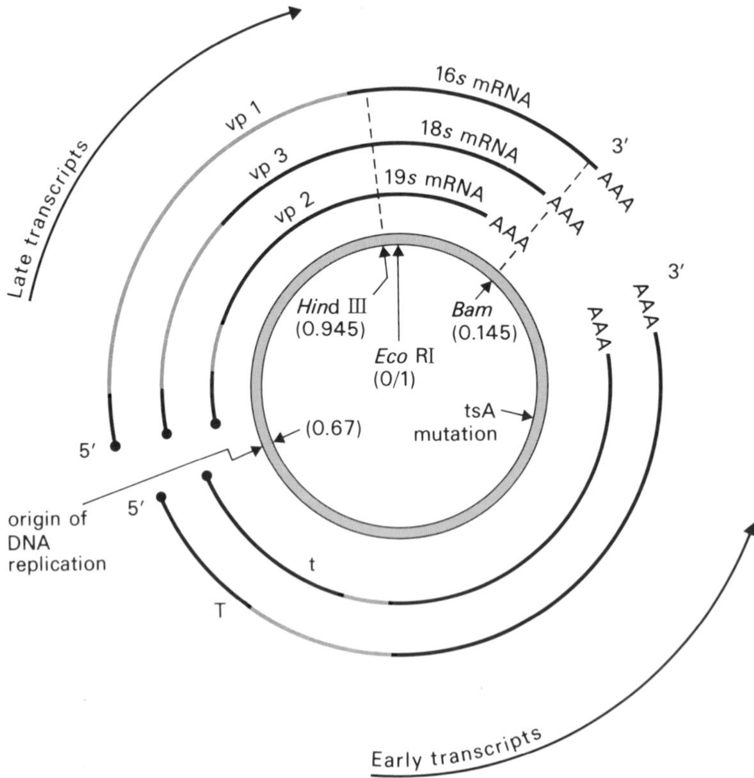

Fig. 11.3 Transcripts and transcript processing of SV40. Intron sequences which are spliced out of the transcripts are shown by red lines.

The 18*s* and 19*s* mRNAs have coding sequences in common and direct the synthesis of VP3 and VP2 such that they contain amino acid sequences in common. More striking, however, is the finding that the VP1 coding region overlaps VP2 and VP3 in a different translational reading frame.

SV40 vectors

The assembly of SV40 virions imposes a strict size limitation on the amount of recombinant DNA which can be packaged. In view of this, two strategies have been employed in developing SV40 vectors. The first is to replace a region of the viral genome with an equivalently sized fragment of foreign DNA, and hence produce a recombinant DNA that can replicate and be packaged into virions in permissive cells. In order to supply genetic functions lost by replacement of virus sequences, a helper SV40 virus must genetically complement the recombinant.

In the alternative strategy, the size limitation of the virion is avoided. Recombinants are constructed which are never packaged into virions and give no lytic infection. These are maintained in host cells transiently as high copy number, unintegrated, plasmid-like DNA molecules.

First experiments with late region replacement

SV40 lacking the entire late region functions can be propagated in *mixed infections* with a temperature-sensitive helper virus that can complement the late region defect. The mixed infection is maintained since the *ts* defect in the helper's early region must be complemented. A suitable helper is a *ts*A mutant, which produces a temperature sensitive T protein (Mertz & Berg 1974).

Based on this observation Goff and Berg (1976) prepared an SV40 vector by excising virtually the entire late region of the viral DNA by cleavage with *Hpa* II and *Bam* HI restriction endonucleases (Fig. 11.4). Cleavage with these two enzymes produces fragments approximately 0.6 and 0.4 of the genome length. These two fragments were separated by electrophoresis and the smaller fragment discarded. The large fragment, called SVGT-1, was then modified by the addition of 5′ poly (dA) tails using deoxynucleotidyl terminal transferase.

A suitable insert was prepared by cleaving λ DNA with *Eco* RI and *Hin*d III restriction endonucleases. This treatment yielded 12 DNA fragments which were separated by electrophoresis in agarose. Fragment 8 is 1.48 kb in length and contains *ori*, the origin of λ DNA replication, two structural genes, cII and *cro*, as well as four transcriptional promoters.

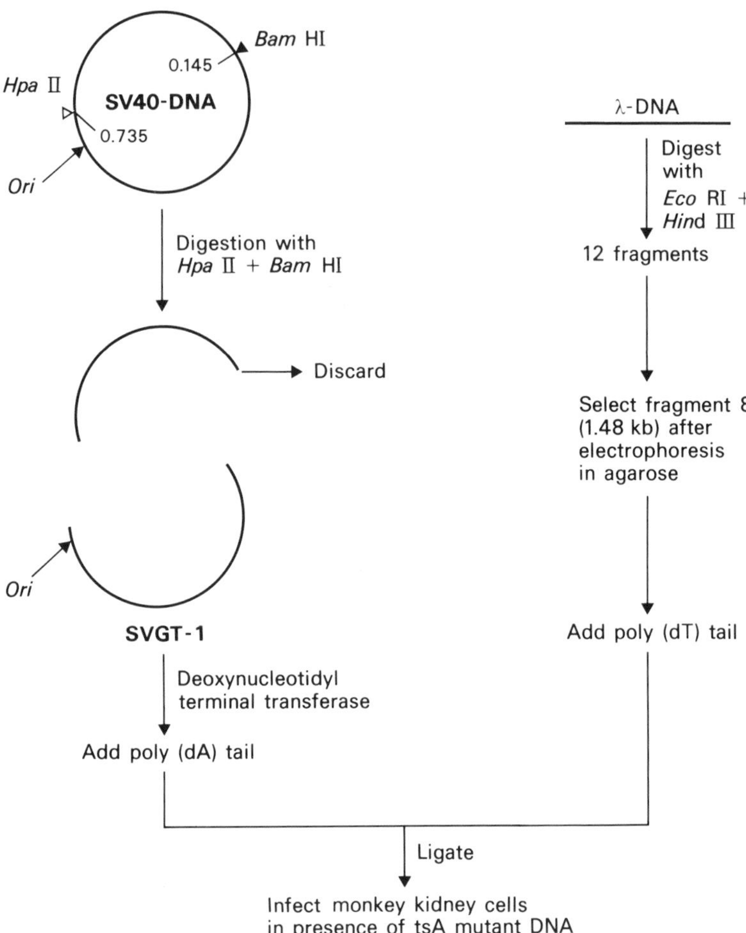

Fig. 11.4 Construction of an SV40-λ DNA hybrid as described in the text. The bold lines represent duplex DNA molecules.

After elution of fragment 8 from the agarose gel it was modified by the addition of poly (dT) tails (Fig. 11.4). SVGT-1 containing the poly (dA) termini and the poly-(dT)-tailed fragment 8 were mixed and annealed. The annealed DNA was then used to transfect monkey kidney cells in the presence or absence of *ts*A helper virus DNA. At the restrictive temperature (41 °C) *ts*A DNA alone gave no plaques, the annealed DNA alone gave no plaques, but transfection with the two DNAs together produced 2.5×10^3 pfu/μg of annealed DNA.

Thirty-four plaques from the mixedly infected cultures were extracted, mixed with more *ts*A mutant virus and used to infect fresh monkey cells at 41 °C. DNA was extracted from the progeny virus and the presence of the λ phage DNA segment was detected by measuring the reassociation kinetics with ^{3}H-labelled λ fragment 8 DNA. DNA from infections with 9 of the 34 plaques caused a striking increase in the reassociation rate of the labelled λ DNA fragment indicating the presence of high levels of the λ-SVGT hybrid. Heteroduplex analysis and restriction endonuclease mapping of the viral DNA from these 9 infected cultures confirmed that it was the expected mixture of *ts*A DNA and recombinant DNA.

Although the SVGT-λ hybrid DNA replicated in monkey cells in the presence of *ts*A helper DNA, the λ DNA sequences were not transcribed. At a time after infection when substantial amounts of SV40-specific RNA could be detected no λ-specific RNA was found. The search for new polypeptides in the SVGT-hybrid infected cells was also fruitless. These disappointingly negative results were still obtained when hybrids with the λ insert in opposite orientations were used.

Hamer (1977) synthesized SV40-*E. coli* Su$^+$III recombinant DNA by a method similar to that of Goff and Berg (1976). In contrast to the results of the latter workers, Hamer (1977) found that monkey cells infected with the SV40-Su$^+$III recombinant DNA synthesized surprisingly large amounts of RNA complementary to the insert. The normal product of the *E. coli* Su$^+$III gene is a suppressor RNA but all attempts to isolate a free or charged tRNA were unsuccessful.

It is now clear that the lack of proper transcription of the foreign DNA was due to the disruption of the late region post-transcriptional processing in these constructs.

Construction of an improved SV40 vector

From the information presented above on transcription of SV40 it is clear that the ideal vector would retain all the regions implicated in transcriptional initiation and termination, splicing and polyadenylation. Inspection of the SV40 map (Fig. 11.3) reveals two restriction endonuclease sites that could be used to generate a suitable vector. First, there is a *Hin*d III site at the map position 0.945 which is 6 nucleotides *proximal* to the initiation codon for VP1 (Fig. 11.5) and 50 nucleotides *distal* to the site at which the leader sequence is joined to the body of 16*s* mRNA. Second, the *Bam* HI site at map position 0.145 is 50 nucleotides proximal to the termination codon for VP1 translation and 150 nucleotides before the poly A sequence at the 3′ end of 16*s* RNA (Fig. 11.5). If the DNA between coordi-

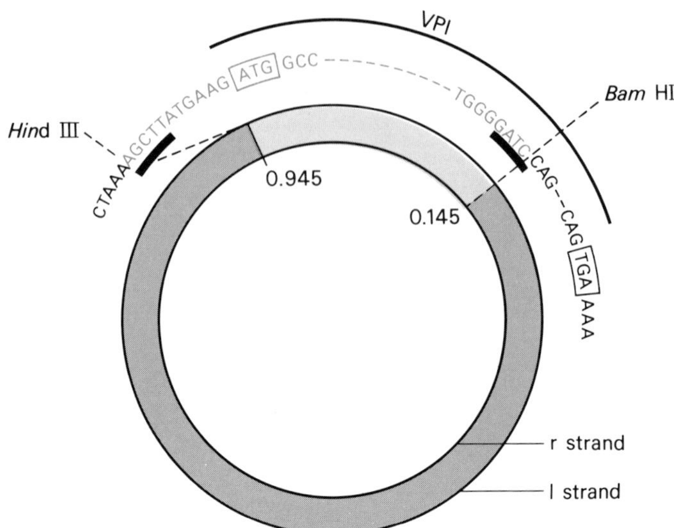

Fig. 11.5 Location of the *Hin*d III and *Bam* HI cleavage sites in relation to the coding sequence for VP1. The triplets enclosed in boxes show the initiation and termination signals for translation of VP1. The sequences underlined are the recognition sites for the *Hin*d III and *Bam* HI restriction endonucleases.

nates 0.945 and 0.145 were removed, the remaining molecule could be used as a vector for it would retain:

1 the origin of replication;

2 the regions at which splicing and polyadenylation occur;

3 the entire early region and hence could be complemented by a *ts*A mutant.

Such a vector (SVGT-5) has been constructed and used successfully to clone the rabbit β-globin gene in monkey kidney cells (Mulligan *et al*. 1979).

The first step in constructing SVGT-5 was partially to digest SV40 DNA with restriction endonuclease *Hin*d III. The digestion products were separated by electrophoresis and full-length molecules, i.e. those with a single cut, selected. Clearly these full-length molecules could have a cut at any one of the six *Hin*d III sites on SV40 DNA. The mixture of different full-length molecules was then digested with a mixture of restriction endonucleases *Bam* HI and *Eco* RI and fragments of 4.2 kb (SVGT-5) selected by electrophoresis on an agarose gel. Examination of the SV40 restriction map (Fig. 11.6) shows that cleavage of the *Hin*d III site at position 0.945 and at the *Bam* HI site produces the desired SVGT-5 fragment which is 4.2 kb long. However, cleavage at the *Hin*d III site at position 0.325 and at the *Bam* HI site

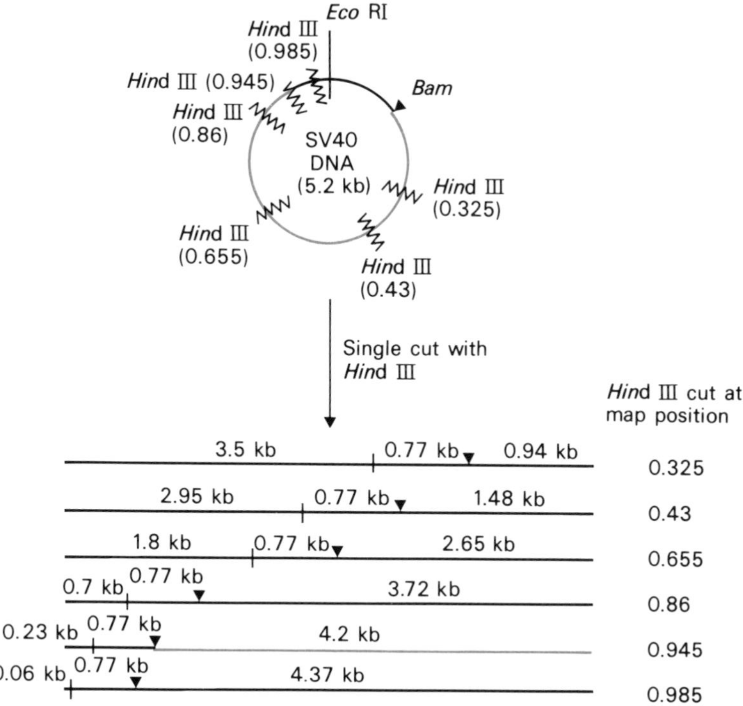

Fig. 11.6 Molecules produced by cutting SV40 DNA once with restriction endonuclease *Hin*d III. Note that when these six different molecules are cleaved with a mixture of *Eco* RI and *Bam* HI restriction endonucleases only one of them will produce a fragment 4.2 kb in length. For convenience only one strand of the DNA is shown.

would produce a fragment of similar size. Since only the latter fragment contains an *Eco* RI site, it can be selectively eliminated by including endonuclease *Eco* RI in the digestion mixture.

Example use of SVGT-5: Cloning of rabbit β-globin gene in monkey kidney cells

The starting point for this experiment was the recombinant plasmid pβG1 which contains the rabbit β-globin coding sequence. pβG1 was constructed by Maniatis *et al.* (1976) by inserting a cDNA copy of purified β-globin mRNA at the *Eco* RI site of plasmid pMB9 using the homopolymer tailing method. Nucleotide sequence analysis confirmed that pβG1 contained the entire β-globin coding sequence, all of the 3′ and 80% of the 5′ non-coding sequences in β-globin mRNA (Efstratiadis *et al.* 1976). Basically, the experiment was conducted in two stages:

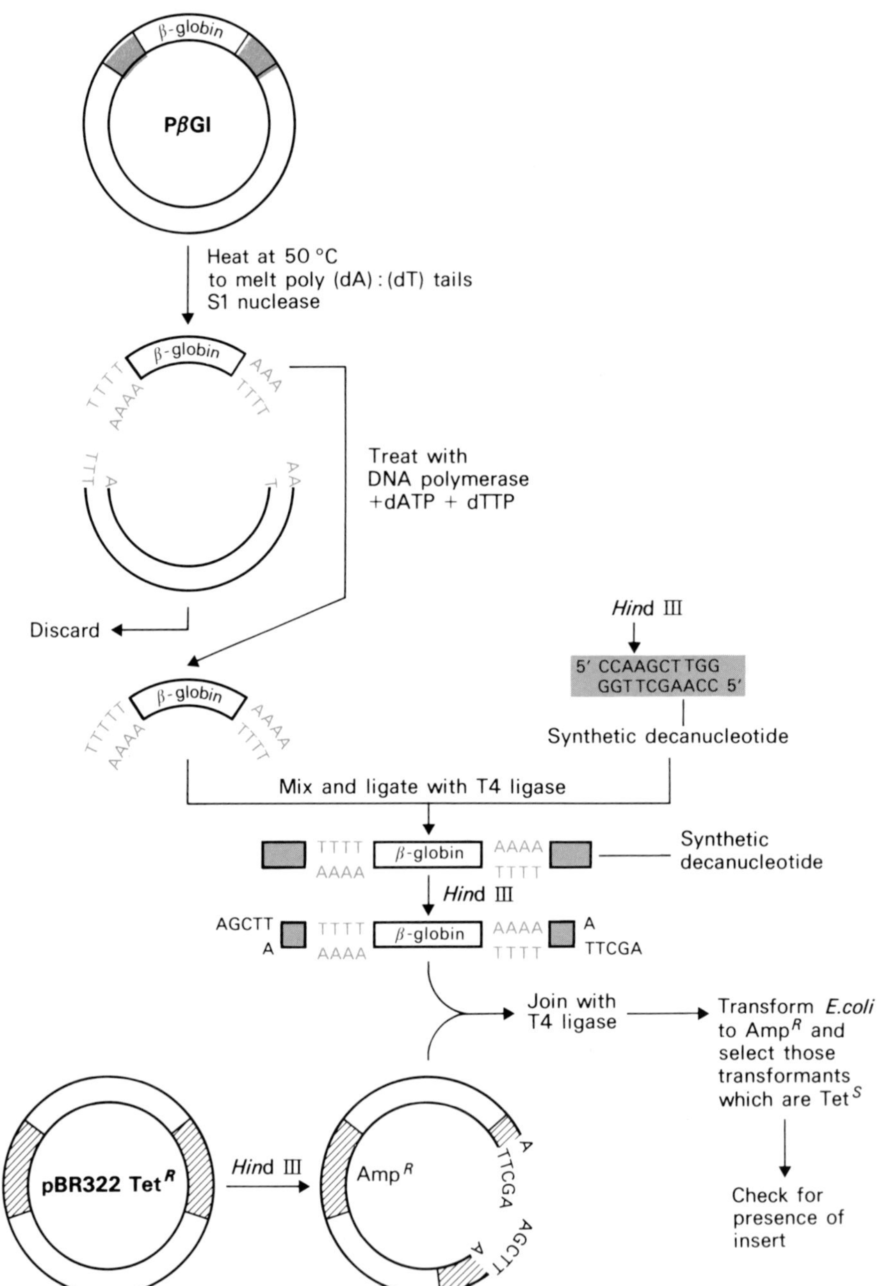

Fig. 11.7 Cloning of the β-globin cDNA in pBR322. See text for details.

1 excision of the β-globin cDNA from cDNA from pβG1 and its modification to produce *Hin*d III cohesive sites at either end;
2 insertion of the altered cDNA into SVGT-5.

Alteration of β-globin cDNA (Fig. 11.7)

pβG1 was incubated at 50 °C in the presence of S1 nuclease. At this temperature the poly (dA.dT) joints melt and the resulting single strands are digested with nuclease S1. The two fragments were separated by electrophoresis and the β-globin cDNA fragment isolated and treated with DNA polymerase I to convert the 'ragged' single-stranded ends to 'blunt' ends. A synthetic decanucleotide was then attached to either end of the cDNA with the aid of T4 DNA ligase. The resultant molecules were digested with endonuclease *Hin*d III and cloned in the *E. coli* vector pBR322. One of the recombinants between pBR322 and the modified β-globin cDNA was selected and designated pBR322-βG2.

Construction of SVGT-5-RaβG (Fig. 11.8)

The function of the cloning step in pBR322 was the enrichment of the altered cDNA prior to its insertion into SVGT-5. This altered cDNA was excised from pBR322-βG2 by sequential digestion with *Hin*d III and *Bgl* II endonucleases. This yielded a fragment having the codon for initiating translation of β-globin 37 nucleotides downstream from the *Hin*d III endonuclease-generated cohesive end and the translation termination codon just proximal to the *Bgl* II endonuclease-generated end. Even though the *Bam* HI (GGATCC) and *Bgl* II (AGATCT) endonuclease recognition sites differ, the cohesive ends generated by the two endonucleases are identical (GATC). Accordingly, the 0.485 kb β-globin cDNA fragment was ligated to SVGT-5 to produce SVGT-5-RaβG.

To propagate the SVGT-5-RaβG genome, monkey kidney cells were transfected at 41 °C with a mixture of SVGT-5-RaβG DNA and the DNA from a *ts*A mutant of SV40.

Expression of SVGT-5-RaβG

In SVGT-5-RaβG the coding sequence of VP1 has been precisely replaced by a cDNA copy of rabbit β-globin mRNA. SV40 late region leader and splice junctions are linked to the

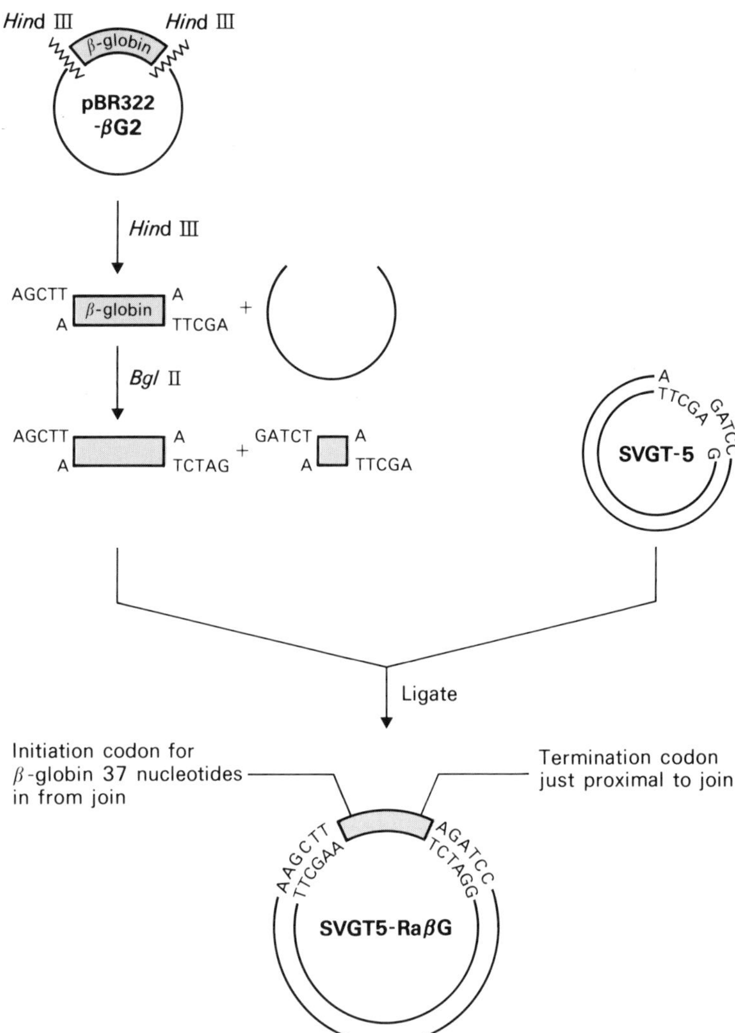

Fig. 11.8 The construction of SVGT-5-RaβG. See text for details.

cDNA sequence, and the downstream polyadenylation signal is specified by viral sequences. Following infection of monkey cells with the mixed recombinant/helper virus stock, the synthesis of the appropriate recombinant late mRNA was detected. The hybrid derivative of 16*s* late mRNA (it is actually 0.5 kb smaller than the normal 16*s* late mRNA) directed the synthesis of a protein indistinguishable from authentic rabbit β-globin. The protein was synthesized in large amounts, comparable to normal amounts of VP1, but appeared to be unstable presumably because it was not assembled into haemoglobin.

Application of late-region replacement vectors

Vectors of this type have been effective in obtaining expression of many foreign genes in monkey cells: mouse β-globin (Hamer & Leder 1979a); rat preproinsulin (Gruss & Khoury 1981, Gruss *et al.* 1981a); the p21 protein of Harvey murine sarcoma virus (Gruss *et al.* 1981b); influenza virus haemagglutinin (Gething & Sambrook 1981; Sveda & Lai, 1981), and hepatitis B virus surface antigen (Moriarty *et al.* 1981). Late-region replacement vectors have been used to study post-translational RNA processing and stability. Some of these studies have demonstrated a requirement for an intron in the late-region primary transcript if a cytoplasmic mRNA is to be produced (Gruss *et al.* 1979, Hamer & Leder 1979b, Lai & Khoury 1979). However, this has not proved consistently to be the case for all constructs (Gething & Sambrook 1981, Gruss *et al.* 1981a, Sveda & Lai 1981) so that the reasons for the requirement remain unclear.

Early region replacement vectors

When the early region of SV40 is replaced, the essential T function is lost and must be provided by complementation. A major breakthrough in the development of SV40 vectors came when it was found that the T protein can be complemented in the COS cell line (this must not be confused with *cos,* the cohesive end site of phage lambda). This cell line is a derivative of the permissive CV-1 monkey cell line. When CV-1 cells are infected with SV40, the normal lytic cycle ensues and no growth transformants are recovered. Gluzman (1981) found that CV-1 cells could be transformed by a segment of SV40 early-region DNA (cloned in an *E. coli* plasmid vector) in which the SV40 origin of DNA replication had been inactivated. The resulting COS cell line (CV-1, *o*rigin of *S*V40) expresses T protein from integrated SV40 sequences and does so in a cellular background which is permissive for SV40 DNA replication. Infection of these cells with SV40 lacking a functional early region therefore leads to a normal lytic cycle. Since mixed infections are not necessary there is no contaminating helper virus. Therefore SV40 recombinants in which early DNA sequences have been replaced with foreign DNA can be constructed and propagated. Gething and Sambrook (1981) have made recombinants of this type which express influenza virus haemagglutinin in COS cells.

Transient vectors in COS cells

In COS cells *any* circular DNA, with no definite size limitation, containing a functional SV40 origin of replication should be replicated independently of the cellular DNA as a plasmid-like episome. Many laboratories have constructed vectors based on this principle (Myers & Tjian 1980, Lusky & Botchan 1981, Mellon *et al.* 1981). In general, these vectors consist of a small SV40 DNA fragment containing the SV40 origin cloned in an *E. coli* plasmid vector. They are thus shuttle vectors from which recombinant derivatives can be constructed and grown in *E. coli*. These are then transfected into COS cells where very high copy numbers of the recombinants are obtained. It is important that the *E. coli* plasmid sequence does not contain the so-called 'poison' sequence. This has been identified as a small region of pBR322 which in mammalian cells causes inefficient replication of any DNA in which it is included (Lusky & Botchan 1981). Plasmid pAT153 (see Chapter 3) and other related deletion derivatives of pBR322 have lost this site and are, therefore, suitable for constructing shuttle vectors for mammalian cells. The poison sequence is coincident with, or very near to, the *nic/bom* site necessary for conjugative mobilization of pBR322 but it is not known if this coincidence is significant.

Permanent cell lines are not established when transient vectors are transfected into COS cells, but even though only a low proportion of cells are transfected, the high copy number is compensatory and the *transient* expression of cloned genes can be analysed. The major application of these vectors is in providing a rapid means of screening the effects of in-vitro manipulations upon transcriptional and post-transcriptional control sequences. In such studies the important transcriptional stimulation by the SV40 enhancer sequences was discovered (Banerji *et al.* 1981, Khoury & Gruss 1983).

The enhancer sequences are a pair of 72 nucleotide, tandemly repeated sequences which are located near the origin of DNA replication and which are necessary for efficient early transcription. By linking the enhancer sequences to other genes it was found that they stimulate or *enhance* transcription from virtually any promoter that is placed near them. This effect operates on both sides of the enhancer, i.e. it is independent of enhancer orientation and extends to a promoter placed several kilobases away. The SV40 enhancer sequences are important to gene manipulators because of their natural stimulatory effect on the SV40 early transcription unit, to which foreign genes may be fused and hence efficiently expressed. Alternatively transcription from foreign promoters may be enhanced by the presence of the enhancer sequences nearby. Subsequent to the discovery of enhancer sequences in SV40, cellular sequences with similar properties have been discovered. Some of these are cell-type specific, i.e. their enhancement effect operates in only certain cell types. How enhancers work is currently unclear. They may be involved in generating regions of DNA which are free of nucleosomes and thus are good access points for RNA polymerase.

Vectors incorporating SV40 transcription units and dominant selectable markers

A large series of vectors has been constructed by Berg and his co-workers (Berg 1981, Mulligan & Berg 1981b). These comprise an *E. coli* plasmid replicon derived from pBR322, an SV40 origin and functional early region providing T protein, and an SV40 transcription unit into which foreign genes can be inserted and expressed. These vectors are grown and manipulated using *E. coli* as the host and then transferred into permissive monkey cells.

Analogous vectors contain a virus origin of replication and early region derived from polyoma virus, whose structural organization closely resembles that of SV40. Mouse cells, but not monkey cells, are permissive for polyoma virus and therefore are the host for these polyoma-derived vectors.

Such vectors have been used in constructs designed to express the bacterial genes encoding neomycin phosphotransferase and XGPRT, which provide dominant selectable markers in animal cells (see p. 236). These vectors have the ability to replicate episomally in permissive hosts. However, when *stable* transfected lines were selected on the basis of drug resistance, it was found that the complete DNA had integrated into the transfected cell genome, with only a few (1–5 range) copies per cell (Mulligan & Berg 1981a).

Bovine papilloma virus

Bovine papilloma virus (BPV) causes warts in cattle. The genome is a covalently closed circular, duplex DNA of about 8.0 kb. The unique feature of this virus DNA is its ability to propagate as a multicopy plasmid in mouse cells, in which it causes growth transformation.

Shuttle vectors have been developed which comprise a large subgenomic transforming fragment of BPV linked to a deletion derivative of pBR322. Such vectors have been used to introduce a rat preproinsulin genomic DNA into cultured mouse cells, using transformation as the basis for selection (Sarver *et al.* 1981a,b). The recombinant replicated as a plasmid at about 100 copies per cell, and expression of the rat gene from its own promoter was readily detected at the mRNA and protein level. Similar experiments have been performed in which a 7.6 kb, human β-globin gene cluster was maintained at about 10–30 copies per mouse cell (Di Maio *et al.* 1982). In related experiments with a BPV vector, transformed cells contained as many as 200 copies of the recombinant DNA (Ostrowski *et al.* 1983). Law *et al.* (1983) have developed an advanced shuttle vector in which a BPV hybrid plasmid contains the neomycin phosphotransferase gene inserted into a mammalian cell transcription unit. The plasmid is maintained as a stable extrachromosomal DNA in mouse cells selected either for resistance to G418 or for the transformed cell phenotype.

BPV-derived vectors show great promise because *permanent*, clonal cell lines can be obtained which carry the recombinant DNA at a relatively high copy number. An additional potential expected advantage is the ability to carry large foreign DNA inserts. The detaiied molecular biology of this virus, including its transcription pattern, will be worked out soon.

Vectors based on other viruses

Experience has accumulated at such a pace that virtually any virus that has been studied in detail can now be rapidly subjugated as a vector. Adenoviruses and retroviruses are attractive potential vectors because of their wide range of avian, mammalian and other animal hosts. Retroviruses offer additional advantages.

1 The production of virus does not lead to cell death.

2 Integrated proviral DNA has strong promoters.

3 In the case of murine mammary tumour virus the promoter function can be switched on and off experimentally by glucocorticoid hormones (Lee *et al.* 1981, Ostrowski *et al.* 1983, Scheidereith *et al.* 1983). An approach that is being actively pursued involves replacement of regions of cloned proviral DNA with foreign DNA, combined with complementation of the defective, recombinant retrovirus by a helper virus

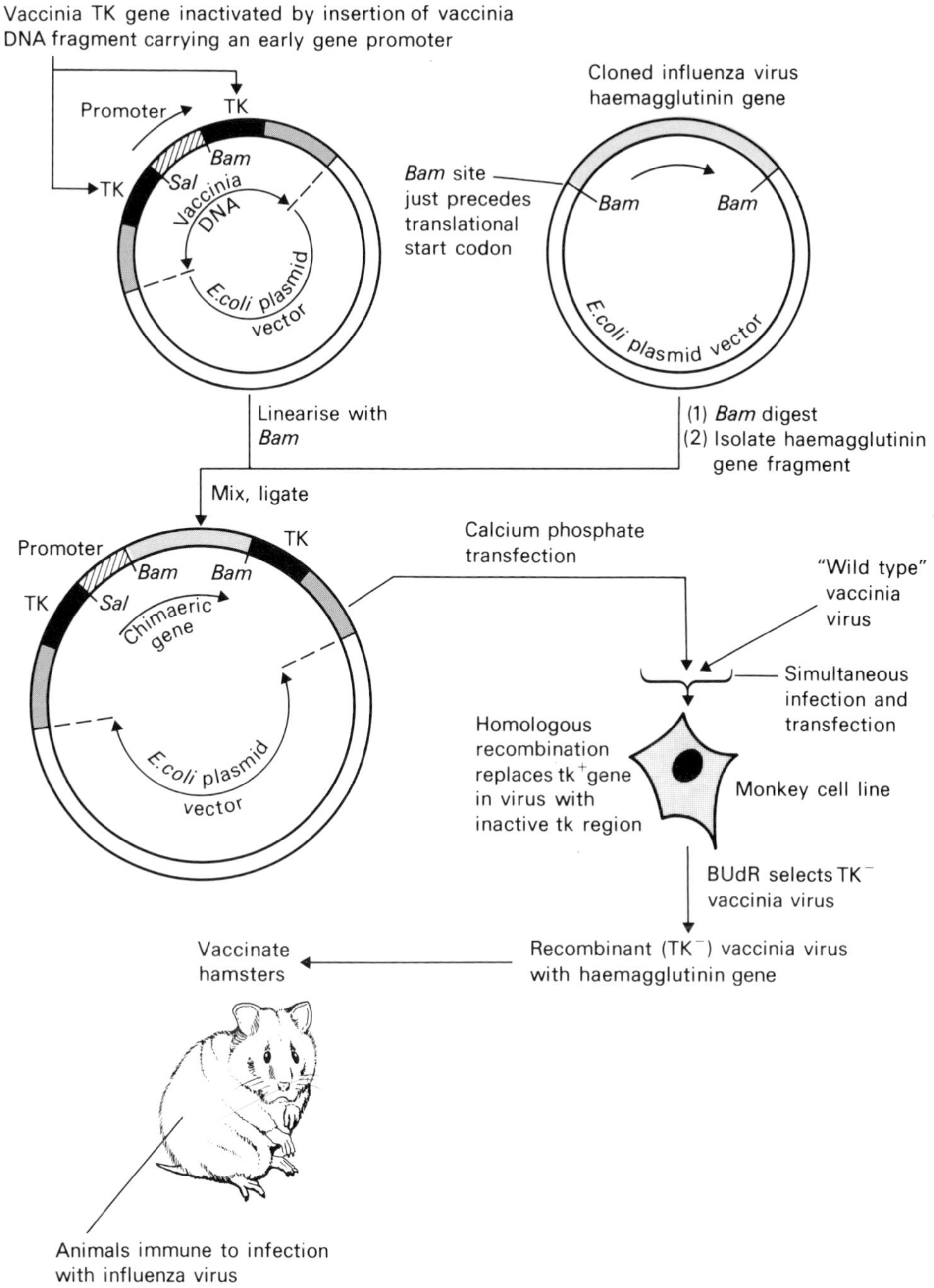

Vaccinia TK gene inactivated by insertion of vaccinia DNA fragment carrying an early gene promoter

Promoter
TK
TK
Bam
Sal
Vaccinia DNA
E.coli plasmid vector

Cloned influenza virus haemagglutinin gene

Bam site just precedes translational start codon

Bam
Bam
E.coli plasmid vector

Linearise with *Bam*

(1) *Bam* digest
(2) Isolate haemagglutinin gene fragment

Mix, ligate

Promoter
TK
TK
Bam
Bam
Sal
Chimaeric gene
E.coli plasmid vector

Calcium phosphate transfection

"Wild type" vaccinia virus

Simultaneous infection and transfection

Homologous recombination replaces tk^{+} gene in virus with inactive tk region

Monkey cell line

BUdR selects TK^{-} vaccinia virus

Recombinant (TK^{-}) vaccinia virus with haemagglutinin gene

Vaccinate hamsters

Animals immune to infection with influenza virus

Fig. 11.9 Construction of an infectious vaccinia virus recombinant expressing influenza virus haemagglutinin.

(Shimotohno & Temin 1981, 1982, Tabin *et al.* 1981, Wei *et al.* 1981, Sorge & Hughes 1982). Most of the principles of this experimentation have been encountered in the SV40 systems described in this chapter.

One final and elegant example of the use of a viral vector deserves mention. This is the exploitation of vaccinia virus. Vaccinia virus is closely related to variola virus, which causes smallpox, and inoculation with vaccinia virus provides a high degree of immunity to smallpox. There has been a proposal to construct vaccinia virus recombinants which express antigens of unrelated pathogens and use them as live vaccines against those pathogens. Smith *et al.* (1983) adopted a clever strategy for expressing the hepatitis B virus surface antigen (HBsAg) in vaccinia which took into account:

1 the large size (180 kb) of vaccinia DNA;
2 the lack of infectivity of isolated viral DNA;
3 the packaging of viral enzymes necessary for transcription within the virion; and
4 the probability that vaccinia virus has evolved in its own transcriptional regulatory sequences operative in the cytoplasm where viral transcription and replication occur.

Briefly, fragments of vaccinia DNA were cloned in an *E. coli* plasmid vector which contained a non-functional vaccinia thymidine kinase gene. This gene had been rendered inactive owing to the insertion of a vaccinia DNA fragment containing a promoter derived from another early vaccinia gene. The HBsAg gene was inserted next to the vaccinia promoter in the correct orientation. This chimaeric HBSAg gene was then inserted into vaccinia DNA by homologous recombination as follows. Monkey cells were infected with wild-type vaccinia and simultaneously transfected with the recombinant *E. coli* plasmid. Homologous recombination could then replace the functional thymidine kinase gene of the wild-type virus with the non-functional tk gene sequence which included the HBsAG chimaeric gene. Such virus would be Tk^- and would be selectable on the basis of resistance to BUdR (Fig. 11.1). When cells were infected with such TK^- virus, they were found to synthesize HBsAg and secrete it into the culture medium. Vaccinated rabbits rapidly produced high-titre antibodies to HBsAg. A similar strategy (Fig. 11.9) was then used to construct an infectious vaccinia virus recombinant which expresses the influenza haemagglutinin gene and induces resistance to influenza virus infection in hamsters (Smith *et al.* 1983).

Chapter 12 Microinjecting Genes into Animal Oocytes, Eggs and Embryos

The successful introduction of exogenous DNA into cultured mammalian cells has led to new experimentation on gene regulation in animals. We saw in Chapter 11 that by using differentiated cell types in culture it has been possible to analyse DNA sequences essential for cell-specific gene expression. A drawback of this approach is that cultured cell lines are incapable of development into organisms *in vitro*. However, the microinjection systems described in this chapter allow the study of transferred gene sequences in the context of normal embryonic development. These are based upon direct microinjection of DNA into fertilized eggs or embryos. The experimental systems which are most advanced in this context involve the frog *Xenopus laevis*, the mouse and the fly *Drosophila melanogaster*. In addition, the oocytes (which are precursor cells of the mature egg cells) of *Xenopus* have been exploited very usefully as an assay for gene expression quite apart from the developmental context.

Each of the experimental systems has its particular advantages and limitations. Frog oocytes are readily available in large numbers, as are the eggs and embryos. The molecular biology and biochemistry of early development of this vertebrate are, therefore, relatively well studied since material is not limiting. The fact that amphibian embryogenesis occurs outside the mother means that access is not a problem. However, there is very little genetics of *Xenopus*, and breeding experiments requiring the raising of more than one generation to adulthood are tedious. The mouse has a much shorter generation time, but by comparison, access to and manipulation of the mammalian embryo is more complicated. *Drosophila* has all the advantages of an extremely powerful genetic system. Particularly useful is the ability to locate gene sequences simply and precisely by hybridization *in situ* to polytene chromosomes. The relatively small genome size (1.4×10^5 kb, compared with 3×10^6 kb and 2.7×10^6 kb for *Xenopus* and mouse) is also a distinct advantage for molecular genetics.

EXPRESSION OF DNA MICROINJECTED INTO *XENOPUS* OOCYTES AND UNFERTILIZED EGGS

Oocytes can be obtained in large numbers by removal of the ovary of adult female *Xenopus*. Each fully grown oocyte is a large cell (0.8 to 1.2 mm in diameter) arrested at first meiotic prophase (Fig. 12.1). This large cell has a correspondingly large nucleus (also called the germinal vesicle) which is located in the darkly pigmented, animal hemisphere of the oocyte; because of the large size, microinjection of DNA into the nucleus is technically easy. Typically 20–40 nanolitres is injected through a finely drawn glass capillary. The oocyte nucleus contains a store of the three eukaryotic RNA polymerases; enough to furnish the needs of the developing embryo at least until the 60 000 cell stage (Roeder 1974). This store is available for transcription of the exogenous DNA. The oocyte also contains large stores of histones and other essential components for early embryonic development (reviewed by Davidson 1976). The exogenous DNA is assembled into chromatin (Gargiulo & Worcel 1983; Ryoji & Worcel 1984), but there is no replication of injected duplex DNA in the oocyte. Single-stranded DNA, such as that obtained by cloning in M13 vectors, is converted to duplex form efficiently in the oocyte nucleus. Recently, genetic recombination between two genetically marked, co-injected DNAs has been reported (Carroll 1983).

Expression of microinjected genes

Since the ability of oocytes to transcribe exogenous DNAs was first demonstrated by Mertz and Gurdon (1977) a wide variety of cloned genes from diverse sources has been microinjected. Table 12.1 gives an indication of the range of such genes. Note that genes transcribed by each of the three RNA polymerases, I, II and III, have been expressed in this system. When protein-coding genes transcribed by RNA polymerase II have been injected correct transcriptional initiation has often been observed, giving rise ultimately to stable transcripts which are exported to the cytoplasm and translated.

This ready ability to introduce a manipulated DNA into a living nucleus where its expression can be monitored has proved very powerful. The effects of directed changes in the DNA sequence, changes in the RNA transcribed from it, and changes in proteins encoded by it can all be assayed. This approach has been given the name 'surrogate genetics' by Birnstiel (Birnstiel & Chipchase 1977). He and his co-workers

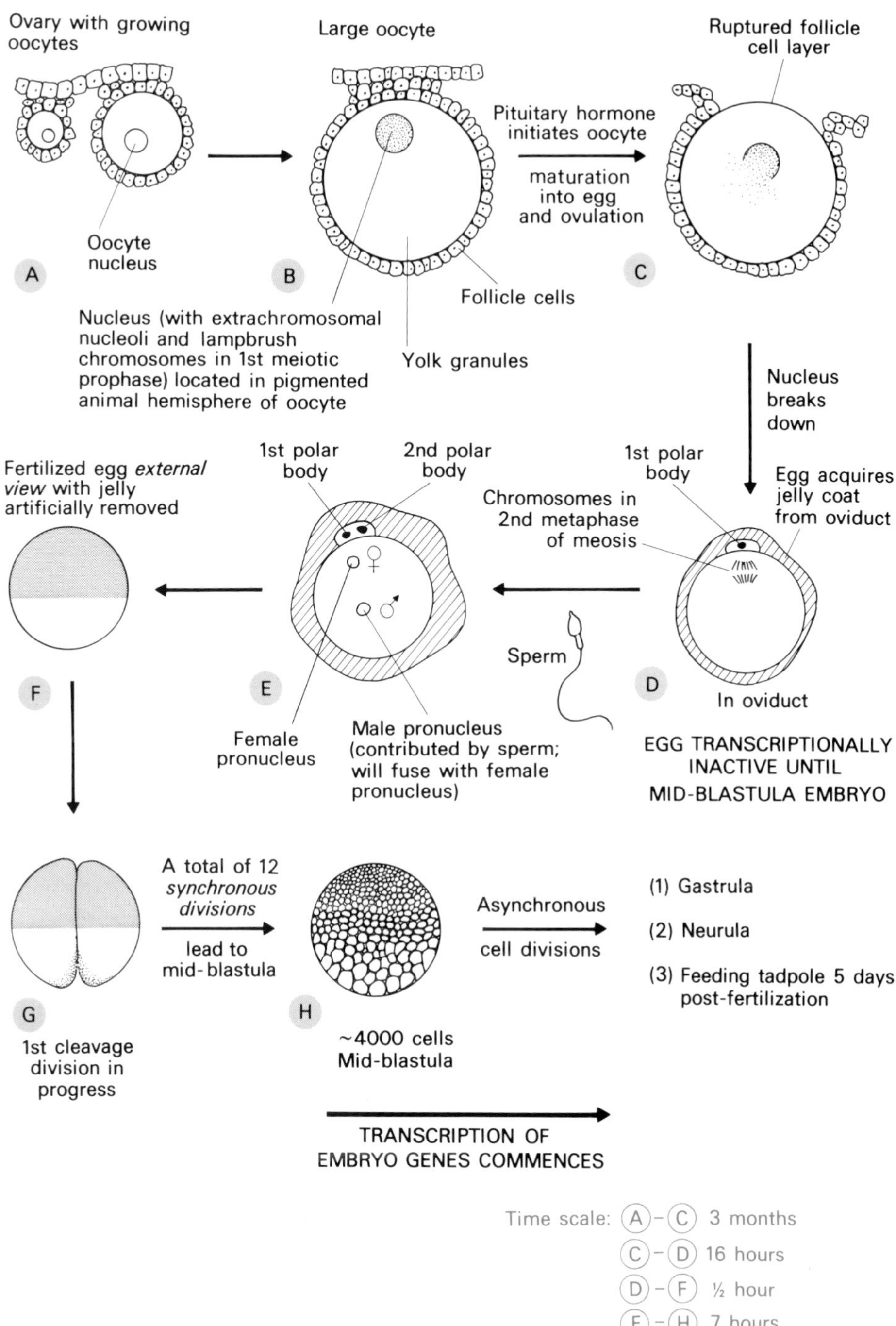

Fig.12.1 Events in oogenesis and early embryogenesis of *Xenopus*. (Adapted from Gurdon (1974) and Newport & Kirschner (1982a,b).)

Table 12.1 Expression of genes injected into the *Xenopus* oocyte nucleus.

Source of gene	Gene	RNA polymerase	Correct transcripts	Proteins synthesized	Reference
Xenopus	rDNA	I	some	—	Moss (1982)
SV40	viral DNA	II	some	T antigen and late proteins	Mertz & Gurdon (1977), De Robertis & Mertz (1977), Rungger & Turler (1978)
Sea-urchin	histones H1, H2A, H2B, H3, H4	II	H1, H2A, H2B, H4, H3 (H3 mainly with 3′ extension)	H2A, H2B, H3 (H1 not analysed)	Hentschel *et al.* (1980), Probst *et al.* (1979)
Xenopus	histones H1, H2A, H2B, H3, H4	II	—	H1, H2A, H2B, H3, H4	Old *et al.* (1982)
Chicken	ovalbumin	II	No correct length transcripts detected	ovalbumin	Wickens *et al.* (1980)
Xenopus	5*s*	III	Yes	—	Brown & Gurdon (1977)
Yeast	$tRNA^{leu}$	III	Yes	—	Newport & Kirschner (1982)

have applied the surrogate genetic approach to the analysis of promoter elements and related regions of sea-urchin histone genes which affect their transcription in *Xenopus* oocyte nuclei (Grosschedl & Birnstiel 1980a,b, Grosschedl *et al.* 1983). Surrogate genetics in the oocyte system has also been employed in studying sequences necessary for polyadenylation at the 3′-terminus of globin mRNA (Williams *et al.* 1984). Other workers have examined the effects of DNA conformation on transcription. In general, circular molecules are transcribed much more efficiently than linear molecules (Harland *et al.* 1983). An especially noteworthy application of the oocyte as an assay has exploited the *failure* of the oocyte to produce a correct transcript from an injected gene. It was observed that a sea-urchin histone H3 gene was transcribed from the correct initiation site but failed to generate substantial amounts of mRNA because transcripts extended beyond the correct 3′-terminus. Birnstiel and his co-workers reasoned that the *Xenopus* oocyte may be deficient in an activity necessary for the generation of the correct 3′-terminus of the transcript. In accord with this idea they made extracts of sea-urchin cells and co-injected them with the H3 gene. They were able to identify and purify a component of the sea-urchin cells which complemented the *Xenopus* deficiency (Stunnenberg & Birnstiel 1982). This component was a nuclear RNA-protein complex. Even the pure 60 nucleotide nuclear RNA alone would stimulate correct 3′-terminus formation when coinjected with the H3 gene (Galli *et al.* 1983).

The above experiments suggested that endonucleolytic cleavage is necessary to generate the correct 3′-terminus of the transcript, and that the *Xenopus* oocyte is deficient in an RNA component necessary for the endonucleolytic cleavage of certain, but not all, exogenous DNA transcripts. Krieg and Melton (1984) have extended this approach by transcribing a chicken histone gene *in vitro* so as to create transcripts which extend beyond the normal 3′ end of the mRNA. These transcripts were then injected into oocytes where they were processed into molecules with the correct 3′-terminus. Taken together, these experiments demonstrate that the 3′-termini can be, and presumably normally are, generated by cleavage of a larger primary transcript and that a nuclear RNA-protein complex is involved. The enzymatic machinery involved in this processing step resembles the splicase necessary for intron removal, in having the requirement for a small nuclear RNA.

As can be seen from Table 12.1, several genes that are transcribed by RNA polymerase II and which contain introns have been found to direct the synthesis of the correct protein product in oocytes. Therefore it is certain that some correct splicing must occur in these cases. The fidelity of the splicing of the RNA polymerase II transcripts from only one exogenous DNA, SV40 DNA, has been examined in detail so far (Wickens & Gurdon 1983). About 50% of the stable transcripts were spliced. The splicing of tRNA precursors has also been investigated. This splicing activity was shown to be confined to the oocyte nucleus (De Robertis *et al.* 1981).

In addition to such investigations of transcription and post-transcriptional processing of RNA in oocytes, it is possible to apply the surrogate genetic system to other cell biological processes; for example, cloned *Xenopus* histone genes have been microinjected and are expressed so as to result in a very large stimulation of histone synthesis by the oocyte. As would be expected, these histones are concentrated in the nucleus just like the normal store of histones accumulated in the oocyte for the early phase of embryogenesis (Old, R. W. *et al.* 1982). This was easily demonstrated because it is simple to isolate the germinal vesicle separate from oocyte cytoplasm by manual dissection. It is, therefore, possible to manipulate such genes and study those features of a protein that specify its 'karyophilicity' or concentration in the nucleus (De Robertis 1983).

A further example of the power of the oocyte system follows from the observation that secretory proteins encoded by DNAs injected into the nucleus, or mRNAs injected cytoplasmically, are secreted by the oocyte into the surrounding incubation medium. It has been found that the oocyte recognizes the hydrophobic signal sequence of secretory proteins from a variety of phylogenetically diverse organisms, making this process amenable to surrogate genetics (Lane *et al.* 1981).

Injection of DNA into fertilized eggs of *Xenopus*

DNA may be injected into de-jellied fertilized eggs at the one- or two-cell stage. The fate of this DNA has been the subject of recent experiments. In experiments with unfertilized eggs, nucleus-like bodies were formed in which the injected DNA is surrounded by an envelope similar to the normal nuclear envelope (Forbes *et al.* 1983). It is reasonable to assume that fertilized eggs respond in a similar way. The DNA is also replicated. There is presently some controversy about whether specific origins of DNA replication are operative in *Xenopus* eggs, or in higher eukaryotes in general for that matter. However, it is clear that all circular, duplex DNAs tested so far are replicated in unfertilized *Xenopus* eggs (reviewed in Laskey *et al.* 1983). In eggs the injected DNA is immediately subject to replication, and thereafter is replicated further only in synchronization with the cell cycle of the activated egg (Harland & Laskey 1980). As an embryo develops, exogenous DNA persists, and at least some of it continues to be replicated. In a typical experiment in which a recombinant plasmid carrying *Xenopus* globin genes was injected, the amount of plasmid DNA increased 50- to 100-fold by the gastrula stage. At subsequent stages, the amount of DNA per embryo decreased and most of the persisting DNA comigrated with high molecular weight chromosomal DNA (Bendig & Williams 1983). Possibly it had become integrated into the *Xenopus* genome. Alternatively, it may have formed large concatemers. Concatenation *is* known to occur when linear DNA is injected into oocyte nuclei (Bayne *et al.* 1984) and linear phage lambda DNA circularizes when injected into eggs (Mechali *et al.* 1983). Although there is no direct evidence that any of the persisting DNA has become chromosomally integrated, exogenous DNA sequences can be detected in tissues of adult animals (Rusconi & Schaffner 1981). The ability of adult frogs to transmit such DNA to their progeny has not been thoroughly investigated. The time scale of breeding experiments in *Xenopus* is not prohibitive, but it is long (at least 6 months from egg to adult). The mouse has been much more actively studied in this regard.

Transcription of DNA injected into fertilized *Xenopus* eggs

Transcription of a variety of exogenous genes has been observed following injection into fertilized eggs. These include: a yeast $tRNA_{leu}$ gene (Newport & Kirschner 1982b); rabbit β-globin gene (Rusconi & Schaffner 1981) and *Xenopus* adult globin genes (Bendig & Williams 1983). Transcription is low or absent immediately following injection, but subsequently becomes activated in accord with the mid-blastula transition in the developmental programme of *Xenopus* embryos (Newport & Kirschner 1982a,b).

While these experiments show that expression of DNA injected into *Xenopus* embryos is feasible, it is at present too early to assess the importance of this surrogate genetic system for analysing developmentally regulated or tissue-specific gene expression. With the cloning and characterization of appropriate *Xenopus* genes, such as α-actin genes which are expressed in mesodermal tissue to the embryo, rapid progress in this area is expected.

MICROINJECTION OF CLONED DNA INTO MOUSE EMBRYOS

In early experiments on microinjecting DNA into mouse embryos, SV40 DNA was deposited in embryos at the pre-implantation blastocyst (4–30 cell) stage (Jaenisch & Mintz 1974). These embryos were implanted into the uteri of foster mothers and allowed to develop. Some cells of the embryo incorporated DNA into their chromosomes, but the adult animals which resulted were mosaics, with only a proportion of the cells in a tissue containing integrated DNA. However, integration into some germ line cells did occur and genetically defined substrains could be obtained in the next generation. In later experiments viral DNA (cloned proviral Moloney leukaemia virus DNA) was injected into the cytoplasm of one-cell embryos (zygotes). Such embryos developed into adults carrying a single inserted copy of the viral DNA in every cell (Harbers *et al.* 1981).

In more recent experiments, the usual procedure has been to inject cloned plasmid DNA into one of the *pronuclei* of the newly fertilized egg. The male pronucleus is contributed by the sperm and, being larger than the female pronucleus, is the one usually chosen for microinjection. Typically about 2 picolitres of DNA-containing solution is introduced. The two pronuclei subsequently fuse to form the diploid, zygote nucleus of the fertilized egg. The injected embryos are cultured *in vitro* to morulae or blastocysts and then transferred to pseudopregnant foster mothers (Gordon & Ruddle 1981). In practice, between 3% and 40% of the animals developing from these embryos contain copies of the exogenous DNA (Lacey *et al.* 1983). In such *transgenic* mice the foreign DNA must have been integrated into one of the host chromosomes at an early stage of embryo development. There is usually no mosaicism, and so the foreign DNA is transmitted through the germ line. In different transgenic animals the copy number of integrated plasmid sequences differs, ranging from one copy to several hundred, and the chromosomal location differs (Palmiter *et al.* 1982a, Lacy *et al.* 1983).

Expression of foreign DNA in transgenic mice

The mouse metallothionein gene promoter

The mouse metallothionein-1 (MMT) gene encodes a small cystein-rich polypeptide that binds heavy metals and is thought to be involved in zinc homeostasis and detoxification of heavy metals. The protein is present in many tissues of the mouse, but is most abundant in the liver. Synthesis of the protein is induced by heavy metals and glucocorticoid hormones. This regulation occurs at the transcriptional level (Durnam & Palmiter 1981).

Brinster *et al.* (1981) constructed plasmids in which the MMT gene promoter and upstream sequences had been fused to the coding region of the Herpes simplex virus *tk* gene. The thymidine kinase enzyme can be assayed readily and provides a convenient 'reporter' of MMT promoter function. The fused MK (Metallothionein-thymidine kinase) gene was injected into the male pronucleus of newly fertilized eggs which were then incubated *in vitro* in the presence or absence of cadmium ions (Brinster *et al.* 1982). The thymidine kinase activity was found to be induced by the heavy metal. By making a range of deletions of mouse sequences upstream of the MMT promoter sequences, the minimum region necessary for inducibility was localized to a stretch of DNA 40 to 180 nucleotides upstream of the transcription initiation site. Additional sequences that potentiate both basal and induced activities extended to at least 600 bp upstream of the transcription initiation site.

The same MK fusion gene was injected into embryos which were raised to transgenic adults (Brinster *et al.* 1981). Most of these mice expressed the MK gene and in such mice there were from 1 to 150 copies of the gene. The reporter activity was inducible by cadmium ions and showed a tissue distribution very similar to that of metallothionein itself (Palmiter *et al.* 1982b). Therefore these experiments showed that DNA sequences necessary for heavy metal induction and tissue-specific expression can be functionally dissected in eggs and transgenic mice. For unknown reasons, there was no response to glucocorticoids in either the egg or transgenic mouse experiments.

As expected, the transgenic mice transmitted the MK gene to their progeny. The genes were inherited as though they were integrated into a single chromosome. When reporter activity was assayed in these offspring the amount of expression could be very different from that in the parent. Examples of increased, decreased or even totally extinguished expression were found. In some, but not all, cases the changes in expression correlated with changes in methylation of the gene sequences (Palmiter *et al.* 1982b).

In a dramatic series of experiments, Palmiter *et al.* (1982a) fused the MMT promoter to a rat growth hormone genomic DNA. This hybrid gene (MGH) was constructed using the same principles as the MK fusion. Of 21 mice that developed from microinjected fertilized eggs, 7 carried the MGH fusion gene and 6 of these grew significantly larger than their littermates. The mice were fed zinc to induce transcription of the MGH gene, but this did not appear to be absolutely necessary since they showed an accelerated growth rate before being placed on the zinc diet. Mice containing high copy numbers of the MGH gene (20–40 copies per cell) had very high concentrations of growth hormone in their serum, some 100 to 800 times above normal. Such mice grew to almost double the weight of littermates at 74 days old (Fig. 12.2).

These experiments have subsequently been repeated with the MMT promoter linked to a human growth hormone gene (Palmiter *et al.* 1983). Synthesis of growth hormone was found to be inducible by heavy metals. The gene was expressed in all tissues examined, but the ratio of human growth hormone mRNA to endogenous metallothionein-1 mRNA varied among different tissues and animals suggesting

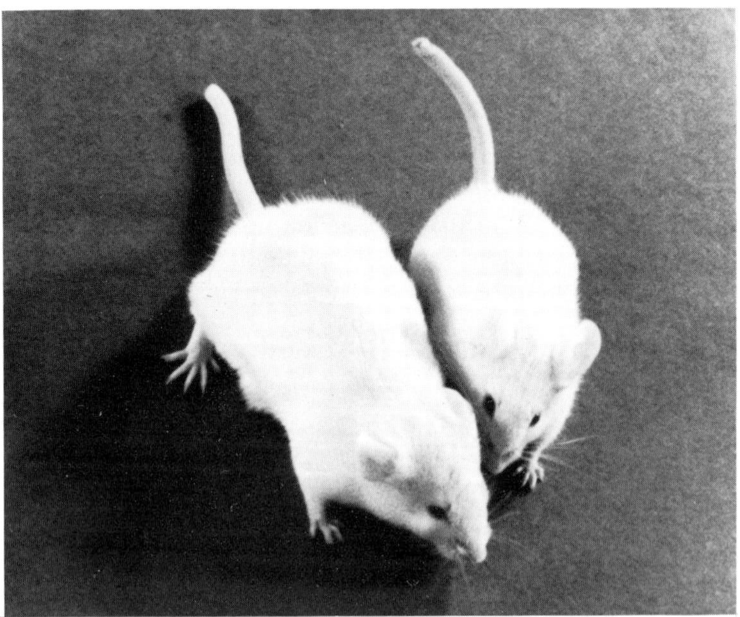

Fig. 12.2 Transgenic mouse containing the mouse metallothionein promoter fused to the rat growth hormone gene. The photograph shows two male mice at about 10 weeks old. The mouse on the left contains the MGH gene and weighs 44 g; his sibling without the gene weighs 29 g. In general, mice that express the gene grow 2–3 times as fast as controls and reach a size up to twice the normal. (Photograph kindly provided by Dr R. L. Brinster.)

that the expression of the fused gene was affected by factors such as cell type and integration site. The pituitaries of the transgenic mice were histologically abnormal, suggesting dysfunction of cells that normally synthesize growth hormone.

These experiments show the power of the MMT promoter in obtaining high levels of expression of any gene to which it is fused. The application of this technology to the production of important polypeptides in farm animals is discussed by Palmiter and his co-authors (1982a). The concentration of growth hormone in the transgenic mice was impressively high, much greater than bacterial or cell cultures genetically engineered for growth hormone production. The genetic farming concept is comparable to the practice of raising valuable antisera in animals, except that a single injection of a gene into a fertilized egg would substitute for multiple somatic injections. An added advantage is the heritable nature of genes, but the variability in expression already encountered in the progeny may be problematical here.

Gene regulation in transgenic mice

The similiarities between the tissue distribution of MK expression and normal MMT expression encouraged the hope that transgenic mice would provide a general assay for functionally dissecting DNA sequences responsible for tissue-specific or developmental regulation of a variety of genes. A further example of appropriate tissue-specific regulation was observed in transgenic mice that incorporated the chicken transferrin gene. Transferrin is an iron-binding protein which has also been called conalbumin. Normally in the chicken the transferrin gene is active in the liver and oestrogen-stimulated oviduct. Transcription of the gene in the liver is constitutive but the rate can be increased by oestrogen or dietary iron deficiency. The cloned gene, together with substantial 5′ and 3′-flanking sequences, was injected into fertilized eggs and six of the seven resulting transgenic mice were found to express chicken transferrin mRNA in several tissues. Chicken transferrin was detectable in the serum (McKnight *et al.* 1983). In five of the six mice expressing the gene the amount of mRNA per cell was significantly higher in the liver than in any other tissue tested. Therefore there appeared to be at least a partial retention of tissue-specific regulation even with this heterologous gene.

Sadly for the developmental biologist, however, incorporation of a cloned rabbit β-globin gene into transgenic mice (Wagner *et al.* 1981) has resulted in abnormal expression (Lacey *et al.* 1983). In seven transgenic lines examined neither rabbit β-globin mRNA nor rabbit β-globin polypeptides were detected in the mouse erythroid cells. In two of the mouse lines, rabbit β-globin transcripts were found at low levels in particular but inappropriate tissues: skeletal muscle in one line, and testis in another. These patterns of transcription were heritable traits in the two lines, and Lacey *et al.* (1983) speculate that they are the result of DNA integration at abnormal chromosomal positions.

These results with a heterologous β-globin gene did not promise well for the application of transgenic mice to analysing gene regulation. However, subsequent experiments in which a rearranged mouse immunoglobulin K gene was introduced were extremely encouraging. In this case high expression of the K gene was restricted to the appropriate cell type, B lymphocytes, in several different transgenic mouse lines. It appeared that the microinjected K gene contained target sequences for gene activation which are specific for B lymphocytes and which can override the influence of different integration sites (Storb *et al.* 1984).

INTRODUCTION OF CLONED GENES INTO THE GERM LINE OF *DROSOPHILA* BY MICROINJECTION

P elements of *Drosophila*

P elements are transposable DNA elements which, in certain circumstances, can be highly mobile in the germ line of *Drosophila melanogaster*. The subjugation of these sequences as specialized vector molecules in *Drosophila* represents a landmark in modern *Drosophila* genetics. Through the use of P element vectors any DNA sequence can be introduced into the genome of the fly.

P elements are the primary cause of a syndrome of related genetic phenomena called P-M hybrid dysgenesis (Bingham *et al.* 1982, Rubin *et al.* 1982). Dysgenesis occurs when males of a P (paternally contributing) strain are mated with females of an M (maternally contributing) strain, but usually not when the reciprocal cross is made. The syndrome is confined mainly to effects of the germ line and includes a high rate of mutation, frequent chromosomal aberrations and, in extreme cases, failure to produce any gametes at all.

P strains contain multiple genetic elements, the P elements, which may be dispersed throughout the genome. These P elements do not produce dysgenesis within P strains because transposition is repressed, probably due to the presence of a P-encoded repressor of a P element-specific *transposase* which is also encoded by the P element. However, when a sperm carrying chromosomes harbouring P elements fertilizes an egg of a strain that does not harbour P elements (i.e. an M strain), the P element transposase is temporarily derepressed owing to the absence of repressor. P element transposition occurs at a high frequency and this leads to the dysgenesis syndrome, the high rate of mutation results from the insertion into and consequent disruption of genetic loci.

Several members of the P transposable element family have been cloned and characterized (O'Hare & Rubin 1983). It appears that the prototype is a 2.9 kb element and that other members of the family have arisen by different internal deletion events within this DNA. The elements are characterized by a perfect 31 bp inverted terminal repeat. It is likely that this repeat is the site of action of the putative transposase. Three long open-reading frames have been identified in the prototype DNA sequences. These are candidates for transposase and repressor genes. Some of the short members of the family are defective. They cannot encode functional transposase but are transposable in *trans* in the presence of a non-defective P element within the same nucleus.

Spradling and Rubin (1982) have devised an approach for introducing the P element DNA into *Drosophila* chromosomes which mimics events taking place during a dysgenic cross. Essentially, a recombinant plasmid which consisted of a 2.9 kb P element together with some flanking *Drosophila* DNA sequences, cloned in the pBR322 vector, was microinjected into the posterior pole of embryos from an M-type strain. The embryos were injected at the syncytial blastoderm stage. This is a stage of insect development in which the cytoplasm of the multinucleate embryo has not yet become partitioned into individual cells (Fig. 12.3). The posterior pole was chosen because it is the site at which the cytoplasm is first partitioned, resulting in cells that will form the germ line. P element DNA introduced in this way became integrated into the genome of one or more posterior pole cells. Because of the multiplicity of such germ line precursor cells the integrated P element DNA was expected to be inherited by only some of the progeny of the resulting adult fly. Therefore the progeny of injected embryos were used to set up genetic lines which could be genetically tested for the presence of incorporated P elements.

A substantial proportion of progeny lines were indeed

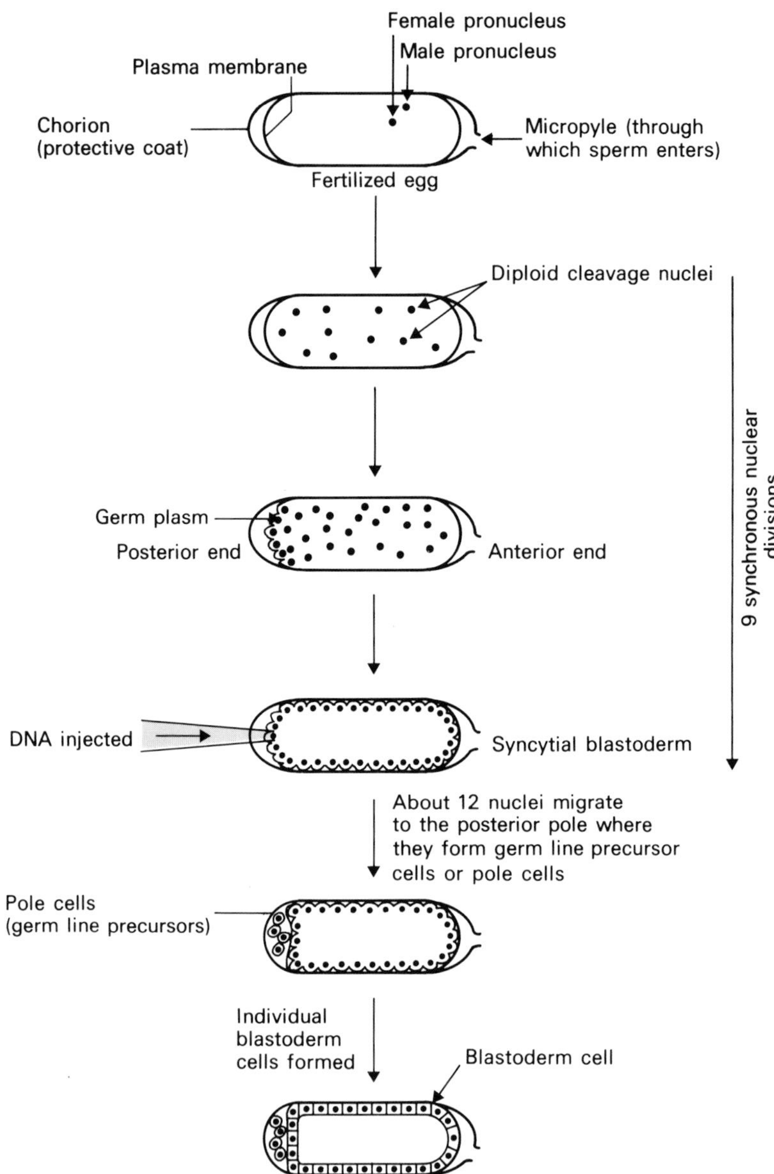

Fig. 12.3 Early embryogenesis of *Drosophila*. DNA injected at the posterior end of the embryo just prior to pole cell formation is incorporated into germ-line cells.

found to contain P elements integrated at a variety of sites in each of the five major chromosomal arms, as revealed by in-situ hybridization to polytene chromosomes. It may be asked whether integration really does mimic normal P element transposition or whether it is simply some non-specific integration of the microinjected plasmid. The answer is that integration occurs by a mechanism analogous to transposition. By probing Southern blots of restricted DNA it was found that the integrated P element was not accompanied by the flanking *Drosophila* or pBR322 DNA sequences which were present in the recombinant plasmid that was microinjected (Spradling & Rubin 1982). Injected plasmid DNA must presumably have been expressed at some level *before* integration so as to provide transposase activity for integration by the transposition mechanism.

These experiments, therefore, showed that P elements can transpose with a high efficiency from injected plasmid into diverse sites in chromosomes of germ line cells. At least one of the integrated P elements in each progeny line remained functional as evidenced by the hypermutability it caused in subsequent crosses to M strain eggs.

P element as a vector

Rubin and Spradling (1982) exploited their finding that P elements can be artificially introduced in the *Drosophila* genome. A possible strategy for using the P element as a vector would be to attempt to identify a suitable site in the 2.9 kb P element sequence where insertion of foreign DNA could be made without disrupting genes essential for transposition. However, an alternative strategy was favoured. A recombinant plasmid was isolated which comprised a short (1.2 kb), internally deleted member of the P element family together with flanking *Drosophila* sequences, cloned in pBR322. This naturally defective P element cannot encode any of the putative protein products of the 2.9 kb prototype element (O'Hare & Rubin 1983). Target DNA was ligated into the defective P element. The aim was to integrate this recombinant P element into the germ line of injected embryos by providing transposase function in *trans*. Two approaches for doing this were tested. In one approach a plasmid carrying the recombinant P element was injected into embryos derived from a P–M dysgenic cross in which transposase activity was therefore expected to be high. This approach does have the disadvantage that frequent mutations and chromosomal aberrations would also be expected. In the other approach the plasmid carrying recombinant P element was co-injected with a plasmid carrying the non-defective 2.9 kb element. This approach is formally similar to the application of bacterial transposons discussed in Chapter 8.

In the first experiments of this kind embryos homozygous for a *rosy* mutation were microinjected with the P element vector containing a wild-type *rosy* gene. Both methods for providing complementing transposase were effective. Rosy$^+$ progeny, recognized by their wild-type eye colour, were obtained from 20% to 50% of injected embryos. The chromosomes of these flies contained one or two copies of the integrated *rosy*$^+$ DNA. Genetically stable progeny lines were established. In a simple extension of this scheme Rubin and Spradling (1982) demonstrated that any DNA, even if unselectable, could be introduced into the genome of *Drosophila* with a high efficiency.

Application of the P element vector

The initial successful demonstration of the P element vector system was rapidly followed by the simultaneous publication of three reports which described the reintroduction of three cloned *Drosophila* genes (Dopa decarboxylase, xanthine dehydrogenase and alcohol dehydrogenase) into the chromosomes of the fly. The genes were accompanied by substantial flanking sequences. These experiments were particularly exciting because in each case the regulation of the gene in question was correct.

Dopa decarboxylase is a gene which is subject to both temporal and tissue-specific regulation during development. Expression of the reintegrated copies of the gene showed the expected pattern of regulation (Scholnick *et al.* 1983). Xanthine dehydrogenase is the product of the *rosy* gene. Again, when the gene was reintroduced into the chromosomes of the fly, the tissue distribution of xanthine dehydrogenase was normal (Spradling & Rubin 1983). Finally, reintroduction of a cloned alcohol dehydrogenase gene resulted in normal expression according to several criteria: quantitative amounts of enzyme in larvae and adults; tissue specificity; and a previously recognized and characteristic developmental switch in transcription initiation site which is a feature of this gene, which has two sets of promoter elements (Goldberg *et al.* 1983).

The above successes with a variety of genes promise rapid progress in understanding DNA control signals which are required for the developmental and tissue-specific regulation of gene expression in *Drosophila*. A further aspect of gene regulation raised by these experiments concerns dosage compensation. *Drosophila* females are XX and males are XY, as in mammals. Both the fly and mammals face the problem of controlling the level of expression of genes on their X chromosome since females have a double dose and the males a single dose, of each. In mammals this is achieved by total inactivation and heterochromatinization of almost all of an X-chromosome in each somatic cell of the female. *Drosophila* has a different mechanism; transcription of each copy of the X-linked genes in the female is less active than in the male. How this is brought about is not known. An insight is given by the finding that when the autosomal genes encoding dopa decarboxylase or xanthine dehydrogenase were integrated into the X-chromosome, expression was at least partially dosage compensated. This should be amenable to further analysis using the P element vector.

A final example of the application of this vector involves the heat shock response of *Drosophila*. High temperatures and other stress-inducing treatments evoke a dramatic change in the pattern of gene expression in *Drosophila*. Among many effects, transcription of certain genes, the heat shock genes, is greatly increased. This can be examined in many ways (see, for example, Bienz & Pelham 1982) but is most striking in the polytene chromosomes where the heat shock rapidly induces large puffs. Lis *et al.* (1983) have constructed a hybrid gene comprising the 5′ flanking region of a *Drosophila* heat shock gene and some accompanying amino-terminal heat shock protein codons, fused in phase with the *E. coli* β-galactosidase gene. The fused gene and a selectable *rosy*$^+$ gene were cloned into the P element vector and then inserted into the *Drosophila* genome. Three lines of flies were established, two contained a single inserted DNA and one contained two copies of the DNA at separate sites. On heat shocking such flies large chromosomal puffs occurred in the polytene chromosomes at the sites of insertion. By dissecting out organs of the larvae and adult flies and then incubating them in the chromogenic β-galactosidase substrate, X-gal, it was evident that β-galactosidase activity was inducible by heat shock and showed the expected widespread distribution throughout the animal's tissues.

SECTION 4

Applications of Recombinant DNA Technology

Chapter 13 Applications of Recombinant DNA Technology in Biology and Medicine

To date the greatest contributions of recombinant DNA research have undoubtedly been within molecular biology itself. In combination with Southern blot-transfer hybridization and DNA sequencing, recombinant DNA technology has made possible the detailed structural analysis of potentially any gene. Those interested enough to read this book will probably know of the recent advances in our understanding of gene structure and will be aware of importance of such knowledge. It is with eukaryotes that the challenge has been greatest, because of their large genomes, but the contribution of recombinant DNA research to prokaryote molecular biology must not be underestimated. The first step in analysing gene structure and gene expression is to isolate the gene in question. The DNA sequence is then analysable and, importantly, specific hybridization probes of absolute sequence purity can be made. The relationship of gene structure to regulated expression and protein function may then be analysed by reintroducing manipulated sequences into living cells using methods described in previous chapters.

With just this basic strategy for research current progress in many fields of molecular biology is very rapid, and similar researches will probably remain a major application of recombinant DNA technology for many years to come.

In previous chapters we have already discussed many important applications of recombinant DNA technology; for example, in Chapter 9 the construction of artificial chromosomes in yeast was discussed. These artificial chromosomes include centromeres, telomeres and origins of DNA replication. Therefore it is now possible to analyse chromosome mechanics in detail by manipulating these sequences and analysing the effects of changing them. As further examples, we have discussed in Chapter 11 how the use of COS cells for transient assays of gene expression has led to the discovery of a new class of genetic elements, enhancer elements; and in Chapter 12 the important application of the *Xenopus* oocyte as a surrogate genetic system has been considered. In this chapter we aim to indicate some selected but illustrative extra applications which extend material discussed in previous chapters.

Chromosome walking

Walking along the chromosome is a term used to describe an approach which allows the isolation of gene sequences whose function is quite unknown but whose genetic location is known. A particularly dramatic application of this approach involves homeotic mutations in *Drosophila*. Homeotic mutations are an especially interesting class of genetic defects which characteristically cause one imaginal disk of the larva to develop in the form of a region of the body derived from an entirely different imaginal disk (for review see Morata & Lawrence 1977). *Antennapedia*, for example, is a mutation which results in adult flies whose antennae are homeotically transformed into leg-like structures. Although the developmental basis of this transformation is unknown, the genetic-locus of the mutation has been accurately mapped by classical genetic methods and assigned to a region comprising just a few bands in the polytene chromosomes of *Drosophila* salivary glands. This allows DNA from this region to be isolated by molecular cloning procedures, which, in turn, enables the mutation to be defined precisely at the DNA sequence level. This provides reagents for analysing the molecular biology of development in normal and mutant imaginal disks. The principle is as follows (see also Fig. 13.1).

1 Cloned genomic sequences can be localized in the genome by radiolabelling the cloned fragment and using it as a probe in a hybridization experiment *in situ* with cytological preparations of polytene chromosomes.

2 A random set of cloned genomic DNA is localized in this way, and one is chosen whose location on the chromosome in question is closest to the map position of the mutation under investigation.

3 The genomic library is then screened with this chosen clone as probe to identify other clones containing DNA with which it reacts and which represent clones overlapping with it. The overlap can be to the left or to the right.

4 Repetition of this single walking step along the chromosome.

Such walking will necessarily occur in both directions along the chromosome unless there is some means of distinguishing

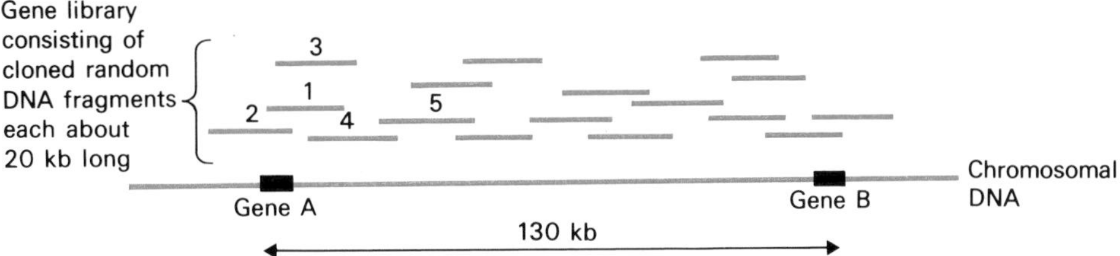

Fig. 13.1 Chromosome walking. It is desired to clone DNA sequences of gene B, which has been identified genetically but for which no probe is available. Sequences of a nearby gene A are available in cloned fragment 1. Alternatively, a sequence close to gene B could be identified by in-situ hybridization to *Drosophila* polytene chromosomes. In a large, random genomic DNA library many overlapping cloned fragments are present. Clone 1 sequences can be used as a probe to identify overlapping clones 2, 3 and 4. Clone 4 can, in turn, be used as a probe to identify clone 5, and so on. It is, therefore, possible to walk along the chromosome until gene B is reached. See text for details.

the direction. In *Drosophila* this can be achieved by hybridization *in situ* to polytene chromosomes.

A possibility that has to be recognized arises from the existence of repeated DNA sequences. These may occur dispersed at several places in the genome and could disrupt the orderly progress of the walk. For this reason the probe used for stepping from one genomic clone to the next must be a unique sequence clone, or a sub-clone which has been shown to contain only a unique sequence. Once the investigator has arrived at, or very near, the locus in question it should be possible to compare mutant and wild-type DNA sequences directly or by restriction mapping with the Southern blot technique. Cloned probes could also be applied to analysing transcripts from the region, hence making inroads into the analysis of the previously unknown molecular biology of the locus.

As outlined here the process of walking is simple in principle. On a small scale it can be applied to almost any eukaryote. If starting with a cloned fragment of a gene, or a cDNA clone, the entire genomic sequence together with substantial flanking sequences may be obtained with a single step in each direction. Walking on a larger scale from a clone chosen to be near another gene of interest in order to walk to the other gene, is very demanding. However, here the advanced genetics of *Drosophila* comes to the rescue, for combined with the inherent usefulness of polytene chromosome hybridization are the numerous inversions and translocations which may be exploited to allow one to make a jump, rather than walk. Such a combination of chromosome walking and jumping has enabled Hogness and his co-workers (Bender *et al.* 1983) to clone DNA from the *Ace* and *rosy* loci and the homeotic *Bithorax* gene complex in *D. melanogaster.*

The existence of the polytene chromosomes in *Drosophila* permits a different, more direct, approach. It has, by a tremendous technical *tour de force*, been found possible to physically excise a region of such a salivary gland chromosome by micromanipulation, and thence to extract its DNA, restrict it and ligate it to a phage λ vector, all within a microdrop under oil, and thereafter obtain clones with reasonable efficiency (Scalenghe *et al.* 1981). With this technique it should be possible to isolate clones from any desired region of the genome of the fruit fly. The main interest in these micromanipulation experiments lies not in the ability to clone sequences from particular regions of the *Drosophila* genome—this can be achieved by other means such as chromosome walking—but rather in demonstrating that minute quantities of DNA can be cloned successfully.

Isolation of genes transferred to animal cells in culture

In Chapter 11 the calcium phosphate method for transferring DNA into cultured animal cells was described. Techniques have been devised for screening libraries specifically for the transferred sequences. Thus any gene which can be selected or recognized by its phenotypic effects in tissue culture can be isolated. There are several variants of this approach. The simplest is 'plasmid rescue' (Hanahan *et al.* 1980, Perucho *et al.* 1980). In the example shown in Fig. 13.2, plasmid rescue is used to isolate a chicken tk^+ gene. First *Hin*d III fragments of non-mutant, total nuclear chicken DNA are ligated into pBR322. This recombinant DNA is used directly to transform mouse tk^- cells, from which tk^+ transformants are selected in HAT medium. DNA from these transformants is prepared and used in a second round of transformation. This second round eliminates most of the non-selected recombinant plasmid DNA which becomes integrated in the first round. DNA from these secondary tk^+ transformants is cut with an appropriate restriction enzyme, circularized and selected in *E. coli* on the basis of plasmid-borne drug resistance. The tk^+ gene accompanies the drug resistance marker in the plasmid which has been rescued. Thus in the plasmid rescue approach the transferred DNA is 'tagged' with the drug resistance marker, which can be selected in *E. coli*.

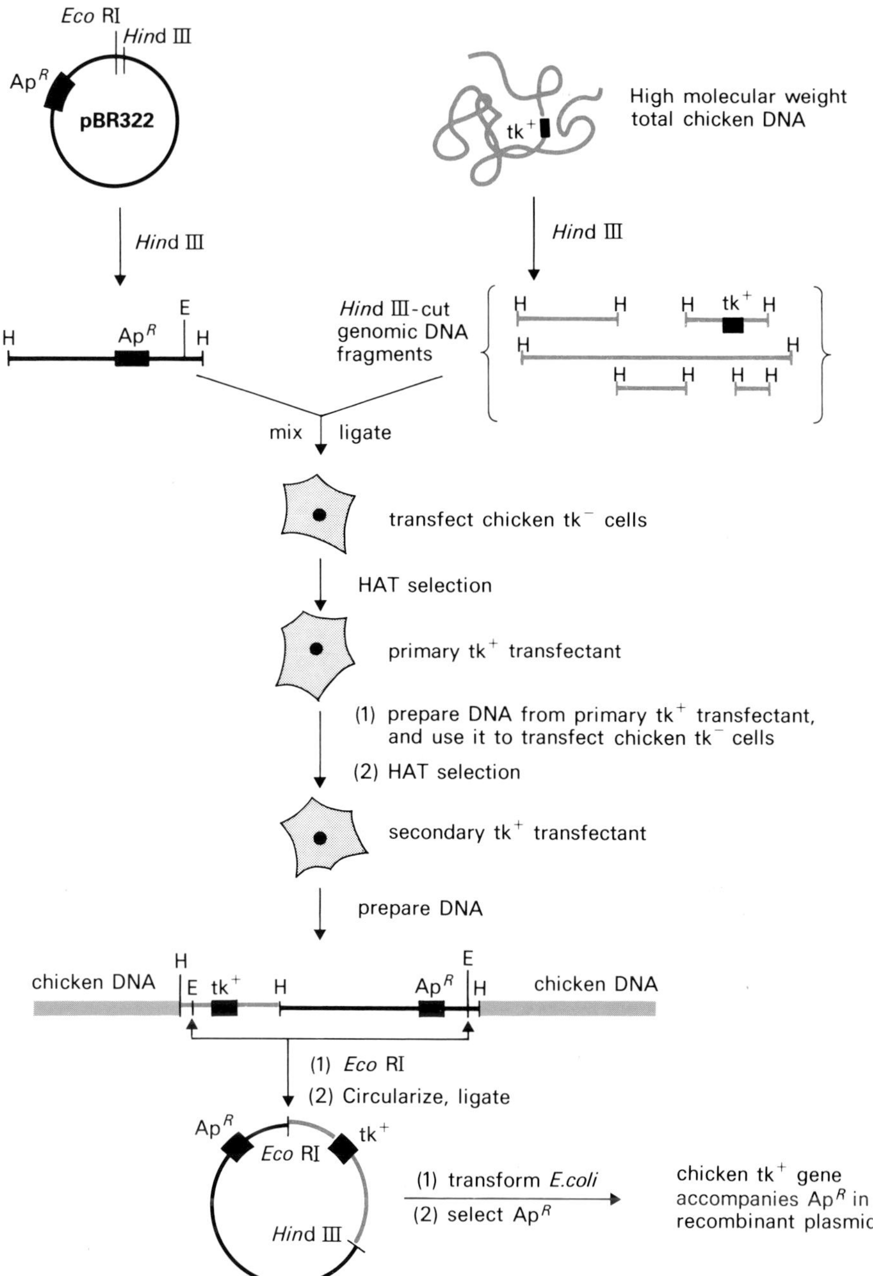

Fig. 13.2 Plasmid rescue, applied to the isolation of the chicken thymidine kinase gene (Perucho *et al.* 1980). See text for details.

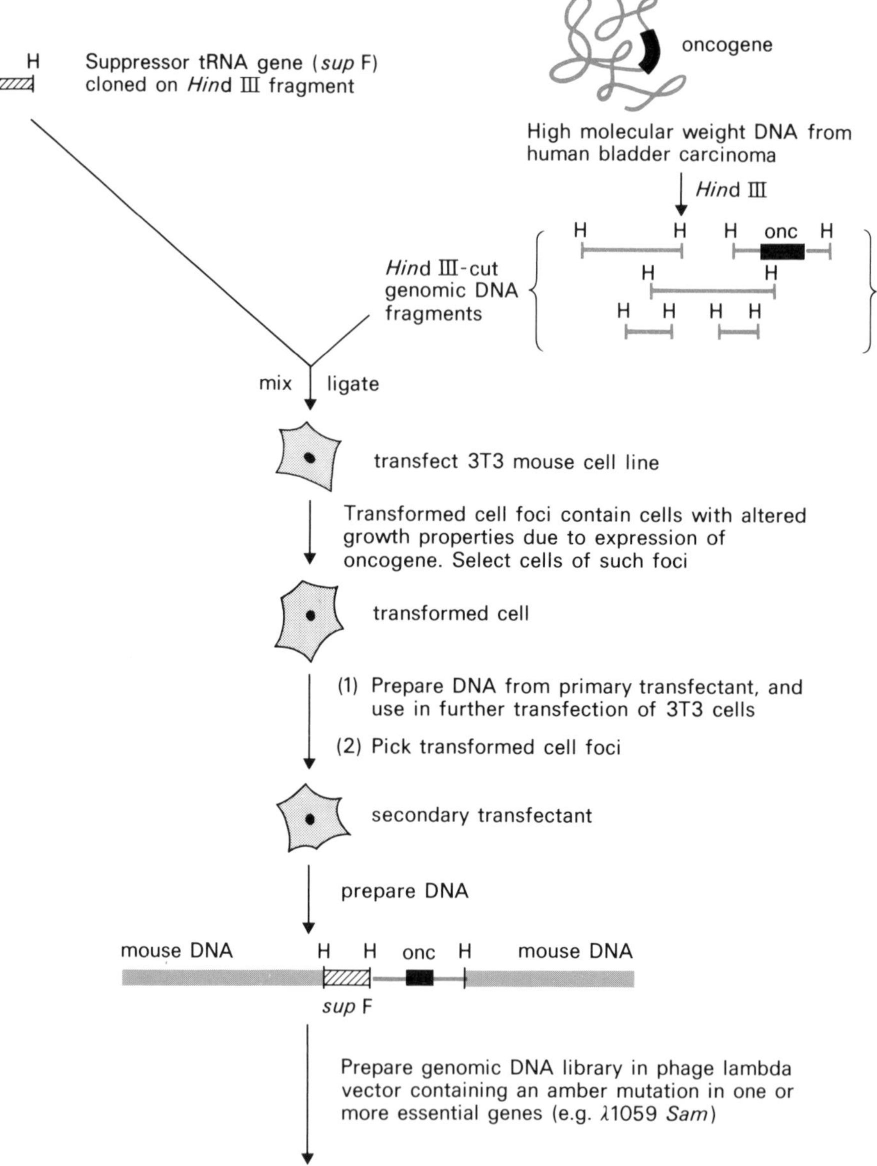

Fig. 13.3 Suppressor rescue, applied to the isolation of the human oncogene in the T24 bladder carcinoma (Goldfarb *et al.* 1981).

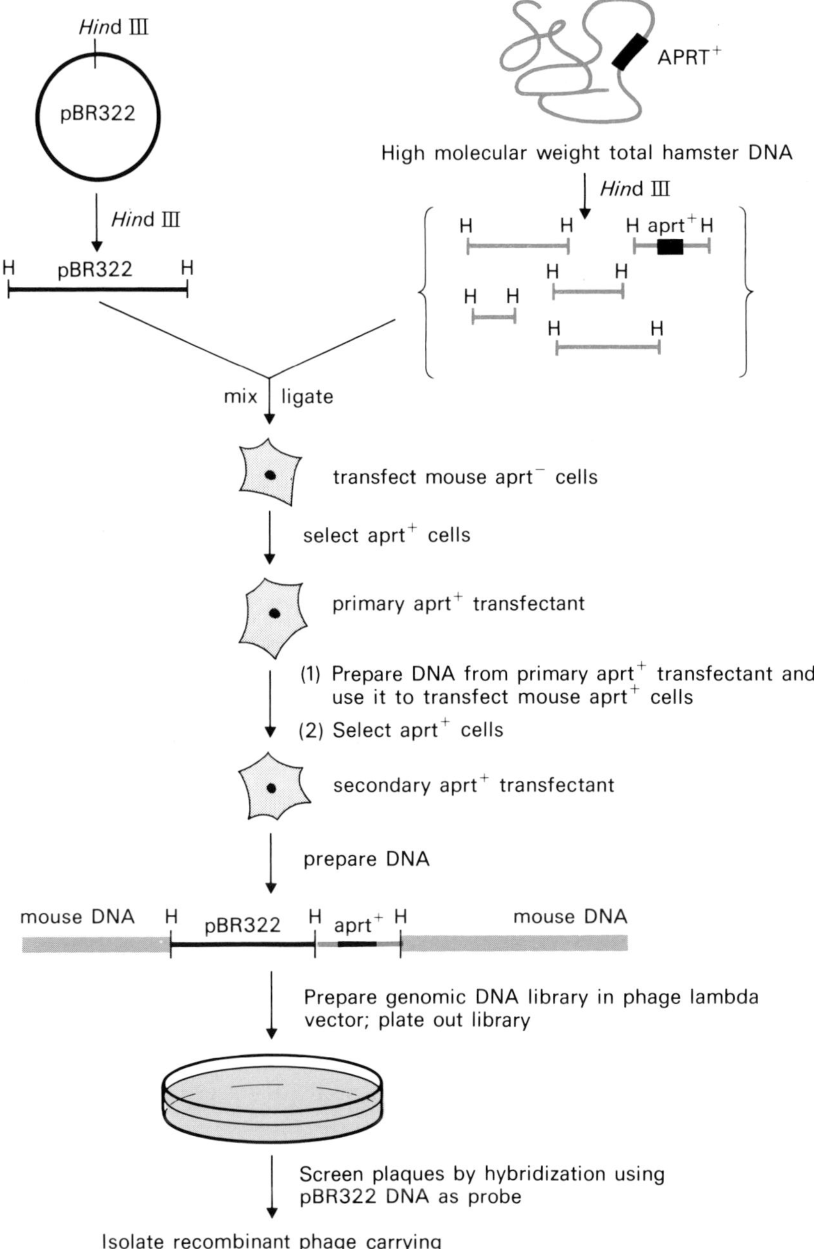

Fig. 13.4 Tag identification by hybridization, applied to the isolation of the hamster adenine phosphoribosyl transferase (aprt) gene (Lowry *et al.* 1980).

In a related approach the transferred DNA is tagged with an amber suppressor gene. This is illustrated in Fig. 13.3, in which the cloning of a human oncogene is used as an example (Goldfarb *et al.* 1982).

In other variants of this approach, the transferred DNA is 'tagged' with DNA sequences that are not selected genetically but which can be detected by molecular hybridization probes. Figure 13.4 shows the procedure of Lowry *et al.* (1980) in which this tag is simply the plasmid pBR322 to which the transferred DNA has been ligated. There is, therefore, a formal resemblance to the plasmid rescue approach, but an advantage of this procedure is the incorporation of the phage lambda cloning step. As we saw in Chapter 4, genomic libraries can be prepared in phage lambda very efficiently when combined with in-vitro packaging. In an alternative variant the 'tag' for the transferred DNA is not provided artificially, but is provided by 'Alu' sequences (Rubin *et al.* 1983). These are members of a highly repetitive family of sequences. Copies are dispersed throughout the human genome such that any substantial fragment of human DNA several kb in size is likely to contain at least one copy of an Alu sequence. It has been demonstrated that under stringent hybridization conditions the presence of Alu sequences can be used to establish the presence of human DNA in mouse or hamster cells (Gusella *et al.* 1980, Murray *et al.* 1981, Perucho *et al.* 1981b).

Recombinant DNA probes for diagnosis of genetic diseases in human medicine

There are several hundred recognized genetic diseases in man which result from single recessive mutations. In some of these a protein product which is defective or absent has been identified. In many others the nature of the mutation is completely unknown. Of all genetic diseases, the inherited haemoglobin disorders have been the most extensively studied at the DNA level. The best-known example is the change from GAG to GTG in the codon encoding the sixth amino acid residue of the β-globin chain. In homozygous form this results in sickle-cell anaemia. This is one of many inherited disorders of haemoglobin. Clinically, the most important of the haemoglobinopathies are the thalassaemias and sickle-cell anaemia and its many variants.

The thalassaemias are a group of disorders in which there is an imbalance in the synthesis of globin chains due to the low, or totally absent, synthesis of one of them. Many α-thalassaemias are caused by gene deletions, although several non-deletion forms have been identified (Weatherall & Clegg 1982). The β-thalassaemias are complex and more than 20 different molecular lesions have been identified. Most are nonsense or frameshift mutations in exons of the β-globin gene. Others are point mutations affecting transcript processing, or point mutations affecting the promoter. Only one form has been identified as due to a major deletion of the β-globin gene (Weatherall & Clegg 1982).

For these disorders, and many other genetic diseases (but fortunately not all), there is no definitive treatment. Their prevention is the current strategy in many countries. Primary prevention, by identifying heterozygous carriers and dissuading them from reproducing in marriages with other carriers, is not practicable. Usually the discovery that both parents are carriers is made only after the birth of an affected child. When a pregnant woman has been identified as at risk, as indicated by a previous birth or other factors, the approach is to offer antenatal diagnosis and the possibility of abortion.

In what follows, haemoglobinopathies will be taken as examples. Their antenatal diagnosis by recombinant DNA techniques has served as a prototype for other genetic disorders (Weatherall & Old 1983). Sadly however, for most of the other disorders suitable cloned probes are not available yet. Their isolation is the subject of very active research.

Currently, antenatal diagnosis of haemoglobin disorders is carried out by fetal blood sampling, followed by estimation of the relative rates of globin chain synthesis by radiolabelling (Alter 1981). It has the disadvantage that fetal blood sampling is not possible until about the 18th week of pregnancy.

Fetal cells can also be obtained by amniocentesis. A sample of amniotic fluid which surrounds the fetus and contains fetal urine and other secretions, is taken at about the 16th week of pregnancy. Fetal cells are recovered from the fluid and can be cultured so that biochemical and chromosomal analyses can be made. Such analyses are useful for detecting disorders other than the haemoglobinopathies, such as certain enzyme deficiencies or Down's syndrome. The fetal cells can also be used to provide DNA for restriction enzyme analysis (Orkin 1982).

Fetal DNA analysis

Clearly, if a mutation either removes or produces a restriction enzyme site in genomic DNA, this can be used as a marker for the presence or absence of the defect. The mutation from GAG to GTG in sickle-cell anaemia eliminates a restriction site for the enzyme *Dde* I (CTNAG) or the enzyme *Mst* II (CCTNAGG) (Chang & Kan 1981, Orkin *et al.* 1982). The mutation can therefore be detected by digesting mutant and normal DNA with the restriction enzyme and performing a Southern-blot hybridization with a cloned β-globin DNA probe. Such an approach is applicable only to those disorders where there is an alteration in a restriction site, or where a major deletion or rearrangement alters the restriction pattern. It is not applicable to most β-thalassaemias.

There are many polymorphic restriction sites scattered throughout the β-globin gene cluster (Weatherall & Old 1983). These are revealed as restriction-fragment-length polymorphisms in Southern-blot experiments. These can be used as linkage markers for antenatal diagnosis, i.e. the close physical linkage with a β-globin gene will mean that the polymorphic site will trace the inheritance of that gene. These polymorphisms can be used in two ways. First, some polymorphisms are linked to specific globin mutations; for example in the US among the black population the sickle mutation is associated 60% of the time with a polymorphic mutation near the gene that eliminates a *Hpa* I site (Kan & Dozy 1978). The linkage could form the basis of diagnosis but is inferior to the direct analysis with *Mst* I. Examples of such associations (so called linkage disequilibrium) are rare. A second approach is required in which it is necessary to establish linkages between polymorphic restriction sites and a particular β-globin gene mutation by carrying out a family study before antenatal diagnosis. This is not always possible, and in any case suitable polymorphic markers may not be present. It has been estimated that this approach is feasible in no more than 50% of β-thalassaemia cases in the UK (Weatherall & Old 1983).

Recently a very powerful and direct approach to analysing point mutations has been devised. Conner *et al.* (1983) synthesized two 19-mer oligonucleotides, one of which was complementary to the amino-terminal region of the normal β-globin (β^A) gene, and one of which was complementary to the sickle cell β-globin gene (β^S). These oligonucleotides were radiolabelled and used to probe Southern blots. Under appropriate conditions, the probes could distinguish the normal and mutant alleles. The DNA from normal homozygotes only hybridized with the β^A probe, and DNA from sickle-cell homozygotes only hybridized with the β^S probe. DNA of heterozygotes hybridized with both probes. These experiments, therefore, showed that oligonucleotide hybridization probes can discriminate between a fully complementary DNA and one containing a single mismatched base (for a related example see p. 101). Similar results have subsequently been obtained with a point mutation in the α-antitrypsin gene which is implicated in pulmonary emphysema (Kidd *et al.* 1983). The generality of this approach is very impressive. It should be applicable to other genetic disorders provided that the nucleotide sequences around the mutation site, which could be a substitution, insertion or deletion, can be established.

The application of such methods to fetal DNA obtained by amniocentesis is not entirely satisfactory because amniocentesis cannot be carried out early in pregnancy, and it is often necessary to grow the fetal cells in culture for several weeks in order to obtain sufficient cells for DNA analysis. Recently techniques have been developed for obtaining fetal DNA from biopsies of trophoblastic villi in the first trimester of pregnancy (Williamson *et al.* 1981). In these techniques some villi of the trophoblast—this is an external part of the human embryo which functions in implantation and becomes part of the placenta—are biopsied with the aid of an endoscope passed through the cervix of the uterus. Up to 100 μg of pure fetal DNA can be obtained between the 6th and 10th weeks of pregnancy (Old, J. M. *et al.* 1982). However the risks of this procedure are being assessed and it may turn out to be an unacceptable alternative to amniocentesis.

Chapter 14 Industrial Applications of Gene Manipulation

Industrial microbiology or biotechnology is not a new field of entrepreneurial activity; its development can be traced back to the production of alcohol and vinegar by the Sumerians before the year 5000 BC. Today biotechnology is a well-established factor in the world economy with an annual value of hundreds of billions of dollars. Despite its long history the major developments have occurred this century, particularly since the Second World War with the introduction of antibiotic fermentations. More recently the industry has been in a ferment, if a pun can be excused, because of the potential of the techniques of gene manipulation which have been described in this book. Using such techniques microorganisms can be engineered to synthesize proteins normally found only in higher eukaryotes thus making many of these proteins commercially available for the first time. Indeed, some human proteins produced in bacteria using recombinant DNA technology are already on the market, e.g. insulin. In addition to enabling microorganisms to produce such foreign proteins, gene manipulation can be used to improve existing conventional industrial fermentations.

There are many ways of classifying industrial fermentations. It will simplify the discussion which follows if we subdivide them as follows:

1 production of cells;
2 production of proteins;
3 production of small molecules;
4 destruction of toxic compounds.

The application of gene manipulation techniques to each of these four classes of fermentation presents unique problems and interesting solutions.

Gene manipulation and the production of single-cell protein (SCP)

A critical factor in determining the suitability of a microorganism for use in the production of single-cell protein (SCP) is the efficiency with which it converts substrate carbon into cellular carbon. In their process for producing SCP from methanol Imperial Chemical Industries use *Methylophilus methylotrophus* because of its high carbon conversion efficiency (greater than 50%). However, they identified a potential source of methanol wastage due to the fact that *M. methylotrophus* assimilates ammonia via the ATP-dependent GS/GOGAT pathway (Fig. 14.1) instead of using the ATP-independent glutamate dehydrogenase (GDH). When the glutamate dehydrogenase gene of *E. coli* was introduced into *M. methylotrophus* on a broad-host-range vector (see p. 158 for details) the efficiency of carbon conversion was 4–7% higher (Windass *et al.* 1980). This represents a considerable cost saving, for *M. methylotrophus* is grown continuously in a reactor with an output of 50 000 l per hour. Since the introduction of the glutamate dehydrogenase gene increases the carbon conversion efficiency of the organisms, maintenance of the plasmid and expression of the *gdh* gene are not an energetic drain on the cell. Thus on prolonged continuous cultivation the *gdh* gene is stably maintained in the absence of antibiotic selection. However, the antibiotic resistance determinants carried by the plasmid are not required and were lost rapidly (Powell & Byron 1983).

Another important consideration in the selection of a microbial strain to be used for the production of SCP is its amino acid profile; for instance, if SCP is to be used to supplement soyabean feed for poultry a high methionine content is desired. In such circumstances the construction of a strain which synthesizes large amounts of a protein enriched in methionine would

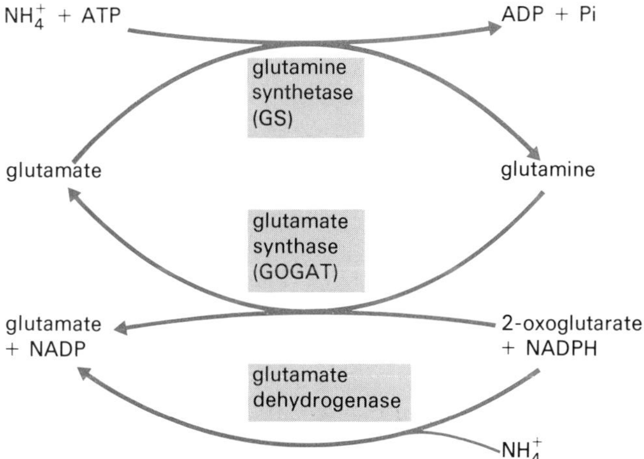

Fig. 14.1 The two routes of ammonia assimilation by bacteria.

be beneficial. As a general approach towards achieving this goal Kangas *et al.* (1982) have reported the construction of a gene whose expression in *E. coli* results in synthesis of a proline-enriched protein.

Gene manipulation and the overproduction of proteins

Microorganisms such as *Bacillus* spp. and filamentous fungi are already used for the commercial production of enzymes, e.g. amylases and proteases. Suitable strains have been isolated by conventional strain improvement involving random mutagenesis and selection. With the advent of in-vitro gene manipulation more rational strain development programmes are underway, particularly with organisms of the genus *Bacillus* (see Chapter 8). Basically, the structural gene for the enzyme is cloned and then expression is maximized. The commercially available enzymes are normal constituents of the cell which produce them. A more glamorous aspect of recombinant DNA technology is its use to construct microorganisms which synthesize proteins such as somatostatin, insulin and β-endorphin (see pp. 60–8) which are not normal constituents of the microbial cell. Other representative examples are shown in Table 14.1. The overproduction in microbes of foreign proteins presents a number of

Table 14.1 Foreign proteins produced in microorganisms.

Protein	Application	Reference
Interferons	Possible treatment of virus infections and cancer	Houghton *et al.* (1980)
Human Growth Hormone	Pituitary dwarfism	Goeddel *et al.* (1979a)
Insulin	Diabetes	Goeddel *et al.* (1979b)
Tissue Plasminogen Activator	Thrombosis	Pennica *et al.* (1983)
β-endorphin	Analgesic	Shine *et al.* (1980)
Interleukin 2	Cancer chemotherapy	Taniguchi *et al.* (1983)
α_1-antitrypsin	Emphysema	Bollen *et al.* (1983)
Relaxin	Facilitate childbirth	Stewart *et al.* (1983)
Chymosin	Cheese production	Emtage *et al.* (1983)
Thaumatin	Sweet-tasting protein	Edens *et al.* (1982)
Poly aspartyl phenylalanine	Preparation of aspartame	Doel *et al.* (1980)

interesting problems, particularly where these proteins have pharmaceutical applications.

Proteolysis

In order to achieve satisfactory overproduction of proteins, the techniques for maximizing gene expression, as outlined in Chapter 7, are used. An important aspect is the role of intracellular proteases in degrading foreign proteins. Their effects can be minimized, in part, by the methods outlined on page 135; for example, by use of the *lon* mutants and the T4 *pin* gene, and partly by increasing the level of expression to such an extent that the proteases are swamped with protein. Even where proteolytic degradation can be minimized the action of proteases can still present problems; for example, partial degradation products which retain at least some biological activity may be produced whose presence may be undesirable if the protein is to be used parenterally.

Aggregate formation

In many instances high expression of recombinant proteins leads to the formation of high molecular weight aggregates, often referred to as inclusions (Fig. 14.2). The inclusions fall into two categories. First, paracrystalline arrays as found in *E. coli*, producing human insulin chains A and B (Williams *et al.* 1982); in this state the protein presumably is in a stable conformation, although not necessarily native. Second, amorphous aggregates, e.g. which contain partially and completely denatured proteins as well as aberrant proteins synthesized as a result of inaccurate translation. Although inclusions probably afford protection against proteases they do present problems of extraction and purification. In most instances denaturants, e.g. SDS or urea, have to be used to extract the protein. For proteins of pharmaceutical interest, particularly if for parenteral administration, the use of detergents is undesirable since it is difficult to remove them completely. SDS and Triton are particularly troublesome. If urea is used as the extractant, chemical modification of some amino acid residues can occur. Regardless of the extraction method used the protein will almost certainly have to be renatured and this may prove difficult, if not impossible.

Presence of fMet

The initiating codon for most genes is ATG and this encodes N-formyl methionine. In their native form the proteins of most

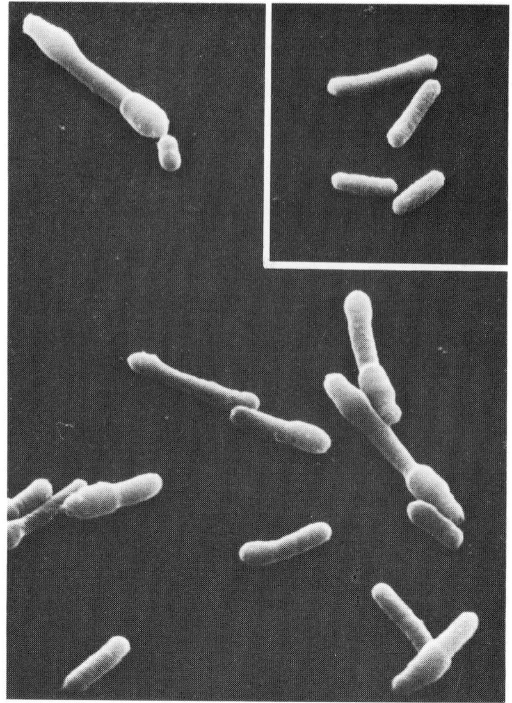

Fig. 14.2 Inclusions of Trp polypeptide-proinsulin fusion protein in *E. coli* (a) scanning electron micrograph of cells fixed in the late logarithmic phase of growth; the inset shows normal *E. coli* cells (magnification ×9200).

cells lack the amino-terminal methionine residue. However, many of the proteins overproduced in *E. coli* as a result of recombinant DNA technology retain the fMet residue; for example, human growth hormone, rabbit β-globin, SV40t-antigen and lambda repressor (Backman & Ptashne 1978, Goeddel *et al.* 1979a, Roberts *et al.* 1979a,b, Guarante *et al.* 1980).

Mistranslation

One of the most poorly understood aspects of high expression systems is the accuracy with which recombinant genes are translated. The normal error rate for translation *in vivo* is about 1 codon in 3000 read (Edelman & Gallant 1977). Depending on the position of the error the resulting protein may retain some activity whilst being more or less susceptible to proteolysis. These translation errors could be of great importance if the proteins are administered parenterally, but of little importance if the protein has a non-pharmaceutical application. It should

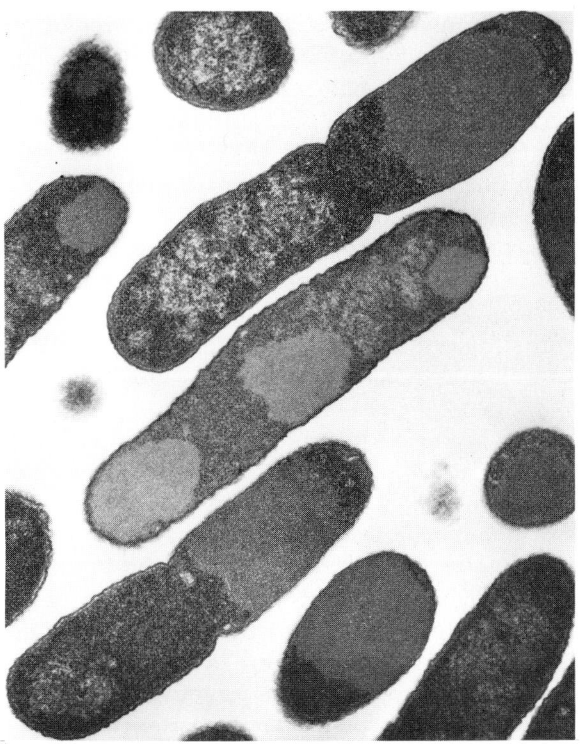

(b) Thin section through *E. coli* cells producing Trp polypeptide-insulin A chain fusion protein (magnification ×53 000). (Photographs reproduced from *Science* courtesy of Dr D. C. Williams (Eli Lilly & Co) and the American Association for the Advancement of Science.)

be noted that under conditions of amino acid limitation the error frequency tends to increase significantly, indeed under even moderate tRNA imbalance the error frequency increases. An artificial tRNA imbalance may be set up if a highly expressing recombinant gene contains a high proportion of infrequently used codons. Minor tRNA species which now are required in much larger quantities may be sequestered on ribosomes to such an extent that they cause a potentially serious starvation of the cognate tRNA. The most obvious solution is the use of artificially synthesized genes with fully optimized codons, but it should be realized that the formation of novel codon contexts may result in undesirable mRNA structures.

Contamination with bacterial proteins

Another problem associated with the pharmaceutical use of recombinant proteins is their contamination with material of microbial origin. Not only must these proteins be free of endotoxin but they must be free of trace amounts of antigenic microbial peptides. Some early batches of recombinant-derived HGH caused the formation of anti-HGH antibodies, most likely because traces of *E. coli* proteins in the HGH preparation acted as an adjuvant. Although these traces of contaminating protein can be reduced to insignificant levels this is not a particularly easy task. For this reason many companies are trying to develop host/vector systems which permit the recombinant protein to be secreted. Although secretion will greatly reduce the starting levels of contaminating protein, in a fermentor the recombinant protein will be subjected to considerable shear forces while in dilute solution.

Despite all these problems in producing recombinant proteins for pharmaceutical use success has been achieved already with insulin (Johnson 1983). However, insulin differs from many proteins of pharmaceutical interest in that it is not glycosylated. Many proteins such as beta- and gamma-interferon retain their pharmacological activities in the absence of the appropriate carbohydrate moieties but their in-vitro and in-vivo stability may be reduced. As yet there is no method which permits *E. coli* to glycosylate recombinant proteins. Yeast almost certainly can glycosylate such proteins but the pattern of glycosylation may be different from that found in the protein isolated from animal tissue.

The production of novel proteins

One of the most exciting aspects of recombinant DNA technology is that its commercial use need not be restricted to introducing intact eukaryotic genes into microbes. Genes for completely novel proteins can also be constructed and some examples have been cited already; for example, the proline enriched protein of Kangas *et al*. (see p. 288). The simplest example of the generation of novel proteins involves redesigning enzyme structure by site-directed mutagenesis. The investigations of Winter and his colleagues have already demonstrated the power of this approach (Winter *et al.* 1982, Wilkinson *et al.* 1984). From a detailed knowledge of the enzyme tyrosyl-tRNA synthetase from *Bacillus stearothermophilus*, including its crystal structure, they were able to predict point mutations in the gene encoding the enzyme which should increase the enzyme's affinity for the substrate ATP. These changes were introduced, and in one case a single amino acid change improved the affinity for ATP by a factor of 100. Oligonucleotide-directed mutagenesis has also been applied to the systematic study of the structure and function of *E. coli* dihydrofolate reductase, another enzyme for which there was available the prerequisite detailed crystal structure (Villafranca *et al.* 1983). These academic investigations have clear industrial and commercial implications for the directed modification of enzyme catalysts with a view to increasing their stability, ease of immobilization by linkage to a solid matrix, changes in substrate specificity and, possibly, fusion of related enzymes into a multi-enzyme complex.

Once the rules governing protein folding and enzyme catalysis are known sufficiently well to permit prediction, it should be possible to design functional protein molecules from scratch. A small but promising start on this major undertaking has been made by Gutte and his co-workers (Gutte *et al.* 1979). Using secondary structure prediction rules and model building, they designed and then synthesized a neutral, artificial 34-residue polypeptide with potential nucleic acid binding activity. The monomeric polypeptide and a covalent dimer of it interacted strongly with single-stranded DNA. Interestingly, the dimer displayed considerable ribonuclease activity. More recently, a polypeptide which binds the insecticide DDT was successfully designed and synthesized (Moser *et al.* 1983).

A standard technique in conventional medicinal chemistry is to identify a chemical entity with a desirable pharmacological effect and then to construct analogues and determine if any of them have improved therapeutic properties. A similar technique can be applied to proteins by means of recombinant DNA technology. Thus by deliberately introducing sequence variations into a protein with multiple biological activities, novel proteins which retain only one or a few of these activities may be obtained. A slightly different approach has been used by Streuli *et al.* (1982) and Weck *et al.* (1982). By combining fragments from different alpha interferon genes they were able to construct genes which encode novel, hybrid interferons, e.g. a hybrid combining the amino-terminal half of IFN-α_2 with the carboxy-terminal half of IFN-α_1. These hybrid interferons had antiviral properties distinct from the two parental interferon molecules. The commercial success of this approach remains to be seen, for such novel proteins may be immunogenic.

The production of vaccines

Whereas proteins for therapeutic use should not be immunogenic, the opposite is true for proteins used as vaccines. For many virus diseases effective vaccines can be made readily by inactivating purified virus preparations but such vaccines are not without their problems as the example of Foot and Mouth Disease virus (FMDV) shows. Several strains of FMDV cannot be grown to a titre high enough to provide sufficient antigenic mass for effective vaccination. In addition, the virus particle is very unstable, particularly below pH 7.0; so it is necessary to store the vaccines under refrigeration. This could be a problem in tropical climates. The recent association of outbreaks of FMDV in Europe with improperly inactivated virus vaccine illustrates another disadvantage of conventional vaccines. Of all the polypeptides of FMDV, only VP1 has been shown to have immunizing activity. Thus bacterially produced VP1 might represent a safer route to a vaccine for FMDV. With this consideration in mind it perhaps was disappointing that polypeptide VP1 produced by recombinant means had low immunogenicity (Kleid *et al.* 1981). Surprisingly, however, it now has been shown that short peptides are more effective immunogens than the complete polypeptide. Bittle *et al.* (1982) found that a peptide corresponding to amino acids 141–160 of VP1 elicited more neutralizing antibody than intact VP1 produced either by disruption of virus particles or by expression in *E. coli* cells.

The work on FMDV has been extended to polypeptide VP1 of poliovirus where an eight amino acid neutralization epitope has been recognized. Emini *et al.* (1983) constructed five peptides varying in size from 11 to 14 amino acids and which corresponded to portions of poliovirus protein. Four of the five were recognized by a poliovirus-neutralizing antibody, but the only one which was able to induce neutralizing antibodies when injected into rabbits was the peptide that incorporated the eight amino acid epitope. As an added bonus, the other four peptides were found to potentiate the antibody response to a subsequent injection of poliovirus.

It must be stated at this point that the peptides used by Bittle *et al.* (1982) and Emini *et al.* (1983) were synthesized chemically. Nevertheless, there is no reason why they could not be made biologically using recombinant DNA technology. Regardless of how the peptides are made, a clear advantage which they have over intact proteins concerns antigenic variability. Variant viruses with altered antigenicity occur with infuriating regularity necessitating the development of new vaccines. The synthesis of new short peptides would be very easy and some might be generated before they are required.

Despite the optimism surrounding recombinant vaccines, numerous development problems remain to be solved. Thus for the protozoal parasites and the lesser-studied viral pathogens, e.g. hepatitis B, identification of a suitable peptide could be a lengthy process. Furthermore, methods are needed for presenting the peptides to animals, and in particular humans, such that the maximal immune response is obtained. Finally, questions of shelf life and stability need to be answered.

Another problem is that whereas vaccines prepared from live, attenuated organisms often give life-long protection, vaccines prepared from killed (inactivated) organisms have to be injected repeatedly according to carefully balanced immunization schedules. Recombinant vaccines almost certainly will fall into the latter category. A safe procedure for attenuating a vaccine organism has been proposed by Kaper *et al.* (1984). They attenuated a pathogenic strain of *Vibrio cholerae* by deletion of DNA sequences encoding the A_1 subunit of the cholera enterotoxin. A restriction endonuclease fragment encoding the A_1, but not the A_2 or B sequences was deleted *in vitro* from cloned cholera toxin genes. The mutation then was recombined into the chromosome of a pathogenic *V. cholerae* strain. The resulting strain produces the immunogenic but non-toxic B subunit of cholera toxin but is incapable of producing the A subunit.

Another method of using recombinant DNA technology for producing live vaccines exploits vaccinia virus as described on p. 254.

Gene manipulation and the production of economically important small molecules

In the above examples of the industrial application of recombinant DNA technology the protein product of the cloned gene has been the desired species. However, the object of most industrial fermentations is the production of small molecules such as amino acids and antibiotics which may be the end-product of lengthy biochemical pathways. Thus the cloning of entire biochemical pathways represents the next degree of complexity of in-vitro gene manipulation. In the case of amino acid production the problems are not too great, for progress is facilitated by the fact that the genetic control of synthesis of most amino acids is well understood, at least in *E. coli*. Not surprisingly, good progress has been made already. By comparison with amino acids very little is known about the biochemical steps in the biosynthesis of most antibiotics, many of which have complex structures. In addition, there is usually no direct selection procedure for antibiotic biosynthetic genes. Fortunately, most of the known antibiotics are synthesized by streptomycetes and hence methods developed for use with one antibiotic producer may be applicable to many others.

Three procedures for cloning antibiotic biosynthetic genes have been described by Hopwood *et al.* (1983). In the case of candicin biosynthesis a key step is the conversion of chorismic acid by the enzyme p-aminobenzoic acid (PABA) synthetase. Cloning of the gene for PABA synthetase was facilitated by the availability of a direct selection method—restoration of prototrophy to a PABA-requiring auxotroph. The second approach that can be used is to isolate mutants which no longer produce

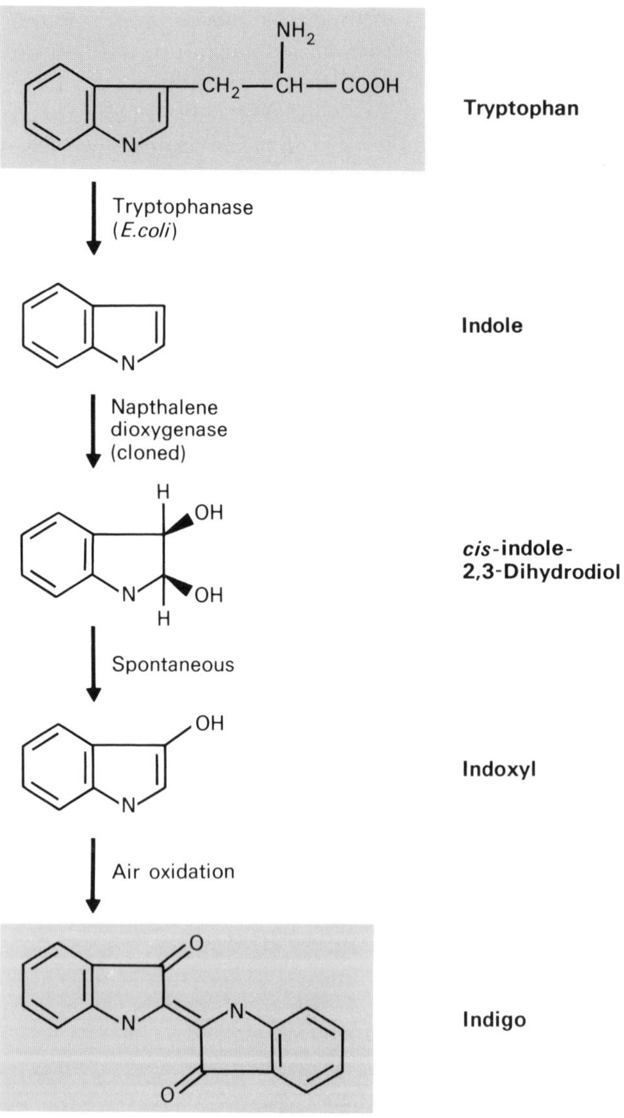

Fig. 14.3 Proposed pathway for indigo biosynthesis by a strain of *E. coli* carrying a cloned gene for naphthalene dioxygenase.

the desired antibiotic and then identify DNA fragments which restore antibiotic synthesis. In this way a gene was identified which encodes an O-methyl transferase involved in undecyl prodigiosin biosynthesis as was a continuous DNA segment carrying the complete genetic information for the synthesis of actinorhodin (Malpartida & Hopwood 1984). The third procedure is to determine if the genes for antibiotic synthesis are plasmid borne, as in the case of methylenomycin, for this facilitates their subsequent manipulation.

Just as novel proteins can be produced by recombinant DNA techniques, so too might novel antibiotics be produced. If libraries of DNA from one antibiotic-producing streptomycete were introduced into a second streptomycete which produces a different antibiotic, new combinations of antibiotic biosynthetic genes will be created; for example, Malpartida and Hopwood (1984) have shown that the genes from *Streptomyces coelicolor* which encode actinorhodin biosynthesis are expressed in *S. parvulus*. Given the lack of specificity of many of the enzymes involved in antibiotic biosynthesis this could result in the formation of novel antibiotics. Evidence which suggests that novel compounds can be formed in this way is provided by the observation that introduction into *E. coli* of a cloned naphthalene dioxygenase gene from *Pseudomonas putida* resulted in the *E. coli* producing the dye indigo (Ensley *et al.* 1983). The postulated mechanism of indigo formation is shown in Fig. 14.3.

Gene manipulation and the production of strains that degrade xenobiotic compounds

In the past quarter of a century the release of various synthetic chemicals, particularly chlorinated aromatics, into the environment has created serious pollution problems. Not only are these compounds toxic and mutagenic/carcinogenic, but they persist for long periods so that they ultimately contaminate man. Examples of such toxic chemicals are DDT and the herbicide 2,4,5-trichlorophenoxyacetic acid (2,4,5-T; Agent Orange). The persistence of these compounds is due to the fact that they do not exist naturally and, consequently, microorganisms have not evolved that degrade them. Rather, they are degraded by co-oxidative metabolism.

Plasmids have been identified which specify the total degradation of chlorinated aromatic compounds such as 3-chlorobenzoate. However, plasmids which encode a complete 3-chlorobenzoate degradative pathway do not permit host cells to metabolize 4-chlorobenzoate unless the cell also carries a

TOL plasmid encoding toluene and xylene decomposition. The latter plasmid provides a benzoate oxygenase with broad specificity which catalyses the first degradative step. By combining plasmids in this way it is possible greatly to extend the range of substrates which bacteria can metabolize (Kellog *et al.* 1981). Thus microorganisms harbouring a variety of plasmids encoding degradation of various aromatic compounds were incubated with 2,4,5-T and after 8–10 months an organism capable of growing on 2,4,5-T as sole carbon source was isolated. Almost certainly a plasmid has evolved by recruitment of genes from other plasmids. This is a very exciting observation. Nevertheless, in future it should be possible to construct such a plasmid *in vitro*; examination of the xenobiotic compound should indicate what degradative enzymes are required and the corresponding genes could be isolated and ligated.

GENETIC ENGINEERING AND PATENT LAW

Patents are rights granted in respect of inventions capable of industrial application. These rights are not automatic but require action to register them—the filing of a patent application. After filing, the application is scrutinized by a patent examiner. If the examiner is convinced that the subject of the patent is both *novel* and *non-obvious*, i.e. involves an inventive step, the patent will be issued. Embodied within the novelty element is the absence of *prior disclosure*. In the USA this means that the subject matter of the patent application should not have been described in public (e.g. publication by the inventor in a scientific journal or presented at a scientific meeting) more than one year before the submission of the patent application. In Europe and most of the world, patent applications must be filed before any publication! In return for the legal protection given to the invention the patentee must provide a written description of the invention which is sufficiently clear and effective to enable a person of *ordinary skill in the art* concerned to practice the invention from its teaching, i.e. a patent must be *enabling*.

In the case of patents involving new strains of mutated microorganisms it is often impossible to write into the patent specification a sufficiently full description of the organism and the method of obtaining it to enable a third party, with certainty, to reproduce the procedure and obtain the strain. The solution to this problem is to deposit the microorganism in a recognized culture collection which, in time, will release the microorganism to the public. A non-mutated microorganism, such as one created by recombinant DNA technology, generally

'And I don't care if they do fix nitrogen, are resistant to acid rain and have a high methionine content—I told you they'd never sell over Christmas!'

(Cartoon redrawn courtesy of University of Birmingham *Biotechnology News*.)

does not need to be deposited if the written patent specification teaches a procedure which 'enables' the exact reconstruction of the microorganism. However, the only way to be certain that the patent application is legally enabling is to deposit the microorganism.

It must be recognized that the acceptability of a patent is governed by the principles of law rather than those of science. This introduces an element of uncertainty for decisions are based on reasoning which, in practice, may mean the forensic abilities of the protagonists. Thus, even if a patent examiner permits a patent to be issued his decision may be challenged in the courts of law by a third party. Alternatively, a third party may deliberately infringe a patent in the belief that the patentee would lose his case if he chose to defend his patent rights in a court of law. Furthermore, each country has its own patent laws and rights granted in respect of a patent in one country might be granted only in part, if at all, in another country. If patent protection in more than one country is desired, separate national applications must be made. These additional applications must be filed within one year of the first patent filing. All the patent applications will usually be given the filing priority date of the first application filed. These and other issues are discussed in detail in two books entitled *Patenting in the Biological Sciences* (Crespi 1982) and *Patenting the Life Forms* (Plant *et al.* 1982) and in an excellent short review (Adler 1984).

Categories of patentable invention

In the biological sciences three categories of patent can be recognized: those dealing with compositions of matter (products), processes and methods of use. Many of the products we expect to produce by recombinant DNA technology are not new. Rather, they are natural products, e.g. insulin or interferon, produced in quantity by an inexpensive procedure made possible by in-vitro genetic manipulation. As these products are not new they are not patentable. As such, they still may be patented in 'substantially pure form' or in a 'pharmaceutical carrier' depending upon prior disclosures. However, if the processes used in making the natural products are unique, the processes *may* be patentable. From the foregoing it should be apparent that most of the patents in the field of genetic engineering will be process patents. A good example is the patenting of an improved process for the manufacture of single-cell protein, namely the incorporation of a *gdh* gene in *M. methylotrophus* as described on page 287.

Generally speaking, process patents are weak patents, in that, given the secrecy under which most companies operate, it is difficult for patentees to know if a third party is infringing their patent rights. Product patents are the most effective form of protection because proof that the claim is being infringed by a competitor is usually apparent from an inspection or analysis of the competitors product without enquiring into the method of manufacture. There are a number of examples in gene manipulation where product patents might be issued; for example, if novel combinations of antibiotic biosynthetic genes resulted in the production of a novel antibiotic, that antibiotic would be patentable. A second example would be a novel microorganism which could degrade a previously biologically recalcitrant molecule, e.g. 2,4,5-T (Agent Orange).

It should be noted that until 1980 it was not possible to patent a living organism other than a plant. However, in that year the US Supreme Court, in a landmark decision (Diamond *vs* Chakrabarty), ruled that the fundamental distinction between living organisms and inert matter was irrelevant for patent purposes. Thus microorganisms *per se* can be patented, not just their use for the production of desirable molecules.

A third example of a patentable product is that of hybrid plasmids. Hybrid plasmids are new substances and, assuming that the problem of accurate definition can be overcome, there appears no reason why product claims should not be allowed for them. If a product claim is allowed for a plasmid then an adequate description of the plasmid must be provided and one way of doing this is to provide the complete nucleotide sequence. This raises the question of whether a competitor could avoid infringing the patent by using the redundancy of the genetic code to construct a plasmid with a different nucleotide sequence but which encodes the same protein(s).

The Cohen and Boyer patents

There are two Cohen and Boyer patents and recently they have been a *cause célèbre*, particularly in the USA. On December 2nd, 1980 the US Patent Office issued patent number 4 237 224 entitled *Process for Producing Biologically Functional Chimeras*. The patent covers a series of steps involved in replicating exogenous genes in microorganisms. A plasmid or viral DNA is cleaved, a new gene is inserted, the replicon is placed in a microorganism and the transformants isolated. In issuing the patent the US Patent Office showed that it considered gene-splicing methods to be patentable. An important facet of this patent is that foreign applications were not filed. This is a result of a prior publication by the inventors which destroyed non-US patent rights. This means that recombinant DNA technology can be used for profit in other countries and that the products of the technology can be imported into the USA unless the now pending changes in the law are accepted and enforced by the US International Trade Commission. Consequently, a second patent application was filed in the USA which laid claim to the products, regardless of country of origin, which resulted from the use of the techniques described in the first patent. The second patent was scheduled for publication on July 13th, 1982 but was not issued because of reconsideration by the US Patent Examiner, prompted, in part, by the growing controversy.*

The controversy surrounding the second patent arose because of unconventional action by the patent assignees, Stanford University. They made details of the patent prosecution in the Patent Office public before the patent was issued. Normally the first indication that a patent application has been made comes when the patent is issued in some country. By their action Stanford University allowed opponents of the patent to voice their criticisms in public thereby influencing the patent examiner. The impetus to prevent the second patent from being issued was, in part, the recompense stipulated by Stanford University—$10 000 per year minimum royalty, plus royalties when products are sold of 1% for sales up to $5 million and falling to 0.5% for sales in excess of $10 million.

Although it was disclosure of the contents of the second patent which caused the furore the criticisms were aimed at the first patent. There were two major criticisms. The first of these was that the patent was *non-enabling*. The first patent covers a key plasmid, pSC101, which was used as the vehicle for inserting new genes in *E. coli* and described the method for producing this plasmid. However, in 1977 Cohen published a paper that admits an error in the original procedure. Deposition of the plasmid in a culture collection would have avoided this problem except that deposition was made too late—after the patent had been filed. The patent assignees claim that Cohen and Boyer made the plasmid available to other scientists but there were restrictions on access to it. Furthermore they argue that the patent is enabling since pSC101 is not vital as the instructions provided in the patent explain how to select, isolate and use other suitable plasmids. However, it should be noted that

*US Patent No. 4 468 464 entitled *Biologically Functional Chimeras* issued August 28th, 1984.

in a submission to the Patent Office in June, 1977 Stanford University took the opposite view; namely, that pSC101 was indeed vital to the patent claims and that this new plasmid found by the inventors was what distinguished their claims from potentially competing work by others. The second objection to the patent was prior disclosure. In an article in *New Scientist* on 25th October 1973, more than a year before the first patent application was filed, a detailed report was given of the work of Cohen and Boyer. Whether the second Cohen and Boyer Patent will issue remains to be seen, and if it does issue, almost certainly that will not be the last of the matter. Most of these present issues will probably reappear as issues in any future legal battles over enforcement of the patent.

SECTION 5
Appendices

Appendix 1 Enzymes Used in Gene Manipulation

See Tables on pages 308–13.

EXONUCLEASES

Enzyme	Activities	Substrate	Product
Exonuclease III (*E. coli*)	(a) $3' \rightarrow 5'$ exonuclease (b) 3′ phosphatase (starts from ends or nicks) (c) endonuclease specific for apurinic/apyrimidinic sites	(a) Double-stranded DNA (b) 3′ OH or 3′ PO_4 (starts from ends or nicks) (c) AP DNA	(a) Single-stranded DNA, 5′ mononucleotides (b) Pi + 3′ OH ends (c) nicked or gapped double-stranded DNA
Lambda exonuclease	$5' \rightarrow 3'$ exonuclease	Double-stranded DNA (single-stranded DNA is degraded at 10 to 1000-fold lower rate depending on size)	Single-stranded DNA + 5′ mononucleotides
Nuclease BAL 31	(a) $5' \rightarrow 3'$ exonuclease $3' \rightarrow 5'$ exonuclease (b) Endonuclease (single-stranded DNA)	(a) Double-stranded DNA (b) Single-stranded DNA	(a) Blunt-ended double-stranded DNA, 5′ mononucleotides (b) Nicked or gapped double-stranded DNA; shortened single-stranded DNA; dNMPs
Exonuclease VII (*E. coli*)	(a) $5' \rightarrow 3'$ exonuclease (b) $5' \rightarrow 3'$ exonuclease $3' \rightarrow 5'$ exonuclease	(a) Single-stranded DNA (b) Double-stranded DNA with single-stranded tails	(a) Dinucleotides to dodecanucleotides (b) Blunt end double-stranded DNA

EXONUCLEASES/POLYMERASES

Enzyme	Activities	Substrate	Product
DNA polymerase I (*E. coli*)	(a) $5' \rightarrow 3'$ exonuclease (b) $5' \rightarrow 3'$ polymerase (c) $3' \rightarrow 5'$ exonuclease (proof-reading function)	(a) Nicked or gapped DNA (3′ OH) (b) Double-stranded DNA (3′ OH) (c) Mismatched nucleotides at ends, nicks or gaps	(a) Double-stranded DNA with single-stranded region (b) Double-stranded DNA (c) Gapped DNA; blunt-ended DNA
T4 DNA polymerase	(a) $3' \rightarrow 5'$ exonuclease. (Much faster on single-stranded DNA than on double-stranded DNA.) (b) $5' \rightarrow 3'$ polymerase	(a) Internal or external 3′ OH ends of double-stranded DNA or single-stranded DNA (b) Double-stranded DNA (3′ recess)	(a) Single-stranded DNA, 5′ mononucleotides (b) Double-stranded DNA (blunt-ended)
Large fragment of DNA polymerase I (Klenow fragment) *E. coli*	(a) $5' \rightarrow 3'$ polymerase (b) $3' \rightarrow 5'$ exonuclease (proof-reading function)	(a) Primed single-stranded DNA (b) Mismatched nucleotides at ends, nicks or gaps	(a) Double-stranded DNA (b) Repaired double-stranded DNA
Reverse transcriptase	Synthesizes complementary DNA	Single-stranded DNA or RNA	Double-stranded DNA or RNA–DNA hybrid
RNA polymerase	Synthesizes RNA	Single-stranded DNA	RNA

DNA AND RNA MODIFYING ENZYMES

Enzyme	Activities	Substrate	Product
Bacterial alkaline phosphatase Calf intestinal phosphatase	Removes 5′ phosphates	RNA or DNA	Polynucleotide lacking 5′-terminal phosphate residues
T4 polynucleotide kinase	Catalyses the transfer of the gamma-phosphate of a nucleoside 5′-triphosphate to the 5′-hydroxyl terminus of a polynucleotide	DNA or RNA	5′-phosphorylated DNA or RNA. If the gamma-phosphate of the donor nucleotide is radiolabelled it produces end-labelled RNA or DNA
T4 RNA ligase	Joins RNA molecules	Single-stranded RNA	Ligated RNA molecules
DNA ligase	Catalyses the formation of a covalent phosphodiester bond from a 5′-phosphoryl group and an adjacent 3′-hydroxyl group	(a) Double-stranded DNA with single-stranded breaks (b) DNA fragments with cohesive or (T4 DNA ligase only) blunt ends	(a) Intact double-stranded DNA (b) Covalently joined DNA fragments
Terminal deoxy-nucleotidyl transferase	Catalyses the polymerization of deoxynucleoside-5′-triphosphates to the 3′-termini of DNA molecules	Single- or double-stranded DNA	DNA with 3′ homopolymer extensions. If alpha-phosphate of deoxynucleotide is radiolabelled then radiolabelled DNA molecule is produced

RecA protein (*E. coli*)	(a) Promotes pairing of single-stranded DNA with homologous duplex DNA (b) Promotes unwinding of duplex DNA by heterologous or homologous single-stranded DNA fragments		
Tobacco acid pyrophosphatase	Removes 5′ cap from eukaryotic mRNA	5′-capped mRNA	Uncapped mRNA with 5′-terminal phosphate
Poly(A) polymerase	Addition of poly(A) at 3′ terminus of RNA	RNA plus ATP	Poly(A)-tailed RNA
dam Methylase	Transfer of methyl group from S-adenosyl methionine to the N6 position of adenine residues in the sequence GATC	Duplex DNA carrying the unmethylated sequence GATC (recognition sequences for endonucleases *Bcl* I and *Mbo* I and some sites for *Cla* I, *Mbo* II, *Taq* I)	Methylated duplex DNA
dcm Methylase (*mec* Methylase)	Transfer of methyl group from S-adenosyl methionine to internal cytosine residues in the sequences CCAGG and CCTGG	Duplex DNA carrying the unmethylated sequences CCAGG and CCTGG (recognition sequences for endonucleases *Eco* RII and *Scr* FI and some sites for *Ava* II and *Sau* 96I)	Methylated duplex DNA

ENDONUCLEASES

Enzyme	Activities	Substrate	Product
S1 Nuclease	Single-stranded specific endonuclease	Single-stranded DNA or RNA	5′ mononucleotides
RNAse H (*E. coli*)	Non-specific endoribonuclease	RNA/DNA hybrids	5′ oligoribonucleotides
RNAse III (*E. coli*)	Non-specific endoribonuclease	Double-stranded RNA	Oligoribonucleotides (15mers or less)
Nuclease P1	(a) Single-stranded specific endonuclease (b) 3′ nucleotidase	Single-stranded DNA or RNA	5′ mononucleotides from DNA or RNA
Ribonuclease T_1	Endoribonuclease (preferentially cleaves 3′ to G)	Single-stranded RNA	Oligonucleotides terminating in 3′ GMP
Ribonuclease T_2	Non-specific endoribonuclease (rate of cleavage is $A > U > G > C$)	Single-stranded RNA	3′ mononucleotides
Ribonuclease U_2	Purine specific endoribonuclease (rate of cleavage is $A >> G$ at pH 3.5)	Single-stranded RNA	Oligonucleotides terminating in 3′ AMP
Ribonuclease CL3	Endoribonuclease (preferentially cleaves 3′ to C)	Single-stranded RNA	Oligonucleotides terminating in 3′ CMP
Ribonuclease *B. cereus*	Pyrimidine specific endoribonuclease	Single-stranded RNA	Oligonucleotides terminating in 3′UMP or 3′CMP
Ribonuclease Phy M	Endoribonuclease (preferentially cleaves 3′ to U and A)	Single-stranded RNA	Oligonucleotides terminating in 3′UMP or 3′AMP

TOPOISOMERASES

Enzyme	Activities	Substrate	Product
Topoisomerase I (from Calf Thymus)	Relaxation of negatively or positively supercoiled DNA	Supercoiled DNA	Relaxed, covalently closed duplex DNA
Topoisomerase II (DNA Gyrase)	(a) Supercoiling of duplex DNA (b) Catenation of duplex DNA rings	(a) Relaxed DNA + ATP (b) Relaxed or supercoiled DNA rings	(a) Supercoiled DNA + ADP (b) Catenated rings

Appendix 2 List of Known Restriction Endonucleases

New restriction endonucleases are discovered almost weekly and at the last count over 400 were known. From time to time Dr R. J. Roberts of the Cold Spring Harbor Laboratory produces an up-to-date list of all known restriction and modification enzymes and their recognition sequences. At the time of going to press the latest compilation was Roberts (1984). In addition, some of the suppliers of restriction endonucleases reproduce this information in the form of useful wallcharts.

Table A2.1 lists all the known restriction endonucleases at the time of going to press. In forming this list all endonucleases cleaving DNA at a specific sequence have been considered to be restriction enzymes although in most cases there is no direct genetic evidence for the presence of a restriction-modification system. Perusal of Table A2.1 will reveal some enzymes with apparently unusual properties; for example, *Eco* pDXI is the first example of a Type I enzyme that recognizes an octanucleotide sequence. By contrast, *Eco* PI, *Eco* P15, *Hin*e I and *Hin*f III have characteristics intermediate between those of the Type I and Type II restriction endonucleases and have been designated Type III. Both *Hin*e I and *Hin*f III cleave about 25 bases 3′ of the recognition sequence. The enzyme *Nci* I leaves termini carrying a 3′-phosphate group.

Table A2.1 Properties of restriction endonucleases. When two enzymes recognize the same sequence, i.e. are isoschizomers, the first example to be described is indicated in parentheses. Recognition sequences are written from 5′ → 3′, only one strand being given, and the point of cleavage is indicated by an arrow (↓). When no arrow appears the precise cleavage site has not been determined. For enzymes such as *Hga* I and *Mbo* II which cleave away from their recognition sequence the sites of cleavage are indicated in parentheses; for example, *Hga* I GACGC (5/10) indicates cleavage as shown below:

5′ GACGCNNNNN↓ 3′
3′ CTGCGNNNNNNNNNN↓ 5′

Bases appearing in parentheses signify that either base may occupy that position and N represents any base. Where known, the base modified by the corresponding specific methylase is indicated by an asterisk.

Microorganism	Enzyme	Sequence
Acetobacter aceti	*Aat* I (*Stu* I)	AGGCCT
	Aat II	GACGT↓C
Acetobacter aceti sub. *liquefaciens*	*Aac* I (*Bam* HI)	GGATCC
Acetobacter aceti sub. *liquefaciens*	*Aae* I (*Bam* HI)	GGATCC
Acetobacter aceti sub. *orleanensis*	*Aor* I (*Eco* RII)	CC↓($\frac{A}{T}$)GG
Acetobacter pasteurianus sub. *pasteurianus*	*Apa* I	GGGCC↓C
Achromobacter immobilis	*Aim* I	?
Acinetobacter calcoaceticus	*Acc* I	GT↓($\frac{A}{C}$)($\frac{G}{T}$)AC
	Acc II (*Fnu* DII)	CGCG
	Acc III	?
Actinomadura madurae	*Ama* I (*Nru* I)	TCGCGA
Agmenellum quadruplicatum	*Aqu* I (*Ava* I)	CPyCGPuG
Agrobacterium tumefaciens	*Atu* AI	?
Agrobacterium tumefaciens B6806	*Atu* BI (*Eco* RII)	CC($\frac{A}{T}$)GG
Agrobacterium tumefaciens IIBV7	*Atu* BVI	?
Agrobacterium tumefaciens ID 135	*Atu* II (*Eco* RII)	CC($\frac{A}{T}$)GG
Agrobacterium tumefaciens C58	*Atu* CI (*Bcl* I)	TGATCA
Alcaligenes species	*Asp* AI (*Bst* EII)	G↓GTNACC
Anabaena catanula	*Aca* I	?
Anabaena cylindrica	*Acy* I	GPu↓CGPyC
Anabaena flos-aquae	*Afl* I (*Ava* II)	G↓G($\frac{A}{T}$)CC
	Afl II	C↓TTAAG
	Afl III	A↓CPuPyGT

Table A2.1 *continued*

Microorganism	Enzyme	Sequence
Anabaena oscillarioides	*Aos* I (*Mst* I)	TGC↓GCA
	Aos II (*Acy* I)	GPu↓CGPyC
Anabaena strain Waterbury	*Ast* WI (*Acy* I)	GPu↓CGPyC
Anabaena subcylindrica	*Asu* I	G↓GNCC
	Asu II	TT↓CGAA
	Asu III (*Acy* I)	GPu↓CGPyC
Anabaena variabilis	*Ava* I	C↓PyCGPuG
	Ava II	G↓G($\frac{A}{T}$)CC
	Ava III	ATGCAT
Anabaena variabilis[uw]	*Avr* I (*Ava* I)	CPyCGPuG
	Avr II	CCTAGG
Aphanothece halophytica	*Aha* I (*Cau* II)	CC($\frac{C}{G}$)GG
	Aha II	?
	Aha III	TTT↓AAA
Arthrobacter luteus	*Alu* I	AG↓CT
Arthrobacter pyridinolis	*Apy* I (*Eco* RII)	CC↓($\frac{A}{T}$)GG
Bacillus acidocaldarius	*Bac* I (*Sac* II)	CCGCGG
Bacillus amyloliquefaciens F	*Bam* FI (*Bam* HI)	GGATCC
Bacillus amyloliquefaciens H	*Bam* HI	G↓GAT$\overset{*}{C}$C
Bacillus amyloliquefaciens K	*Bam* KI (*Bam* HI)	GGATCC
Bacillus amyloliquefaciens N	*Bam* NI (*Bam* HI)	GGATCC
	Bam N_x (*Ava* II)	G↓G($\frac{A}{T}$)CC
Bacillus aneurinolyticus	*Ban* I (*Hgi* CI)	GGPyPuCC
	Ban II (*Hgi* JII)	GPuGCPy↓C
	Ban III (*Cla* I)	AT$\overset{*}{C}$GAT
Bacillus brevis S	*Bbv* SI	G$\overset{*}{C}$($\frac{A}{T}$)GC
Bacillus brevis	*Bbv* I	GCAGC (8/12)
Bacillus caldolyticus	*Bcl* I	T↓GATCA
Bacillus centrosporus	*Bcn* I (*Cau* II)	CC($\frac{C}{G}$)↓GG
Bacillus cereus	*Bce* 14579	?
Bacillus cereus	*Bce* 1229	?
Bacillus cereus	*Bce* 170 (*Pst* I)	CTGCAG
Bacillus cereus Rf sm st	*Bce* R (*Fnu* DII)	CGCG
Bacillus globigii	*Bgl* I	GCCNNNN↓NGGC
	Bgl II	A↓GATCT
Bacillus megaterium 899	*Bme* 899	?
Bacillus megaterium B205-3	*Bme* 205	?
Bacillus megaterium	*Bme* I	?
Bacillus pumilus AHU 1387A	*Bpu* I	?
Bacillus sphaericus	*Bsp* 1286	?
Bacillus sphaericus R	*Bsp* RI (*Hae* III)	GG↓CC
Bacillus stearothermophilus C1	*Bst* CI (*Hae* III)	GGCC
Bacillus stearothermophilus C11	*Bss* CI (*Hae* III)	GGCC
Bacillus stearothermophilus G3	*Bst* GI (*Bcl* I)	TGATCA
	Bst GII (*Eco* RII)	CC($\frac{A}{T}$)GG

Table A2.1 *continued*

Microorganism	Enzyme	Sequence
Bacillus stearothermophilus G6	*Bss* GI (*Bst* XI)	CCANNNNNNNTGG
	Bss GII (*Mbo* I)	GATC
Bacillus stearothermophilus H1	*Bst* HI (*Xho* I)	CTCGAG
Bacillus stearothermophilus H3	*Bss* HI (*Xho* I)	CTCGAG
	Bss HII (*Bse* PI)	GCGCGC
Bacillus stearothermophilus H4	*Bsr* HI (*Bse* PI)	GCGCGC
Bacillus stearothermophilus P1	*Bss* PI	?
Bacillus stearothermophilus P5	*Bsr* PI	?
	Bsr PII (*Mbo* I)	GATC
Bacillus stearothermophilus P6	*Bse* PI	GCGCGC
Bacillus stearothermophilus P8	*Bsa* PI (*Mbo* I)	GATC
Bacillus stearothermophilus P9	*Bso* PI (*Bsr* PI)	?
Bacillus stearothermophilus T12	*Bst* TI (*Bst* XI)	CCANNNNNNNTGG
Bacillus stearothermophilus X1	*Bst* XI	CCANNNNN↓NTGG
	Bst XII (*Mbo* I)	GATC
Bacillus stearothermophilus 1503-4R	*Bst* I (*Bam* HI)	G↓GATCC
Bacillus stearothermophilus 240	*Bst* AI	?
Bacillus stearothermophilus ET	*Bst* EI	?
	Bst EII	G↓GTNACC
	Bst EIII (*Mbo* I)	GATC
Bacillus stearothermophilus	*Bst* PI (*Bst* EII)	G↓GTNACC
Bacillus stearothermophilus	*Bst* NI (*Eco* RII)	CC↓($\frac{A}{T}$)GG
Bacillus stearothermophilus 822	*Bse* I (*Hae* III)	GGCC
	Bse II (*Hpa* I)	GTTAAC
Bacillus subtilis strain R	*Bsu* RI (*Hae* III)	GG↓$\overset{*}{C}$C
Bacillus subtilis Marburg 168	*Bsu* M	?
Bacillus subtilis	*Bsu* 6663	?
Bacillus subtilis	*Bsu* 1076 (*Hae* III)	GGCC
Bacillus subtilis	*Bsu* 1114 (*Hae* III)	GGCC
Bacillus subtilis	*Bsu* 1247 (*Pst* I)	CTGCAG
Bacillus subtilis	*Bsu* 1145	?
Bacillus subtilis	*Bsu* 1192 (*Hpa* II)	CCGG
	Bsu 1192II (*Fnu* DII)	CGCG
Bacillus subtilis	*Bsu* 1193 (*Fnu* DII)	CGCG
Bacillus subtilis	*Bsu* 1231I (*Hpa* II)	CCGG
	Bsu 1231II (*Fnu* DII)	CGCG
Bacillus subtilis	*Bsu* 1259	?
Bifidobacterium bifidum	*Bbi* I (*Pst* I)	CTGCAG
	Bbi II (*Acy* I)	GPuCGPyC
	Bbi III (*Xho* I)	CTCGAG
	Bbi IV	?
Bifidobacterium breve	*Bde* I (*Nar* I)	GGCGC↓C
Bifidobacterium breve S1	?	?
Bifidobacterium breve S50	*Bbe* AI (*Nar* I)	GGCGCC
	Bbe AII	?
Bifidobacterium infantis 659	*Bin* I	GGATC
Bifidobacterium infantis S76e	*Bin* SI (*Eco* RII)	CC($\frac{A}{T}$)GG
	Bin SII (*Nar* I)	GGCGCC

Table A2.1 *continued*

Microorganism	Enzyme	Sequence
Bifidobacterium longum E194b	*Blo* I	?
Bifidobacterium thermophilum RU326	*Bth* I (*Xho* I)	CTCGAG
	Bth II	?
Bordetella bronchiseptica	*Bbr* I (*Hind* III)	AAGCTT
Bordetella pertussis	*Bpe* I (*Hind* III)	AAG$\overset{*}{\text{C}}$TT
Brevibacterium albidum	*Bal* I	TGG↓$\overset{*}{\text{C}}$CA
Brevibacterium luteum	*Blu* I (*Xho* I)	C↓TCGAG
	Blu II (*Hae* III)	GGCC
Calothrix scopulorum	*Csc* I (*Sac* II)	CCGC↓GG
Caryophanon latum L	*Cla* I	AT↓CGAT
Caryophanon latum	*Clm* I (*Hae* III)	GGCC
	Clm II (*Ava* II)	GG($^{A}_{T}$)CC
Caryophanon latum	*Clt* I (*Hae* III)	GG↓CC
Caryophanon latum RII	*Clu* I	?
Caryophanon latum H7	*Cal* I	?
Caulobacter crescentus CB-13	*Ccr* I	?
	Ccr II (*Xho* I)	CTCGAG
Caulobacter fusiformis	*Cfu* I (*Dpn* I)	G$\overset{*}{\text{A}}$TC
Chloroflexus aurantiacus	*Cau* I (*Ava* II)	GG($^{A}_{T}$)CC
	Cau II	CC↓($^{C}_{G}$)GG
Chromatium vinosum	*Cvn* I (*Sau* I)	CC↓TNAGG
Chromobacterium violaceum	*Cvi* I	?
Citrobacter freundii	*Cfr* I	Py↓GGCCPu
Clostridium formicoaceticum	*Cfo* I (*Hha* I)	GCGC
Clostridium pasteurianum	*Cpa* I (*Mbo* I)	GATC
Corynebacterium humiferum	*Chu* I (*Hind* III)	AAGCTT
	Chu II (*Hind* II)	GTPyPuAC
Corynebacterium petrophilum	*Cpe* I (*Bcl* I)	TGATCA
Cystobacter velatus Plv9	*Cve* I	?
Desulfovibrio desulfuricans Norway strain	*Dde* I	C↓TNAG
	Dde II (*Xho* I)	CTCGAG
Desulfovibrio desulfuricans	*Dds* I (*Bam* HI)	GGATCC
Diplococcus pneumoniae	*Dpn* I	G$\overset{*}{\text{A}}$↓TC
Diplococcus pneumoniae	*Dpn* II (*Mbo* I)	GATC
Enterobacter aerogenes	*Eae* I (*Cfr* I)	Py↓GGCCPu
Enterobacter cloacae	*Ecl* I	?
	Ecl II (*Eco* RII)	CC($^{A}_{T}$)GG
Enterobacter cloacae	*Eca* I (*Bst* EII)	G↓GTNACC
	Eca II (*Eco* RII)	CC($^{A}_{T}$)GG
Enterobacter cloacae	*Ecc* I (*Sac* II)	CCGCGG
Escherichia coli pDXI	*Eco* DXI	ATCA$(N)_7$ATTC
Escherichia coli J62 pLG74	*Eco* RV	GATAT↓C

Table A2.1 *continued*

Microorganism	Enzyme	Sequence
Escherichia coli RY13	*Eco* RI	G↓A$\overset{*}{\text{A}}$TTC
	Eco RI′	PuPuA↓TPyPy
Escherichia coli R245	*Eco* RII	↓C$\overset{*}{\text{C}}$$(\frac{\text{A}}{\text{T}})$GG
Escherichia coli B	*Eco* B	TG$\overset{*}{\text{A}}$(N)$_8$TGCT
Escherichia coli K	*Eco* K	AAC(N)$_6$GTGC
Escherichia coli (PI)	*Eco* PI	AG$\overset{*}{\text{A}}$CC
Escherichia coli P15	*Eco* P15	CAGCAG
Flavobacterium okeanokoites	*Fok* I	GGATG (9/13)
Fremyella diplosiphon	*Fdi* I (*Ava* II)	G↓G$(\frac{\text{A}}{\text{T}})$CC
	Fdi II (*Mst* I)	TGC↓GCA
Fusobacterium nucleatum A	*Fnu* AI (*Hin*f I)	G↓ANTC
	Fnu AII (*Mbo* I)	GATC
Fusobacterium nucleatum C	*Fnu* CI (*Mbo* I)	↓GATC
Fusobacterium nucleatum D	*Fnu* DI (*Hae* III)	GG↓CC
	Fnu DII	CG↓CG
	Fnu DIII (*Hha* I)	GCG↓C
Fusobacterium nucleatum E	*Fnu* EI (*Mbo* I)	↓GATC
Fusobacterium nucleatum 48	*Fnu* 48I	?
Fusobacterium nucleatum 4H	*Fnu* 4HI	GC↓NGC
Gluconobacter dioxyacetonicus	*Gdi* I (*Stu* I)	AGG↓CCT
	Gdi II	Py↓GGCCG
Gluconobacter dioxyacetonicus	*Gdo* I (*Bam* HI)	GGATCC
Gluconobacter oxydans sub. *melonogenes*	*Gox* I (*Bam* HI)	GGATCC
Haemophilus aegyptius	*Hae* I	$(\frac{\text{A}}{\text{T}})$GG↓CC$(\frac{\text{A}}{\text{T}})$
	Hae II	PuGCGC↓Py
	Hae III	GG↓$\overset{*}{\text{C}}$C
Haemophilus aphrophilus	*Hap* I	?
	Hap II (*Hpa* II)	C↓CGG
Haemophilus gallinarum	*Hga* I	GACGC (5/10)
Haemophilus haemoglobinophilus	*Hhg* I (*Hae* III)	GGCC
Haemophilus haemolyticus	*Hha* I	G$\overset{*}{\text{C}}$G↓C
	Hha II (*Hin*f I)	GANTC
Haemophilus influenzae GU	*Hin* GUI (*Hha* I)	GCGC
	Hin GUII (*Fok* I)	GGATG
Haemophilus influenzae 173	*Hin* 173 (*Hin*d III)	AAGCTT
Haemophilus influenzae 1056	*Hin* 1056I (*Fnu* DII)	CGCG
	Hin 1056II	?
Haemophilus influenzae serotype b, 1076	*Hin*b III (*Hin*d III)	AAGCTT
Haemophilus influenzae serotype c, 1160	*Hin*c II (*Hin*d II)	GTPyPuAC
Haemophilus influenzae serotype c, 1161	*Hin*c II (*Hin*d II)	GTPyPuAC

Table A2.1 *continued*

Microorganism	Enzyme	Sequence
Haemophilus influenzae serotype e	*Hin*e I (*Hin*f III)	CGAAT
Haemophilus influenzae R_b	*Hin*b III (*Hin*d III)	AAGCTT
Haemophilus influenzae R_c	*Hin*c II (*Hin*d II)	GTPyPuAC
Haemophilus influenzae R_d	*Hin*d I	C$\overset{*}{\text{A}}$C
	*Hin*d II	GTPy↓Pu$\overset{*}{\text{A}}$C
	*Hin*d III	$\overset{*}{\text{A}}$↓AGCTT
	*Hin*d IV	G$\overset{*}{\text{A}}$C
Haemophilus influenzae R_f	*Hin*f I	G↓ANTC
	*Hin*f II (*Hin*d III)	AAGCTT
	*Hin*f III	CGAAT
Haemophilus influenzae H-1	*Hin* HI (*Hae* II)	PuGCGCPy
Haemophilus influenzae P_1	*Hin* P_1I (*Hha* I)	G↓CGC
Haemophilus influenzae S_1	*Hin* S_1 (*Hha* I)	GCGC
Haemophilus influenzae S_2	*Hin* S_2 (*Hha* I)	GCGC
Haemophilus influenzae JC9	*Hin* JCI (*Hin*d II)	GTPy↓PuAC
	Hin JCII (*Hin*d III)	AAGCTT
Haemophilus parahaemolyticus	*Hph* I	GGTGA (8/7)
Haemophilus parainfluenzae	*Hpa* I	GTT↓A$\overset{*}{\text{A}}$C
	Hpa II	C↓$\overset{*}{\text{C}}$GG
Haemophilus suis	*Hsu* I (*Hin*d III)	A↓AGCTT
Halococcus agglomeratus	*Hag* I	?
Herpetosiphon giganteus HP 1023	*Hgi* AI	G$\binom{A}{T}$GC$\binom{A}{T}$↓C
Herpetosiphon giganteus Hpg 5	*Hgi* BI (*Ava* II)	G↓G$\binom{A}{T}$CC
Herpetosiphon giganteus Hpg 9	*Hgi* CI	G↓GPyPuCC
	Hgi CII (*Ava* II)	G↓G$\binom{A}{T}$CC
	Hgi CIII (*Sal* I)	G↓TCGAC
Herpetosiphon giganteus Hpa 2	*Hgi* DI (*Acy* I)	GPu↓CGPyC
	Hgi DII (*Sal* I)	G↓TCGAC
Herpetosiphon giganteus Hpg 24	*Hgi* EI (*Ava* II)	G↓G$\binom{A}{T}$CC
	Hgi EII	ACC$(N)_6$GGT
Herpetosiphon giganteus Hpg 14	*Hgi* FI	?
Herpetosiphon giganteus Hpa 1	*Hgi* GI (*Acy* I)	GPu↓CGPyC
Herpetosiphon giganteus HP 1049	*Hgi* HI (*Hgi* CI)	G↓GPyPuCC
	Hgi HII (*Acy* I)	GPu↓CGPyC
	Hgi HIII (*Ava* II)	G↓G$\binom{A}{T}$CC
Herpetosiphon giganteus HFS 101	*Hgi* JI	?
	Hgi JII	GPuGCPy↓C
Herpetosiphon giganteus Hpg 32	*Hgi* KI	?
Klebsiella pneumoniae OK8	*Kpn* I	GGTAC↓C
Mastigocladus laminosus	*Mla* I (*Asu* II)	TT↓CGAA
Microbacterium thermosphactum	*Mth* I (*Mbo* I)	GATC
Micrococcus luteus	*Mlu* I	A↓CGCGT
Micrococcus radiodurans	*Mra* I (*Sac* II)	CCGCGG
Microcoleus species	*Mst* I	TGC↓GCA
	Mst II (*Sau* I)	CC↓TNAGG

Table A2.1 *continued*

Microorganism	Enzyme	Sequence
Moraxella bovis	*Mbo* I	↓GATC
	Mbo II	GAAGA
Moraxella bovis	*Mbv* I	?
Moraxella glueidi LG1	*Mgl* I	?
Moraxella glueidi LG2	*Mgl* II	?
Moraxella kingae	*Mki* I (*Hin*d III)	AAGCTT
Moraxella nonliquefaciens	*Mno* I (*Hpa* II)	C↓CGG
	Mno II (*Mnn* III)	?
	Mno III (*Mbo* I)	GATC
Moraxella nonliquefaciens	*Mnl* I	CCTC (7/7)
Moraxella nonliquefaciens	*Mnn* I (*Hin*d II)	GTPyPuAC
	Mnn II (*Hae* III)	GGCC
	Mnn III	?
	Mnn IV (*Hha* I)	GCGC
Moraxella nonliquefaciens	*Mni* I (*Hae* III)	GGCC
	Mni II (*Hpa* II)	CCGG
Moraxella osloensis	*Mos* I (*Mbo* I)	GATC
Moraxella phenylpyruvica	*Mph* I (*Eco* RII)	CC($\frac{A}{T}$)GG
Moraxella species	*Msp* I (*Hpa* II)	C↓CGG
Myxococcus stipitatus Mxs2	*Msi* I (*Xho* I)	CTCGAG
	Msi II	?
Myxococcus virescens V-2	*Mvi* I	?
	Mvi II	?
Neisseria caviae	*Nca* I (*Hinf* I)	GANTC
Neisseria cinerea	*Nci* I (*Cau* II)	CC↓($\frac{C}{G}$)GG[g]
Neisseria denitrificans	*Nde* I	CA↓TATG
	Nde II (*Mbo* I)	GATC
Neisseria flavescens	*Nfl* I (*Mbo* I)	GATC
	Nfl II	?
	Nfl III	?
Neisseria gonorrhoea	*Ngo* I (*Hae* II)	PuGCGCPy
Neisseria gonorrhoea	*Ngo* II (*Hae* III)	GGCC
Neisseria gonorrhoea KH 7764-45	*Ngo* III (*Sac* II)	CCGCGG
Neisseria mucosa	*Nmu* I (*Nae* I)	GCCGGC
Neisseria ovis	*Nov* I	?
	Nov II (*Hinf* I)	GANTC
Nocardia aerocolonigenes	*Nae* I	GCC↓GGC
Nocardia amarae	*Nam* I (*Nar* I)	GGCGCC
Nocardia argentinensis	*Nar* I	GG↓CGCC
Nocardia blackwellii	*Nbl* I (*Pvu* I)	CGAT↓CG
Nocardia brasiliensis	*Nbr* I (*Nae* I)	GCCGGC
Nocardia brasiliensis	*Nba* I (*Nae* I)	GCCGGC
Nocardia corallina	*Nco* I	C↓CATGG
Nocardia dassonvillei	*Nda* I (*Nar* I)	GG↓CGCC
Nocardia minima	*Nmi* I (*Kpn* I)	GGTACC
Nocardia opaca	*Nop* I (*Sal* I)	G↓TCGAC
	Nop II	?

Table A2.1 *continued*

Microorganism	Enzyme	Sequence
Nocardia otitidis-caviarum	*Not* I	?
Nocardia otitidis-caviarum	*Noc* I (*Pst* I)	CTGCAG
Nocardia rubra	*Nru* I	TCG↓CGA
Nocardia uniformis	*Nun* I	?
	Nun II (*Nar* I)	GG↓CGCC
Nostoc species	*Nsp* BI (*Asu* II)	TTCGAA
	Nsp BII	C($\substack{A\\C}$)G↓C($\substack{T\\G}$)G
Nostoc species	*Nsp* (7524)I	PuCATG↓Py
	Nsp (7524)II (*Sdu* I)	G($\substack{G\\A\\T}$)GC($\substack{C\\A\\T}$)↓C
	Nsp (7524)III (*Ava* I)	C↓PyCGPuG
	Nsp (7524)IV (*Asu* I)	G↓GNCC
	Nsp (7524)V (*Asu* II)	TTCGAA
Nostoc species	*Nsp* HI (*Nsp* CI)	PuCATG↓Py
	Nsp HII (*Ava* II)	GG($\substack{A\\T}$)CC
Oerskovia xanthineolytica	*Oxa* I (*Alu* I)	AGCT
	Oxa II	?
Proteus vulgaris	*Pvu* I	CGAT↓CG
	Pvu II	CAG↓CTG
Providencia alcalifaciens	*Pal* I (*Hae* III)	GGCC
Providencia stuartii 164	*Pst* I	CTGCA↓G
Pseudoanabaena species	*Psp* I (*Asu* I)	GGNCC
Pseudomonas aeruginosa	*Pae* R7	?
Pseudomonas facilis	*Pfa* I (*Mbo* I)	GATC
Pseudomonas maltophila	*Pma* I (*Pst* I)	CTGCAG
Rhizobium leguminosarum 300	*Rle* I	?
Rhizobium lupini #1	*Rlu* I	?
Rhizobium meliloti	*Rme* I	?
Rhodococcus rhodochrous	*Rrh* I (*Sal* I)	GTCGAC
	Rrh II	?
Rhodococcus rhodochrous	*Rro* I (*Sal* I)	GTCGAC
Rhodococcus species	*Rhs* I (*Bam* HI)	GGATCC
Rhodococcus species	*Rhp* I (*Sal* I)	GTCGAC
	Rhp II	?
Rhodococcus species	*Rhe* I (*Sal* I)	GTCGAC
Rhodospirillum rubrum	*Rrb* I	?
Rhodopseudomonas sphaeroides	*Rsp* I (*Pvu* I)	CGATCG
Rhodopseudomonas sphaeroides	*Rsh* I (*Pvu* I)	CGAT↓CG
Rhodopseudomonas sphaeroides	*Rsa* I	GT↓AC
Rhodopseudomonas sphaeroides	*Rsr* I (*Eco* RI)	GAATTC
Salmonella infantis	*Sin* I (*Ava* II)	GG($\substack{A\\T}$)CC
Serratia marcescens S_b	*Sma* I	CCC↓GGG
Serratia species SAI	*Ssp* I	?
Sphaerotilus natans C	*Sna* I	GTATAC
Spiroplasma citri ASP2	*Sci* NI (*Hha* I)	G↓CGC

Table A2.1 *continued*

Microorganism	Enzyme	Sequence
Staphylococcus aureus 3A	*Sau* 3A (*Mbo* I)	↓GATC
Staphylococcus aureus PS96	*Sau* 96I (*Asu* I)	G↓GNCC
Staphylococcus saprophyticus	*Ssa* I	?
Streptococcus cremoris F	*Scr* FI	CCNGG
Streptococcus durans	*Sdu* I	G C G(A)GC(A)C T T
Streptococcus dysgalactiae	*Sdy* I (*Asu* I)	GGNCC
Streptococcus faecalis var. *zymogenes*	*Sfa* I (*Hae* III)	GG↓CC
Streptococcus faecalis GU	*Sfa* GU I (*Hpa* II)	CCGG
Streptococcus faecalis ND547	*Sfa* NI	GCATC (5/9)
Streptomyces achromogenes	*Sac* I	GAGCT↓C
	Sac II	CCGC↓GG
	Sac III	?
Streptomyces albus	*Sal* PI (*Pst* I)	CTGCA↓G
Streptomyces albus subspecies *pathocidicus*	*Spa* I (*Xho* I)	CTCGAG
Streptomyces albus G	*Sal* I	G↓TCGAC
	Sal II	?
Streptomyces aureofaciens IKA 18/4	*Sau* I	CC↓TNAGG
Streptomyces bobili	*Sbo* I (*Sac* II)	CCGCGG
Streptomyces caespitosus	*Sca* I	AGTACT
Streptomyces cupidosporus	*Scu* I (*Xho* I)	CTCGAG
Streptomyces exfoliatus	*Sex* I (*Xho* I)	CTCGAG
	Sex II	?
Streptomyces fradiae	*Sfr* I (*Sac* II)	CCGCGG
Streptomyces ganmycicus	*Sga* I (*Xho* I)	CTCGAG
Streptomyces goshikiensis	*Sgo* I (*Xho* I)	CTCGAG
Streptomyces griseus	*Sgr* I	?
Streptomyces hygroscopicus	*Shy* TI	?
Streptomyces hygroscopicus	*Shy* I (*Sac* II)	CCGCGG
Streptomyces lavendulae	*Sla* I (*Xho* I)	C↓TCGAG
Streptomyces luteoreticuli	*Slu* I (*Xho* I)	CTCGAG
Streptomyces oderifer	*Sod* I	?
	Sod II	?
Streptomyces phaeochromogenes	*Sph* I	GCATG↓C
Streptomyces stanford	*Sst* I (*Sac* I)	GAGCT↓C
	Sst II (*Sac* II)	CCGC↓GG
	Sst III (*Sac* III)	?
	Sst IV (*Bcl* I)	TGATCA
Streptomyces tubercidicus	*Stu* I	AGG↓CCT
Streptoverticillium flavopersicum	*Sfl* I (*Pst* I)	CTGCA↓G
Thermoplasma acidophilum	*Tha* I (*Fnu* DII)	CG↓CG
Thermopolyspora glauca	*Tgl* I (*Sac* II)	CCGCGG
Thermus aquaticus YTI	*Taq* I	T↓CG$\overset{*}{\text{A}}$
	Taq II	?
Thermus aquaticus	*Taq* XI (*Eco* RII)	C$\overset{*}{\text{C}}$↓AGG
Thermus flavus AT62	*Tfl* I (*Taq* I)	TCGA

Table A2.1 *continued*

Microorganism	Enzyme	Sequence
Thermus thermophilus HB8	*Tth* HB8 I (*Taq* I)	TCG$\overset{*}{\text{A}}$
Thermus thermophilus strain 23	*Ttr* I (*Tth* 111 I)	GACNNNGTC
Thermus thermophilus strain 110	*Tte* I (*Tth* 111 I)	GACNNNGTC
Thermus thermophilus strain 111	*Tth* 111 I	GACN↓NNGTC
	Tth 111 II	CAAPuCA
	Tth 111 III	?
Tolypothrix tenuis	*Ttn* I (*Hae* III)	GGCC
Vibrio narveyi	*Vha* I (*Hae* III)	GGCC
Xanthomonas amaranthicola	*Xam* I (*Sal* I)	GTCGAC
Xanthomonas badrii	*Xba* I	T↓CTAGA
Xanthomonas holcicola	*Xho* I	C↓TCGAG
	Xho II	Pu↓GATCPy
Xanthomonas malvacearum	*Xma* I (*Sma* I)	C↓CCGGG
	Xma II (*Pst* I)	CTGCAG
	Xma III	C↓GGCCG
Xanthomonas manihotis 7AS1	*Xmn* I	GAANN↓NNTTC
Xanthomonas nigromaculans	*Xni* I (*Pvu* I)	CGATCG
Xanthomonas oryzae	*Xor* I (*Pst* I)	CTGCAG
	Xor II (*Pvu* I)	CGAT↓CG
Xanthomonas papavericola	*Xpa* I (*Xho* I)	C↓TCGAG

Appendix 3 Useful Information Derived from the Nucleotide Sequence of Plasmid pBR322

The original nucleotide sequence of pBR322 published by Sutcliffe (1979) has been revised by Peden (1983) and Backman and Boyer (1983) by the inclusion of an additional CG base pair at position 526. The entire genome is 4362 base pairs long and the base pairs have been numbered, arbitrarily, in a clockwise direction from the centre of the unique *Eco* RI site. Thus position zero is between the adjacent A and T of the *Eco* RI recognition sequence (G AATTC). Each cleavage site is assigned a coordinate which represents the base immediately 5′ to the nick made by the particular enzyme in the clockwise 5′ to 3′ strand. Thus the asymmetric cut of *Eco* RI is numbered 4361 (−2). Table A3.1 lists those enzymes which do not cut pBR322 DNA. The cleavage coordinates for a number of enzymes which cleave pBR322 DNA are shown in Table A3.2 and the sizes of the fragments produced by some of them are shown in Table A3.3. Since the exact length of every fragment is known they can serve as very valuable markers for sizing DNA fragments by agarose gel electrophoresis (Sutcliffe 1978). Detailed restriction maps for pBR322 and its derivative pAT153 are given in Fig. A3.1.

Table A3.1 Restriction endonucleases which do not cleave pBR322 DNA.

Afl II	*Apa* I	*Asu* II	*Ava* III
Avr II	*Bcl* I	*Bgl* II	*Bss* HII
Bst EII	*Bst* XI	*Hpa* I	*Kpn* I
Mlu I	*Mst* II	*Nco* I	*Sac* I
Sac II	*Sma* I	*Stu* I	*Xba* I
Xho I			

The Ap^{R} gene of pBR322 codes for a β-lactamase which is 263 amino acids long. This β-lactamase is synthesized as a pre-protein with a 23 amino acid signal sequence at its amino-terminal end (see p. 132). The direction of translation of this 186

Table A3.2 The cleavage coordinates for a selection of enzymes which cut pBR322 DNA. The table gives the coordinate of the base which corresponds to the 5′ nucleotide of each recognition sequence.

Enzyme	No. of sites	Coordinates of sites				
Aat II	1	4286				
Afl III	1	2475				
Ava I	1	1425				
Bal I	1	1444				
Bam HI	1	375				
Cla I	1	23				
Eco RI	1	4361				
Eco RV	1	185				
*Hin*d III	1	29				
Nde I	1	2297				
Nru I	1	972				
Pst I	1	3609				
Pvu I	1	3735				
Pvu II	1	2066				
Sal I	1	651				
Sca I	1	3846				
Sna I	1	2246				
Sph I	1	562				
Tth 111 I	1	2219				
Xma III	1	939				
Acc I	2	651	2246			
Ban II	2	471	485			
Hgi EII	2	2295	3056			
*Hin*c II	2	651	3907			
Xmn I	2	2031	3963			
Aha III	3	3232	3251	3943		
Bgl I	3	929	1163	3482		
Rsa I	3	164	2282	3847		
Mst I	4	260	1356	1454	3588	
Nae I	4	401	769	929	1283	
Nar I	4	413	434	548	1205	
Nsp CI	4	562	1816	2110	2475	
Gdi II	5	295	399	531	939	3756
Tth 111 II	5	7	1922	3049	3082	3088
Aha II	6	413	434	548	1205	3904
		4286				
Bst NI	6	130	1059	1442	2502	2623
		2636				
Cfr I	6	295	399	531	939	1444
		3756				
Nsp BII	6	1139	2066	2185	2815	3060
		4001				

Table A3.2 *continued*

Enzyme	No. of sites	Coordinates of sites				
Hae I	7	918	990	1047	1444	2488
		2499	2951			
Taq I	7	24	339	652	1127	1268
		2575	4019			
Ava II	8	799	887	1136	1439	1481
		1760	3506	3728		
Dde I	8	1581	1743	2285	2750	3159
		3325	3865	4291		
Hgi AI	8	276	587	1174	1465	2291
		2789	3950	4035		
Xho II	8	375	1667	3116	3127	3213
		3225	3993	4010		
Ban I	9	76	119	413	434	548
		766	1205	1289	3316	
Bsp 1286	10	276	471	485	587	1174
		1465	2291	2789	3950	4035
Hinf I	10	632	852	1006	1304	1525
		2031	2375	2450	2846	3363
Nci I	10	170	534	1258	1484	1812
		2120	2155	2854	3550	3901
Hae II	11	232	413	434	494	548
		775	1205	1644	1727	2349
		2719				
Hga I	11	390	649	944	976	1240
		1390	2004	2181	2577	3155
		3905				
Mbo II	11	464	738	1009	1601	2354
		3125	3216	3971	4049	4158
		4354				
Bin I	12	375	376	1097	1667	3042
		3116	3128	3213	3226	3690
		3993	4011			
Fok I	12	112	133	987	1032	1681
		1770	1848	2009	2150	3348
		3529	3816			
Hph I	12	126	408	453	1307	1528
		2085	2094	3219	3446	3842
		4068	4083			
Sau 96I	15	172	524	799	887	1136
		1260	1439	1481	1760	1949
		3410	3489	3506	3728	4344

Table A3.2 *continued*

Enzyme	No. of sites	Coordinates of sites				
Alu I	16	15	30	686	1089	1999
		2056	2067	2116	2135	2416
		2642	2778	3035	3556	3656
		3719				
Scr FI	16	130	170	534	1059	1258
		1422	1484	1812	2120	2155
		2502	2623	2636	2854	3550
		3901				
Bbv I	21	226	615	773	1406	1430
		1559	1562	1685	2065	2068
		2114	2211	2380	2398	2817
		2882	2885	3091	3419	3608
		3785				
Hae III	22	173	296	400	524	532
		596	830	919	940	991
		1048	1261	1445	1949	2489
		2500	2518	2952	3410	3490
		3757	4344			
Mbo I	22	349	376	467	826	1098
		1129	1144	1461	1668	3042
		3117	3128	3136	3214	3226
		3331	3672	3690	3736	3994
		4011	4047			
Sfa NI	22	134	204	247	393	405
		658	1026	1033	1421	1673
		1682	1769	1847	1910	2151
		2267	2322	2343	2563	3615
		3825	4055			
Fnu DII	23	346	702	817	946	973
		978	1039	1105	1234	1244
		1389	1415	1537	1634	2006
		2075	2077	2180	2521	3102
		3432	3925	4257		
Hpa II	26	161	170	387	402	411
		534	694	770	930	1020
		1258	1284	1485	1665	1812
		2121	2155	2682	2829	2855
		3045	3449	3483	3550	3660
		3902				
Mnl I	26	115	175	379	598	797
		865	981	1167	1228	1266
		1293	1479	1793	1851	1907
		2072	2102	2364	2590	2647
		2914	3314	3395	3525	3731
		4342				

Table A3.2 *continued*

Enzyme	No. of sites	Coordinates of sites				
Hha I	31	101	233	261	414	435
		495	549	701	776	816
		947	1206	1357	1419	1455
		1645	1728	2076	2179	2209
		2350	2383	2653	2720	2820
		2994	3103	3496	3589	3926
		4258				
*Fnu*4 HI	42	226	297	300	578	581
		615	722	773	938	1023
		1106	1163	1208	1287	1406
		1409	1416	1430	1559	1562
		1685	1766	2065	2068	2114
		2211	2264	2380	2398	2401
		2519	2674	2817	2882	2885
		3091	3419	3608	3758	3785
		3880	4109			

amino acid pre-protein is counter-clockwise and starts at coordinate 4146 and terminates at coordinate 3297. The origin of this β-lactamase gene was Tn3 present originally on R7268. During the construction of pBR322 one of the terminal inverted repeats of Tn3 was lost. The remaining one is located in the region 3147 to 3297. The β-lactamase promoter has been characterized by Russell and Bennett (1981). The transcription initiation site is located 35 base pairs from the translation initiation codon. The mRNA produced *in vitro* has a 5′ pppGpA terminus. RNA polymerase bound at this start site protects a region from about −50 to +20 from DNase I cleavage.

By sequencing mRNA complementary to pBR322 DNA a second promoter for the β-lactamase gene has been identified (Brosius *et al.* 1982). Transcription from this second promoter is initiated 244 or 245 base pairs upstream close to the Tc^R gene. Although this second promoter can function *in vivo* this is an unnatural role created by the ligation of two different DNA fragments via an *Eco* RI site in the construction of pBR322 (see Fig. 3.4).

The Tc^R gene of pBR322 encodes a polypeptide 396 amino acids long. The direction of translation is clockwise, starts at coordinate 86 and terminates at nucleotide 1273. The purine initiation nucleotide of the promoter for the Tc^R gene is at coordinate 44. By means of electron microscopic analysis of in-vitro transcriptional complexes of pBR322, Stüber and Bujard

Table A3.3 The sizes, in base pairs, of the restriction fragments of pBR322. These sizes do not include any extension (3′ or 5′) which may be left by a particular enzyme. Such differences will affect gel mobility slightly, particularly with short fragments, and this should be taken into account when they are used as markers.

Eco RII	*Hae* III	*Hpa* II	*Alu* I	*Hinf* I	*Taq* I	*Tha* I	*Hha* I	*Hae* II	*Mbo* I	*Ava* II
1857	587	622	910	1631	1444	581	393	1875	1374	1743
1060	540	527	659	517	1307	493	347	622	665	1433
928	504	404	655	506	475	452	337	433	358	303
383	458	309	521	396	368	372	332	430	341	279
121	434	242	403	344	315	355	270	370	317	249
13	267	238	281	298	312	341	259	227	272	225
	234	217	257	221	141	332	206	181	258	88
	213	201	226	220		330	190	83	207	42
	192	190	136	154		145	174	60	105	
	184	180	100	75		129	153	53	91	
	124	160	63			129	152	21	78	
	123	160	57			122	151		75	
	104	147	49			115	141		46	
	89	147	19			104	132		36	
	80	124	15			97	131		31	
	64	110	11			68	109		27	
	57	90				66	104		18	
	51	76				61	100		17	
	21	67				27	93		15	
	18	34				26	83		12	
	11	34				10	75		11	
	7	26				5	67		8	
		26				2	62			
		15					60			
		9					53			
		9					40			
							36			
							33			
							30			
							28			
							21			

(1981) detected two partially overlapping promoters at the beginning of the tetracycline resistance region which initiate transcription crosswise in opposite directions. Clockwise transcription is from the Tc^R promoter and counter-clockwise transcription is from the second Ap^R promoter referred to above.

The origin of pBR322 replication is at coordinate 2534, i.e. this is the nucleotide at which the switch from RNA primer to DNA occurs. DNA replication proceeds anti-clockwise from this point. The DNA sequences around the origins of pBR322 and Col E1 have been compared (Sutcliffe 1979). There are 10 single base differences between pBR322 and Col E1 in the first

50 base pairs anti-clockwise from the origin of replication, whereas there is only 1 base difference in the first 190 base pairs in the clockwise direction.

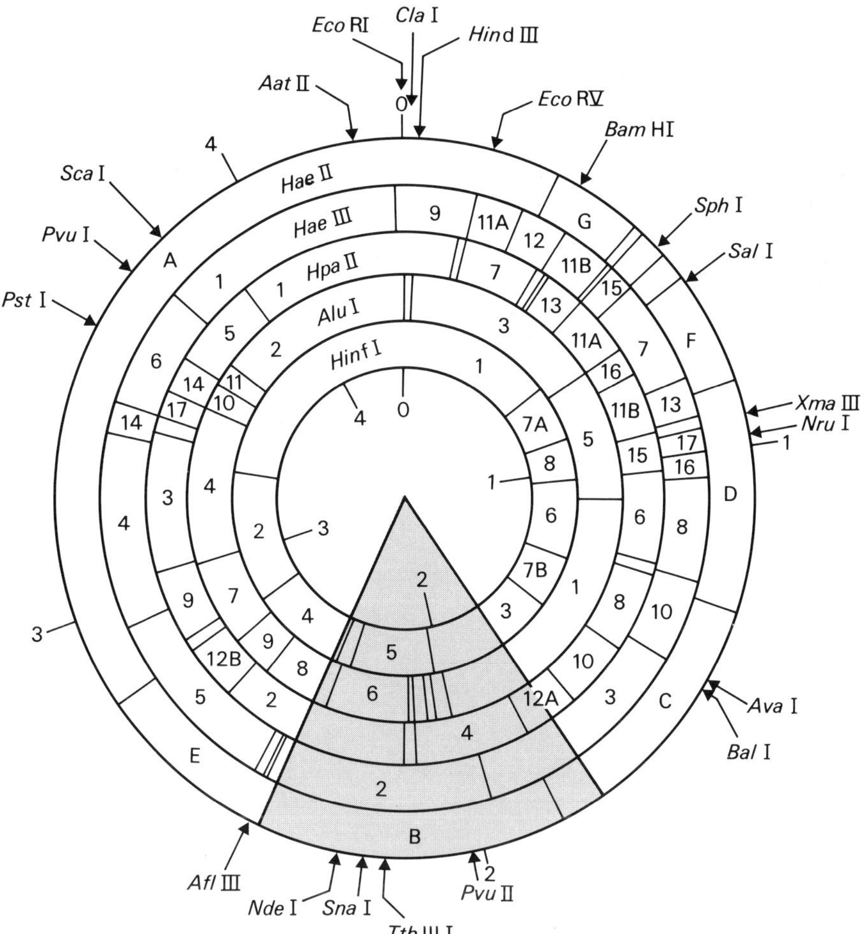

Fig. A3.1 The pBR322 restriction cleavage maps for the enzymes are indicated on the concentric rings. The red tinted area represents the portion of pBR322 missing from pAT153 (i.e. the *Hae* II B fragment plus the minor adjacent fragment as indicated in the legend to Fig. 3.16). The fragments are numbered by size and these sizes are listed in Table A3.3 In the case of the *Hae* II restriction map the fragments are assigned letters in alphabetical order on the basis of sizes to correspond with the map shown in Fig. 3.16. The precise coordinates of the restriction cuts are listed in Table A3.2. Around the perimeter of the circular map appear the sites for enzymes which cut pBR322 and pAT153 once. The numbers inside and outside the circles are the distances in kilobases from the unique *Eco* RI site. (Adapted from Sutcliffe 1978.) (Figure A3.1 continues overleaf.)

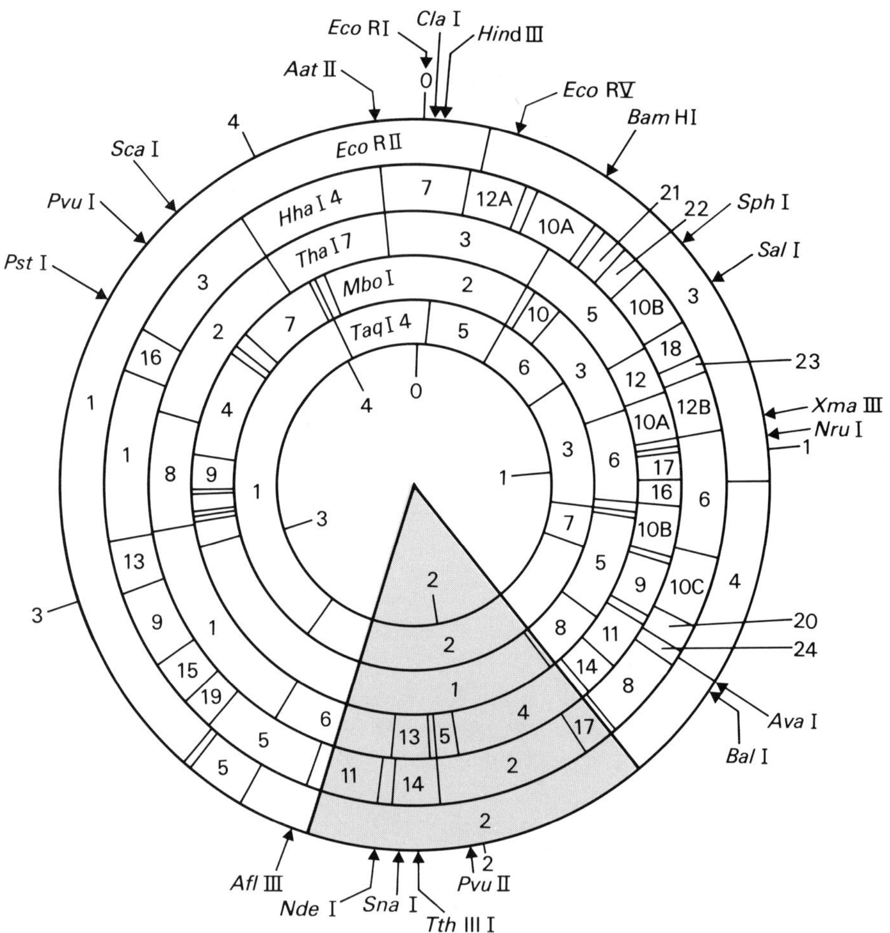

Fig. A3.1 (*contd*)

Queen and Rosenberg (1981) have demonstrated *in vitro* the existence of a promoter signal that is strictly dependent on cAMP and its receptor protein CRP. Transcription initiates at nucleotide 2271 and proceeds counter-clockwise. DNase protection studies have shown that CRP binds to the −35 region of the promoter *in vitro*. The function of this promoter is not known. However, the DNA region which contains it is derived from pMBI which in turn is related to Col E1. Examination of the sequence of Col E1 shows that a similar promoter element is present suggesting that it has a useful function.

Appendix 4 Restriction Map of Bacteriophage Lambda

DNA isolated from phage lambda is linear and, except for the extreme left and right ends, double stranded. At each end the 5′ strand overhangs the 3′ strand by 12 bases giving rise to a short single-stranded terminus. The nucleotide sequences of the single-stranded ends are complementary. They anneal *in vitro* and *in vivo* leading to the formation of circular, completely double-stranded molecules. DNA from one isolate of phage lambda has been sequenced completely (Sanger *et al.* 1982) and the circular form of the molecule is 48502 base pairs in length. By convention, numbering of the nucleotide sequence begins with the first base of the left cohesive end: GGGCGGCGACCT and proceeds along the L strand in the direction late genes to early genes. Numbering of the nucleotide sequence terminates at 48501, the 3′ end of the L strand and does not include the right cohesive end which extends 12 bases further on the complementary strand.

Restriction endonuclease digests of lambda DNA are frequently used as markers for sizing DNA fragments by agarose gel electrophoresis. Figure A4.1 shows restriction maps for the lambda DNA which has been sequenced and Table A4.1 gives the coordinates of the restriction sites. Note that the strains of lambda in current use are *not* descended from a single phage isolate. Rather, these phage strains have been derived as independent isolates from a series of *E. coli* K12 strains and consequently may have different restriction maps. Restriction maps for many of the cloning vectors in current use are given by Williams and Blattner (1980). Figure A4.1 also shows a simplified genetic map of phage lambda. If a more detailed map is required the reader is advised to consult the paper of Szybalski and Szybalski (1979).

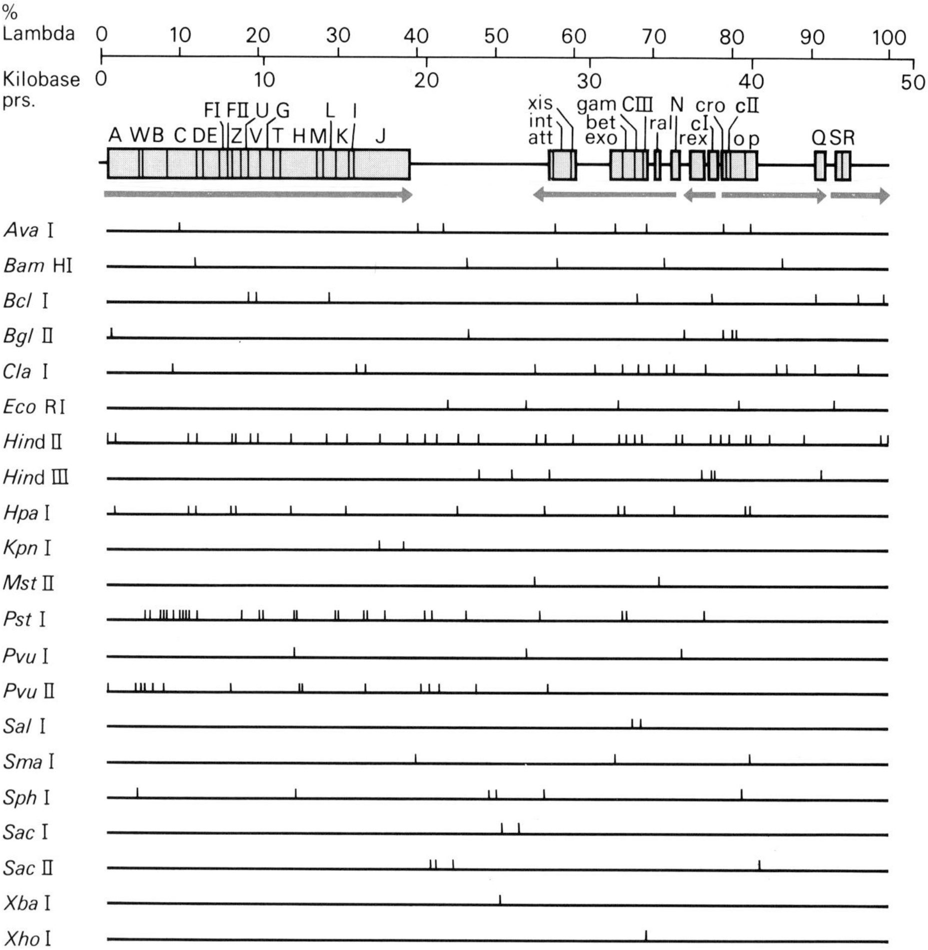

% Lambda 0 10 20 30 40 50 60 70 80 90 100
Kilobase prs. 0 10 20 30 40 50

FI FII U G L I
A WB C DE Z V T H M K J
xis int att
gam bet exo
CIII ral N cro cI rex cII o p
QSR

Ava I
Bam HI
Bcl I
Bgl II
Cla I
Eco RI
Hind II
Hind III
Hpa I
Kpn I
Mst II
Pst I
Pvu I
Pvu II
Sal I
Sma I
Sph I
Sac I
Sac II
Xba I
Xho I

Fig. A4.1 Restriction map and genetic map of bacteriophage λCI_{857}Sam7. (Reproduced courtesy of New England Biolabs.) The red arrows below the genetic map show the different transcriptional units.

Table A4.1 Locations of some of the restriction sites in DNA from λCI_{857}Sam7. (Reproduced courtesy of New England Biolabs.) The entries give the coordinate of the base which corresponds to the 5′ nucleotide of each recognition sequence.

Enzyme	No. of sites	Coordinates of sites				
Apa I	1	10086				
Nae I	1	20040				
Nar I	1	45679				
Xba I	1	24508				
Xho I	1	33498				
Avr II	2	24322	24396			
Kpn I	2	17053	18556			
Mst II	2	26717	34318			
Sac I	2	24772	25877			
Sal I	2	32745	33244			
*Tth*111 I	2	11202	36120			
Xma III	2	19944	36654			
Afl II	3	6540	12618	42630		
Pvu I	3	11933	26254	35787		
Sma I	3	19397	31617	39888		
Sna I	3	15260	18834	19473		
Nco I	4	19329	23901	27868	44248	
Sac II	4	20320	20530	21606	40386	
Bam HI	5	5505	22346	27972	34499	41732
Eco RI	5	21226	26104	31747	39168	44972
Nru I	5	4590	28050	31703	32407	41808
Sca I	5	16421	18684	25685	27263	32802
Bgl II	6	415	22425	35711	38103	38754
		38814				
Bss HII	6	3522	4126	5627	14815	16649
		28008				
*Hin*d III	6	23130	25157	27479	36895	37459
		44141				
Sph I	6	2212	12002	23942	24371	27374
		39418				
Stu I	6	12434	31478	32997	39992	40596
		40614				
Asu II	7	18048	25884	27980	29150	30396
		34331	42637			
Ban II	7	581	10086	19763	21570	24772
		25877	39453			
Mlu I	7	458	5548	15372	17791	19996
		20952	22220			
Nde I	7	27630	29883	33679	36112	36668
		38357	40131			

Table A4.1 *continued*

Enzyme	No. of sites	Coordinates of sites				
Ava I	8	4720	19397	20999	27887	31617
		33498	38214	39888		
Bcl I	8	8844	9361	13820	32729	37352
		43682	46366	47942		
Acc I	9	2190	15260	18834	19473	31301
		32745	33244	40201	42921	
Aat II	10	5105	9394	11243	14974	29036
		40806	41113	42247	45563	45592
Aha III	13	90	8460	16294	23110	23284
		25436	26132	26665	32703	36302
		36530	38833	47429		
Bst EII	13	5687	7058	8322	9024	13348
		13572	13689	16012	17941	25183
		30005	36374	40049		
Bst XI	13	2855	6706	8413	8850	10915
		13263	14338	18029	19741	21622
		34596	38292	46434		
Ava III	14	10325	27206	27372	28432	30342
		30989	32967	33682	34208	35868
		36665	36671	37769	38307	
Hgi EII	14	1785	2250	5903	6555	12513
		13954	15877	17433	20244	26435
		35595	35639	37999	42048	
Hpa I	14	732	5267	5708	7948	8199
		11583	14991	21902	27316	31807
		32217	35259	39606	39834	
Cla I	15	4198	15583	16120	26616	30289
		31990	32963	33584	34696	35050
		36965	41363	42020	43824	46438
Mst I	15	463	2503	4270	5155	6979
		11563	11690	13355	16046	21805
		21826	27949	32683	34821	42380
Pvu II	15	209	1917	2385	2526	3058
		3637	7831	12099	12162	16078
		19716	20059	20695	22991	27412
Bal I	18	1326	2206	3260	4193	6496
		6877	7584	7978	8056	8859
		10609	10777	13934	14903	21260
		26623	28618	36040		
Afl III	20	458	628	5548	11281	15372
		17791	18284	19996	20952	22220
		24133	24168	26528	32764	39395
		42086	42363	43762	44501	46982

Table A4.1 *continued*

Enzyme	No. of sites	Coordinates of sites				
Eco RV	21	650	2084	6681	8084	8822
		13435	14023	17767	18385	21269
		22948	26821	28198	28211	33587
		39352	41273	41541	41576	42231
		45826				
Gdi II	21	2739	5601	6008	8366	10588
		13481	14575	16416	18547	19284
		19332	19944	20239	20323	20928
		20988	22025	35465	36654	39458
		45214				
Xho II	21	415	1606	2531	5505	6422
		22346	22425	24511	27027	27972
		29593	30426	34499	35711	38103
		38664	38754	38814	39576	41732
		47773				
Xmn I	24	33	1151	2319	8490	10111
		13102	16909	22852	22871	23808
		23828	24228	24578	25485	27252
		29015	29993	31085	33811	34185
		42477	44727	45741	47564	
Ban I	25	1180	1365	2331	5407	5665
		5671	5900	8036	8043	8441
		8764	8988	10221	13038	13642
		14623	15199	15237	16236	17053
		18556	21545	39907	42797	45679
Hgi AI	28	5619	6002	9485	10295	11950
		13289	13492	14474	15211	16516
		21612	21798	21852	24772	25877
		26469	27173	33467	35583	37933
		40216	40489	42371	42512	44177
		44846	46698	47660		
Pst I	28	2556	2820	3625	3640	3856
		4370	4709	4909	5120	5214
		5682	8520	9613	9777	11763
		11835	14294	14381	16081	16231
		17390	19833	20281	22421	26928
		32005	32252	37001		
Bgl I	29	404	2660	3798	4360	4451
		4577	5246	5432	6053	6104
		7550	8049	11058	12708	12717
		12832	13198	14401	14890	15157
		17638	18085	19334	20124	20250
		20460	21233	30882	32323	

Table A4.1 *continued*

Enzyme	No. of sites	Coordinates of sites				
Nsp CI	32	628	2212	6478	8375	12002
		17274	18758	21802	23425	23942
		24371	25099	25659	25868	27374
		29170	30738	31542	32493	33717
		34649	35082	38021	38996	39395
		39418	39646	40069	42346	43068
		46123	47840			
Ava II	35	1612	1922	2816	3801	4314
		4622	6042	6440	8995	11000
		11045	12996	13147	13737	13952
		13984	14329	15613	16587	16610
		16683	19289	19356	19867	22001
		22243	28798	32474	32562	39004
		39437	39479	47605	48202	48474
Hinc II	35	197	732	5267	5708	7948
		8199	9054	9624	11583	13783
		14991	17074	18754	19839	20567
		21902	23145	26742	27316	28926
		31807	32217	32745	33244	35259
		35613	37431	37987	38546	39606
		39834	40940	43181	47936	48296
Bsp 1286	38	581	5619	5664	6002	9485
		10086	10295	11414	11950	13039
		13289	13492	14474	14897	15211
		16516	19763	21570	21612	21798
		21852	24772	25877	26469	27173
		32330	33467	35583	37933	39453
		40216	40489	42371	42512	44177
		44846	46698	47660		
Cfr I	39	1326	2206	2739	3260	4193
		5601	6008	6496	6877	7584
		7978	8056	8366	8859	10588
		10609	10777	13481	13934	14575
		14903	16416	18547	19284	19332
		19944	20239	20323	20928	20988
		21260	22025	26623	28618	35465
		36040	36654	39458	45214	
Aha II	40	1475	1496	2303	4947	4985
		5105	6915	8096	8263	9089
		9394	9452	9861	10080	10621
		11243	11768	12929	13318	14799
		14974	16056	17616	17670	28467
		29036	30472	30727	31765	31936
		35072	40806	41113	42247	44221
		44330	44912	45563	45592	45679

Appendix 5 Restriction Endonuclease Cleavage Maps for Some Natural Plasmids Used in the Construction of Cloning Vectors

Restriction maps for the broad host-range plasmids RSF 1010, pSa and RP4 (RP1, RK2) are given on pages 157, 160 and 159 respectively. Figure A5.1 shows the restriction map for pSC101.

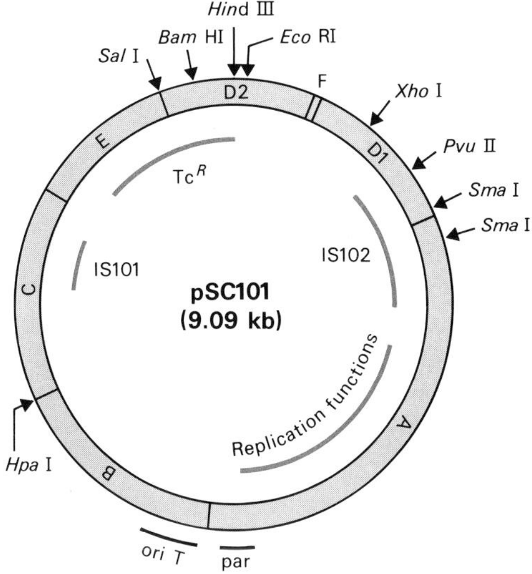

Fig. A5.1 Restriction endonuclease cleavage map for plasmid pSC101. The enzyme *Hinc* II produces 7 fragments which are labelled alphabetically. IS101 and IS102 are two insertion sequences, *ori*T is a sequence required for mobilization of pSC101 by conjugative plasmids, and *par* is a region of DNA required for stable partitioning of pSC101 (see p. 149).

Most of the cloning vectors for use with *B. subtilis* are derived from one or more of the following *Staph. aureus* plasmids: pC194 encoding resistance to chloramphenicol, pE194 specifying erythromycin-induced resistance to marcrolide, lincosamide and streptogramin type B (MLS) antibiotics, and pUB110 encoding resistance to kanamycin. The restriction map of pUB110 is shown in Fig. A5.2 and more details can be found in Jalanko *et al.* (1981). The restriction maps

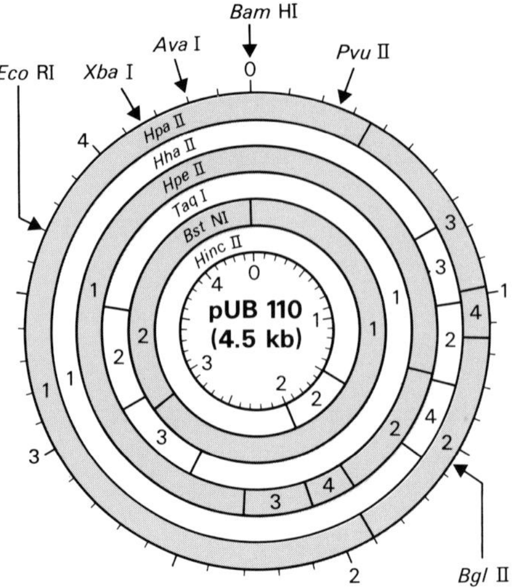

Fig. A5.2 Restriction endonuclease cleavage map for plasmid pUB110.

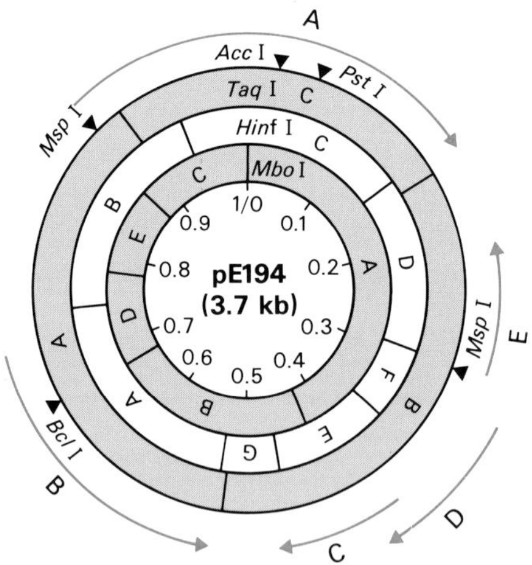

Fig. A5.3 Restriction endonuclease cleavage map for pE194. The bold arrows indicate the open-reading frames, and their respective 5′ to 3′ orientations, as determined from the nucleotide sequence of the plasmid. Frame B specifies the inducible MLS resistance.

for pE194 and pC194 are shown in Figs A5.3 and A5.4 and their complete nucleotide sequences can be found in Horinouchi and Weisblum (1982a,b).

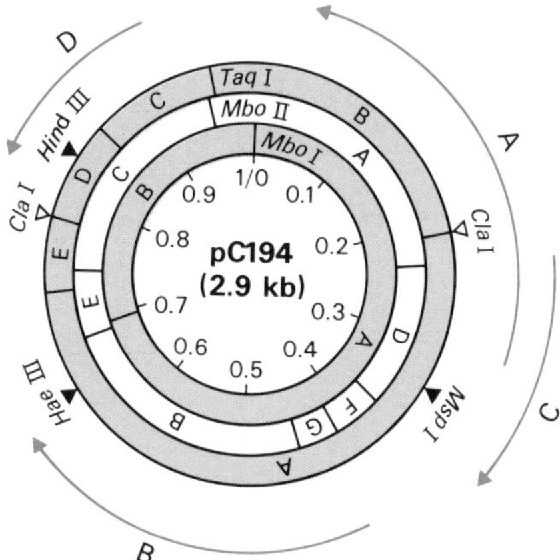

Fig. A5.4 Restriction endonuclease cleavage map plasmid pC194. The bold arrows indicate the open-reading frames and their respective 5′ to 3′ orientations as determined from the nucleotide sequence of the plasmid. Frame B specifies chloramphenicol resistance.

Appendix 6 Properties of the M13-Derived Vectors

The DNA from bacteriophage M13 is a single-stranded circle 6407 bases in length. Double-stranded forms of the molecule arise as intermediates during DNA replication. The DNA strand in the virion is termed the plus strand. By convention, numbering of the nucleotide sequence begins at the unique *Hpa* I site: the first A in the sequence ...GTTAAC.. is designated nucleotide number 1. Nucleotide numbering proceeds around the plus strand in the 3′ to 5′ direction. Figure A6.1 shows the restriction endonuclease map for wild-type phage M13 double-stranded DNA. The complete nucleotide sequence has been presented by van Wezenbeck *et al.* (1980).

Numerous derivatives of M13 have been constructed for use as cloning vehicles, particularly as a prelude to sequencing of DNA by the dideoxy procedure. These cloning vectors all contain a portion of the *E. coli lac* operon inserted in the intergenic region between genes IV and II (see p. 92) and differ from M13 in having a number of single base changes which destroy particular endonuclease cleavage sites. Within the remnant of the *lac*Z gene oligonucleotides comprising multiple cloning sites have been inserted and these are shown in Fig. A6.2. The lengths of the various derivatives are: M13mp7, 7237 nucleotides; M13mp8, 7229 nucleotides; M13mp9, 7599 nucleotides; M13mp10 and M13mp11, 7245 nucleotides. The longer length of M13mp9 is due to the insertion of a *Hae* II DNA fragment from pBR322 (position 2352–2722) at position 6002 in the *lac*i gene.

In addition to the M13 cloning and sequencing vectors, Vieira and Messing (1982, 1983) have developed a number of plasmid vectors, the pUC series of vectors, which also can be used for shotgun cloning and sequencing. These plasmids contain the same cloning sites as the M13mp7–M13mp13 series of vectors. Details of these vectors are given by Messing and Vieira (1982), Vieira and Messing (1982) and Norrander *et al.* (1983).

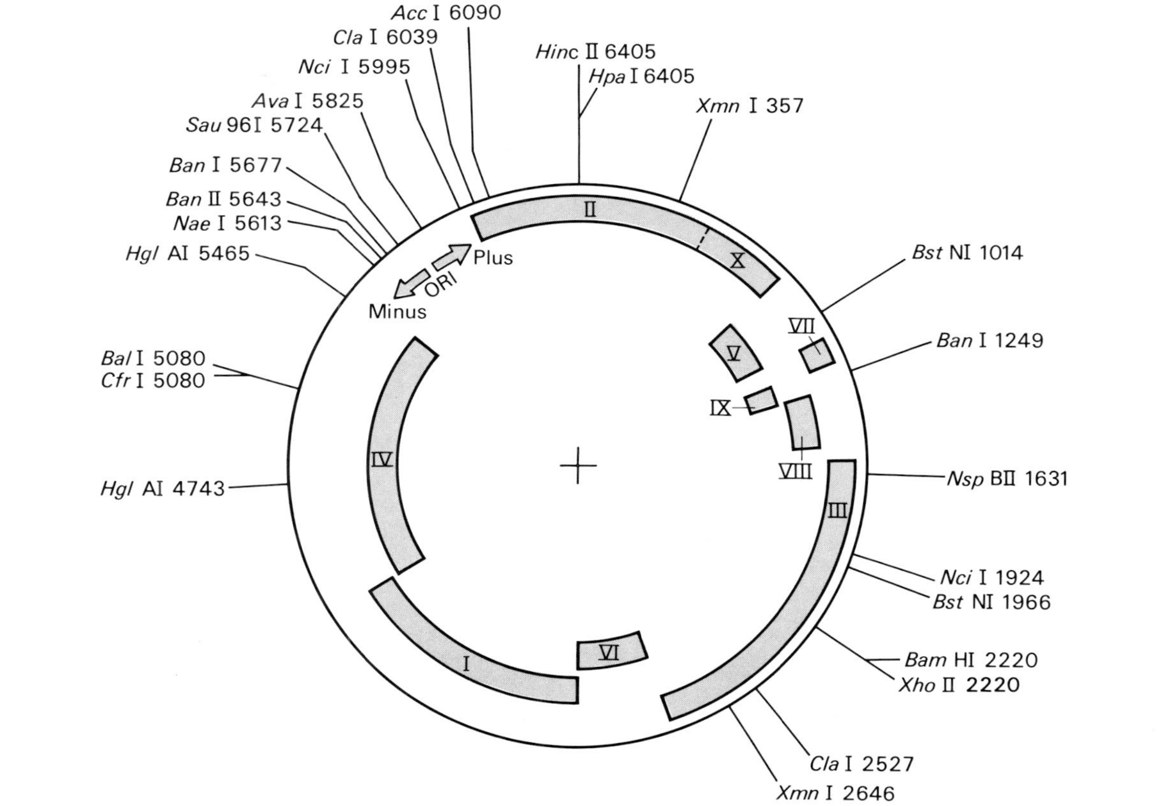

Fig. A6.1 Restriction endonuclease cleavage map of wild-type phage M13 replication form DNA. Genes I to IX are shown in the red-tinted boxes within the circle. The origin of DNA replication is located in the intergenic region between genes IV and II.

```
               1   2   3   4   1   2   3   4   5   6   7   8   9   10  11  12  13  14  15  16  17  18  5   6   7   8
M13mp19/pUC19      THR MET ILE THR pro ser leu his ala cys arg ser thr leu glu asp pro arg val pro ser ser ASN SER LEU ALA
               ATG ACC ATG ATT ACG CCA AGC TTG CAT GCC TGC AGG TCG ACT CTA GAG GAT CCC CGG GTA CCG AGC TCG AAT TCA CTG GCC
                         Hind III / Sph I / Pst I / Sal I, Acc I, Hinc II / Xba I / Bam HI / Xma I, Sma I / Kpn I / Sst I / Eco RI / Hae III

               1   2   3   4   5   6   1   2   3   4   5   6   7   8   9   10  11  12  13  14  15  16  17  18  7   8
M13mp18/pUC18      THR MET ILE THR ASN SER ser ser val pro gly asp pro leu glu ser thr cys arg his ala ser leu ala LEU ALA
               ATG ACC ATG ATT ACG AAT TCG AGC TCG GTA CCC GGG GAT CCT CTA GAG TCG ACC TGC AGG CAT GCA AGC TTG GCA CTG GCC
                         Eco RI / Sst I / Kpn I / Xma I, Sma I / Bam HI / Xba I / Sal I, Acc I, Hinc II / Pst I / Sph I / Hind III / Hae III

               1   2   3   4   1   2   3   4   5   6   7   8   9   10  11  12  13  14  15  16  5   6   7   8
M13mp11/pUC13      THR MET ILE THR pro ser leu gly cys arg ser thr leu glu asp pro arg ala ser ser ASN SER LEU ALA
               ATG ACC ATG ATT ACG CCA AGC TTG GGC TGC AGG TCG ACT CTA GAG GAT CCC CGG GCG AGC TCG AAT TCA CTG GCC
                         Hind III / Pst I / Sal I, Acc I, Hinc II / Xba I / Bam HI / Xma I, Sma I / Sst I / Eco RI / Hae III

               1   2   3   4   5   6   1   2   3   4   5   6   7   8   9   10  11  12  13  14  15  16  7   8
M13mp10/pUC12      THR MET ILE THR ASN SER ser ser pro gly asp pro leu glu ser thr cys ser pro ser leu ala LEU ALA
               ATG ACC ATG ATT ACG ATT TCG AGC TCG CCC GGG GAT CCT CTA GAG TCG ACC TGC AGC CCA AGC TTG GCA CTG GCC
                         Eco RI / Sst I / Xma I, Sma I / Bam HI / Xba I / Sal I, Acc I, Hinc II / Pst I / Hind III / Hae III

               1   2   3   4   1   2   3   4   5   6   7   8   9   10  11  5   6   7   8
M13mp9/pUC9        THR MET ILE THR pro ser leu ala ala gly arg arg ile pro gly ASN SER LEU ALA
               ATG ACC ATG ATT ACG CCA AGC TTG GCT GCA GGT CGA CGG ATC CCC GGG AAT TCA CTG GCC
                         Hind III / Pst I / Sal I, Acc I, Hinc II / Bam HI / Xma I, Sma I / Eco RI / Hae III
```

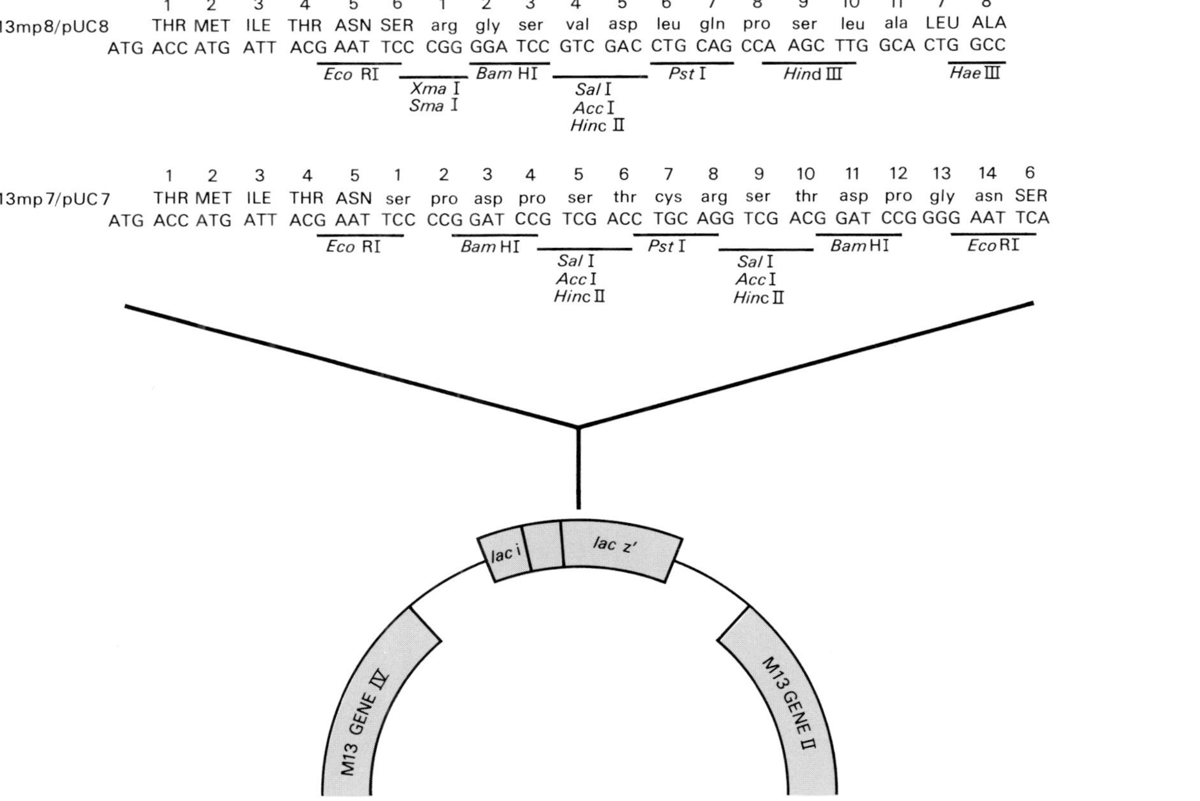

Fig. A6.2 Sequences of the multiple cloning sites within the *lac* region of the M13mp series of vectors. The unique insert within the *lac*Z gene DNA used for DNA cloning/sequencing is shown as an expanded scale with the additional codons in the *lac*Z gene labelled with lower case letters.

Appendix 7 The Genetic Code and Single Letter Amino Acid Designations

5′-OH terminal base	Middle base				3′-OH terminal base
	U	C	A	G	
U	Phe	Ser	Tyr	Cys	U
	Phe	Ser	Tyr	Cys	C
	Leu	Ser	STOP	STOP	A
	Leu	Ser	STOP	Trp	G
C	Leu	Pro	His	Arg	U
	Leu	Pro	His	Arg	C
	Leu	Pro	Gln	Arg	A
	Leu	Pro	Gln	Arg	G
A	Ile	Thr	Asn	Ser	U
	Ile	Thr	Asn	Ser	C
	Ile	Thr	Lys	Arg	A
	Met	Thr	Lys	Arg	G
G	Val	Ala	Asp	Gly	U
	Val	Ala	Asp	Gly	C
	Val	Ala	Glu	Gly	A
	Val[a]	Ala	Glu	Gly	G

[a]Codes for Met if in the initiator position.

Alanine	A	Leucine	L
Arginine	R	Lysine	K
Asparagine	N	Methionine	M
Aspartic Acid	D	Phenylalanine	F
Cysteine	C	Proline	P
Cystine	C	Serine	S
Glycine	G	Threonine	T
Glutamic Acid	E	Tryptophan	W
Glutamine	Q	Tyrosine	Y
Histidine	H	Valine	V
Isoleucine	I		

Appendix 8 Further Reading Material

There are a number of sources of more detailed information and these are listed below.

Molecular Cloning—A Laboratory Manual by T. Maniatis *et al.* and published in 1982 by Cold Spring Harbor Laboratory, New York, is an excellent source of detailed methodology.

Methods in Enzymology, published by Academic Press, New York and London. A number of volumes (68, 100 and 101) are devoted solely to practical details of recombinant DNA technology and others (e.g. volume 65) deal with the biochemistry of nucleic acids.

Genetic Engineering: Principles and Methods, edited by J. K. Setlow and A. Hollaender and published by Plenum Press, New York and London. This is a continuing series of volumes and at the time of going to press 4 volumes had been published.

Genetic Engineering, edited by R. Williamson and published by Academic Press, New York and London. This is a continuing series and at the time of going to press 4 volumes had been published.

Recombinant DNA Technical Bulletin. This is published quarterly by the National Institutes of Health (USA) and the articles cover all aspects of recombinant DNA technology. Details can be had from the Editor, Recombinant DNA Technical Bulletin, Office of Recombinant DNA Activities, Building 31, Room 4A52, National Institutes of Health, Bethesda, MD 20205, USA.

Science, the official journal of the American Association for the Advancement of Science, occasionally devotes an entire issue to recombinant DNA technology and to date there have been 3 special issues. The first of these, volume 196 number 4286, concentrated on methodology; the second, volume 209 number

4463, concentrated on the application of the techniques to the study of the arrangement, expression and regulation of genes; the third, volume 219 number 4585, is devoted to biotechnological applications. All the articles in each issue are written by acknowledged experts and are highly recommended.

Focus. This is a newsletter/information sheet which is published free of charge by Bethesda Research Laboratories Inc. (address given below). Each issue contains at least one short but informative review on a particular aspect of recombinant DNA technology.

Other literature. Much useful information including practical tips is to be found in the catalogues issued by the suppliers of restriction endonucleases, etc., who are listed below.

Bethesda Research
Laboratories Inc.
411 North Stonestree Avenue
P.O. Box 6010
Rockville
MD 20850, USA

and
P.O. Box 145
Science Park
Cambridge
CB4 4BE, England

Amersham International
Amersham
Bucks, England

New England Biolabs
32 Tozer Road
Beverly
MA 01915, USA

and
P.O. Box 22
Bishops Stortford
Herts, England

P.L. Biochemicals
1037 W. McKinley Avenue
Milwaukee
WI 53205, USA

References

Aaij C. & Borst P. (1972) The gel electrophoresis of DNA. *Biochim. biophys. Acta* **269**, 192–200.

Abelson J. (1979) RNA processing. *Annu. Rev. Biochem.* **48**, 1035–69.

Adler R. G. (1984) Biotechnology as an intellectual property. *Science* **224**, 357–63.

Al Ani R., Pfeiffer P. & Lebeurier G. (1979). The structure of cauliflower mosaic virus, II. Identity and location of the viral polypeptides. *Virology* **93**, 188–97.

Al Ani R., Pfeiffer P., Whitechurch O., Lesot A., Lebeurier G. & Hirth L. (1980) A virus specified protein produced upon infection by cauliflower mosaic virus (CaMV). *Ann. Virol.* (Inst. Pasteur) **131**, 33–53.

Alter B. (1981) Prenatal diagnosis of haemoglobinopathies: a status report. *Lancet* **2**, 151–5.

Alwine J. C., Kemp D. J., Parker B. A., Reiser J., Renart J., Stark G. R. & Wahl G. M. (1979) Detection of specific RNAs or specific fragments of DNA by fractionation in gels and transfer to diazobenzyloxymethyl paper. *Methods in Enzymology* **68**, 220–42.

Anderson S., Gait M. J., Mayol L. & Young I. G. (1980) A short primer for sequencing DNA cloned in the single-stranded phage vector M13mp2. *Nucleic Acids Res.* **8**, 1731–43.

Amann E., Brosius J. & Ptashne M. (1983) Vectors bearing a hybrid *trp–lac* promoter useful for regulated expression of cloned genes in *Escherichia coli*. *Gene* **25**, 167–78.

Axel R., Fiegelson P. & Schutz G. (1976) Analysis of the complexity and diversity of mRNA from chicken oviduct and liver. *Cell* **17**, 247–54.

Backman K. & Boyer H. W. (1983) Tetracycline resistance determined by pBR322 is mediated by one polypeptide. *Gene* **26**, 197–203.

Backmann K. & Ptashne M. (1978) Maximizing gene expression on a plasmid using recombination *in vitro*. *Cell* **13**, 65–71.

Backmann K., Ptashne M. & Gilbert W. (1976) Construction of plasmids carrying the cI gene of bacteriophage λ. *Proc. nat. Acad. Sci. U.S.A.* **73**, 4174–8.

Bagdasarian M. M., Amann E., Lurz R., Ruckert B. & Bagdasarian M. (1983) Activity of the hybrid *trp–lac* (*tac*) promoter of *Escherichia coli* in *Pseudomonas putida*. Construction of broad-host-range, controlled-expression vectors. *Gene* **26**, 273–82.

Bagdasarian M., Bagdasarian M. M., Coleman S. & Timmis K. N. (1979) New vector plasmids for gene cloning in *Pseudomonas*. In *Plasmids of Medical, Environmental and Commercial Importance*, eds Timmis

K. N. & Pühler A., pp. 411–22. Elsevier/North Holland Biomedical Press, Amsterdam.

Bagdasarian M., Lurz R., Rückert B., Franklin F. C. H., Bagdasarian M. M., Frey J. & Timmis K. N. (1981) Specific-purpose plasmid cloning vectors. II. Broad host range, high copy number, RSF1010-derived vectors, and a host-vector system for gene cloning in *Pseudomonas*. *Gene* **16**, 237–47.

Bahl C. P. & Wu R. (1978) Cloned seventeen-nucleotide-long synthetic lactose operator is biologically active. *Gene* **3**, 123–34.

Balàzs E., Guilley H., Jonard G. & Richards K. (1982) Nucleotide sequence of DNA from an altered-virulence isolate D/H of the cauliflower mosaic virus. *Gene* **19**, 239–49.

Ballance D. J., Buxton F. P. & Turner G. (1983) Transformation of *Aspergillus nidulans* by the orotidine-5′-phosphate decarboxylase gene of *Neurospora crassa*. *Biochem. Biophys. Res. Commun.* **112**, 284–9.

Band L. & Henner D. J. (1984) *Bacillus subtilis* requires a 'stringent' Shine-Dalgarno region for gene expression. *DNA* **3**, 17–21.

Banerji J., Rusconi S. & Schaffner W. (1981) Expression of a β-globin gene is enhanced by remote SV40 DNA sequences. *Cell* **27**, 299–308.

Barclay S. L. & Meller E. (1983) Efficient transformation of *Dictyostelium discoideum* amoebae. *Molec. cell. Biol.* **3**, 2117–30.

Barnes W. M. (1977) Plasmid detection and sizing in single colony lysates. *Science* **195**, 393–4.

Barnes W. M. (1980) DNA cloning with single-stranded phage vectors. In *Genetic Engineering*, vol. 2, ed. Setlow J. K. & Hollaender A., pp. 185–200. Plenum Press, New York.

Barton K. A., Binns A. N., Matzke A. J. M. & Chilton M. D. (1983) Regeneration of intact tobacco plants containing full length copies of genetically engineered T-DNA, transmission of T-DNA to RI progeny. *Cell* **32**, 1033–43.

Bates P. F. & Swift R. A. (1983) Double *cos* site vectors: simplified cosmid cloning. *Gene* **26**, 137–46.

Bayne M. L., Alexander R. F. & Benbow R. M. (1984) DNA binding protein from ovaries of the frog, *Xenopus laevis* which promotes concatenation of linear DNA. *J. mol. Biol.* **172**, 87–108.

Beck E. & Bremer E. (1980) Nucleotide sequence of the gene *ompA* encoding the outer membrane protein II of *Escherichia coli* K-12. Nucleic Acids Res. **8**, 3011–24.

Beggs J. D. (1978) Transformation of yeast by a replicating hybrid plasmid. *Nature, London* **275**, 104–9.

Beggs J. D., van den Berg J., van Ooyen A. & Weissmann C. (1980) Abnormal expression of chromosomal rabbit β-globin gene in *Saccharomyces cerevisiae*. *Nature, London* **283**, 835–40.

Bender W., Spierer P. & Hogness D. S. (1983) Chromosomal walking and jumping to isolate DNA from the *Ace* and *rosy* loci and the bithorax complex in *Drosophila melanogaster*. *J. molec. Biol.* **168**, 17–33.

Bendig M. M. & Williams J. G. (1983) Replication and expression of *Xenopus laevis* globin genes injected into fertilized *Xenopus* eggs. *Proc. nat. Acad. Sci. U.S.A.* **80**, 6197–201.

Bennetzen J. L. & Hall B. D. (1982) Codon selection in yeast. *J. biol. Chem.* **257**, 3026–31.

Benton W. D. & Davis R. W. (1977) Screening λgt recombinant clones by hybridization to single plaques *in situ*. *Science* **196**, 180–2.

Berg P. (1981) Dissections and reconstructions of genes and chromosomes. *Science* **213**, 296–303.

Bernhard K., Schrempf H. & Goebel W. (1978) Bacteriocin and antibiotic resistance plasmids in *Bacillus cereus* and *Bacillus subtilis*. *J. Bacteriol.* **133**, 897–903.

Bevan M., Barnes W. & Chilton M. D. (1983) Structure and transcription of the nopaline synthase gene region of T-DNA. *Nucleic Acids Res.* **11**, 369–85.

Bevan M. W. & Chilton M. D. (1982) T-DNA of the Agrobacterium Ti and Ri plasmids. *Ann. Rev. Genet.* **16**, 357–84.

Bevan M. W., Flavell R. B. & Chilton M. D. (1983) A chimaeric antibiotic resistance gene as a selectable marker for plant cell transformation. *Nature* **304**, 184–87.

Bibb M. J., Schottel J. L. & Cohen S. N. (1980) A DNA cloning system for inter-species gene transfer in antibiotic-producing *Streptomyces. Nature, London* **284**, 526–31.

Bickle T. A., Brack C. & Yuan R. (1978) ATP-induced conformational changes in restriction endonuclease from *Escherichia coli* K12. *Proc. nat. Acad. Sci. U.S.A.* **75**, 3099–103.

Bienz M. & Pelham H. R. B. (1982) Expression of a *Drosophila* heat-shock protein in *Xenopus* oocytes: conserved and divergent regulatory signals. *EMBO J.* **1**, 1583–88.

Bingham A. H. A., Bruton C. J. & Atkinson T. (1980) Characterization of *Bacillus stearothermophilus* plasmid pAB124 and construction of deletion variants. *J. gen. Microbiol.* **119**, 109–115.

Bingham P. M., Kidwell M. G. & Rubin G. M. (1982) The molecular basis of P-M hybrid dysgenesis: the role of the P element, a P-strain-specific transposon family. *Cell* **29**, 995–1004.

Binns A., Wood H. N. & Brown A. C. (1981) Suppression of the tumorous state in crown gall teratomas of tobacco: a clonal analysis. *Differentiation* **19**, 97–102.

Bird A. P. & Southern E. M. (1978) Use of restriction enzymes to study eukaryotic DNA methylation: I. The methylation pattern in ribosomal DNA from *Xenopus laevis*. *J. molec. Biol.* **118**, 27–47.

Birnboim H. C. & Doly J. (1979) A rapid alkaline extraction procedure for screening recombinant plasmid DNA. *Nucleic Acids Res.* **7**, 1513–23.

Birnstiel M. L. & Chipchase M. (1977) Current work on the histone operon. *Trends biochem. Sci.* **2**, 149–52.

Bittle J. L., Houghton R. A., Alexander H., Shinnick T. M., Sutcliffe J. G. & Lerner R. A. (1982) Protection against foot-and-mouth disease by immunization with a chemically synthesized peptide predicted from the viral nucleotide sequence. *Nature, London* **298**, 30–3.

Blackburn E. H. & Gall J. G. (1978) A tandemly repeated sequence at the termini of the extrachromosomal ribosomal RNA genes in *Tetrahymena*. *J. molec. Biol.* **120**, 33–53.

Blattner F. R., Williams B. G., Blechl A. E., Deniston-Thompson K., Faber H. E., Furlong L. A., Grunwald D. J., Kiefer D. O., Moore D. D., Schumm J. W., Sheldon E. L. & Smithies O. (1977) Charon phages: safer derivatives of bacteriophage lambda for DNA cloning. *Science* **196**, 161–9.

Blobel G. & Dobberstein B. (1975) Transfer of proteins across membranes. I. Presence of proteolytically processed and unprocessed nascent immunoglobulin light chains on membrane-bound ribosomes of murine myeloma. *J. Cell Biol.* **67**, 835–51.

Bloom K. S. & Carbon J. (1982) Yeast centromere DNA is a unique and highly ordered structure in chromosomes and small circular minichromosomes. *Cell* **29**, 305–17.

Boeke J. D., Vovis G. F. & Zinder N. D. (1979) Insertion mutant of bacteriophage f1 sensitive to *Eco* RI. *Proc. nat. Acad. Sci. U.S.A.* **76**, 2699–702.

Bolivar F. (1978) Construction and characterization of new cloning vehicles. III. Derivatives of plasmid pBR322 carrying unique *Eco* RI sites for selection of *Eco* RI generated recombinant DNA molecules. *Gene* **4**, 121–36.

Bolivar F., Rodriguez R. L., Betlach M. C. & Boyer H. W. (1977a) Construction and characterization of new cloning vehicles. I. Ampicillin-resistant derivatives of the plasmid pMB9. *Gene* **2**, 75–93.

Bolivar F., Rodriguez R. L., Greene P. J., Betlach M. V., Heynecker H. L., Boyer H. W., Crosa J. H. & Falkow S. (1977b) Construction and characterization of new cloning vehicles. II. A multipurpose cloning system. *Gene* **2**, 95–113.

Bollen A., Herzog A., Cravador A., Herion P., Chuchana P., Straten A. V., Loriau R., Jacobs P. & Van Elsen A. (1983) Cloning and expression in *Escherichia coli* of full length complementary DNA coding for α_1-antitrypsin. *DNA* **2**, 255–65.

Bomhoff G. H., Klapwijk F. M., Kester M. C. M., Schilperoort R. A., Hernalsteens J. P. & Schell J. (1976) Octopine and nopaline synthesis and breakdown genetically controlled by a plasmid of *Agrobacterium tumefaciens*. *Molec. gen. Genet.* **145**, 177–81.

Botstein D. & Davis R. W. (1982) Principles and practice of recombinant DNA research with yeast. In *The Molecular Biology of the Yeast Saccharomyces*, eds Strathem J. N., Jones E. W. & Broach J. R. Cold Spring Harbor Press, Cold Spring Harbor, New York.

Boyko W. L. & Ganschow R. E. (1982) Rapid identification of *Escherichia coli* transformed by pBR322 carrying inserts at the *Pst* I site. *Analytical Biochem.* **122**, 85–8.

Braun A. C. & Wood H. N. (1976) Suppression of the neoplastic state with the acquisition of specialised functions in cells, tissues and organs of crown gall teratoma of tobacco. *Proc. nat. Acad. Sci. U.S.A.* **73**, 496–500.

Brenner S., Cesareni G. & Karn J. (1982) Phasmids: hybrids between Col E1 plasmids and *E. coli* bacteriophage lambda. *Gene* **17**, 27–44.

Breunig K. D., Mackedonski V. & Hollenberg C. P. (1982) Transcription of the bacterial β-lactamase gene in *Saccharomyces cerevisiae. Gene* **20**, 1–10.

Brinster R. L., Chen H. Y., Trumbauer M., Senear A. W., Warren R. & Palmiter R. D. (1981) Somatic expression of Herpes thymidine kinase in mice following injection of a fusion gene into eggs. *Cell* **27**, 223–31.

Brinster R. L., Chen H. Y., Warren R., Sarthy A. & Palmiter R. D. (1982) Regulation of metallothionein-thymidine kinase fusion plasmids injected into mouse eggs. *Nature, London* **296**, 39–42.

Broda P. (1979) *Plasmids*. W. H. Freeman & Co., San Francisco.

Broglie R., Coruzzi G., Fraley R. T., Rogers S. G., Horsch R. B., Niedermeyer J. G., Fink C. L., Flick J. S. & Chua N. H. (1984) Light-regulated expression of a pea ribulose-1,5-bisphosphate carboxylase small subunit gene in transformed plant cells. *Science* **224**, 838–43.

Broome S. & Gilbert W. (1978) Immunological screening method to detect specific translation products. *Proc. nat. Acad. Sci. U.S.A.* **75**, 2746–9.

Brosius J., Cate R. L. & Perlmutter A. P. (1982) Precise location of two promoters for the β-lactamase gene of pBR322. *J. biol. Chem.* **257**, 9205–10.

Brown A. C. & Wood H. N. (1976) Suppression of the neoplastic state with the acquisition of specialized functions in cells, tissues and organs of crown gall teratomas of tobacco. *Proc. nat. Acad. Sci. U.S.A.* **73**, 496–500.

Brown D. D. & Gurdon J. B. (1977) High fidelity transcription of 5S DNA injected into *Xenopus* oocytes. *Proc. nat. Acad. Sci. U.S.A.* **74**, 2064–8.

Cameron J. R., Panasenko S. M., Lehman I. R. & Davis R. W. (1975) *In vitro* construction of bacteriophage λ carrying segments of the *Escherichia coli* chromosome: selection of hybrids containing the gene for DNA ligase. *Proc. nat. Acad. Sci. U.S.A.* **72**, 3416–20.

Canosi U., Iglesias A. & Trautner T. A. (1981) Plasmid transformation in *Bacillus subtilis*: effects of insertion of *Bacillus subtilis* DNA into plasmid pC194. *Molec. gen. Genet.* **181**, 434–40.

Canosi U., Morelli G. & Trautner T. A. (1978) The relationship between molecular structure and transformation efficiency of some *S. aureus* plasmids isolated from *B. subtilis. Molec. gen. Genet.* **166**, 259–67.

Carlton B. C. & Helinski D. R. (1969) Heterogeneous circular DNA elements in vegetative cultures of *Bacillus megaterium. Proc. nat. Acad. Sci. U.S.A.* **64**, 592–9.

Carroll D. (1983) Genetic recombination of bacteriophage lambda DNAs in *Xenopus* oocytes. *Proc. nat. Acad. Sci. U.S.A.* **80**, 6902–6.

Caruthers M. H., Beaucage S. L., Becker C., Efcavitch W., Fisher E. F., Galluppi G., Goldman R., Dettaseth P., Martin F., Matteucci M. & Stabinsky Y. (1982) New methods for synthesizing deoxyoligonucleotides. In *Genetic Engineering*, eds Setlow J. K. & Hollaender A., pp. 119–45. Plenum Press, New York & London.

Case M. E., Schweizer M., Kushner S. R. & Giles N. H. (1979) Efficient transformation of *Neurospora crassa* by utilizing hybrid plasmid DNA. *Proc. nat. Acad. Sci. U.S.A.* **76**, 5259–63.

Cesarini G., Muesing M. A. & Polisky B. (1982) Control of Col E1 DNA replication: the *rop* gene product negatively affects transcription from the replication primer promoter. *Proc. nat. Acad. Sci. U.S.A.* **79**, 6313–7.

Chang A. C. Y. & Cohen S. N. (1974) Genome construction between bacterial species *in vitro*: replication and expression of *Staphylococcus* plasmid genes in *Escherichia coli. Proc. nat. Acad. Sci. U.S.A.* **71**, 1030–4.

Chang A. C. Y., Ehrlich H. A., Gunsalus R. P., Nunberg J. H., Kaufman R. J., Schimke R. T. & Cohen S. N. (1980) Initiation of protein synthesis in bacteria at a translational start codon of mammalian cDNA: effects of the preceding nucleotide sequence. *Proc. nat. Acad. Sci. U.S.A.* **77**, 1442–6.

Chang A. C. Y., Nunberg J. H., Kaufman R. K., Ehrlich H. A. Schimke R. T. & Cohen S. N. (1978) Phenotypic expression in *E. coli* of a DNA sequence coding for mouse dihydrofolate reductase. *Nature* **275**, 617–24.

Chang J. C. & Kan Y. W. (1981) Antenatal diagnosis of sickle-cell anaemia by direct analysis of the sickle mutation. *Lancet* **2**, 1127–9.

Chang L. M. S. & Bollum F. J. (1971) Enzymatic synthesis of oligodeoxynucleotides. *Biochemistry* **10**, 536–42.

Chang S. & Cohen S. N. (1979) High-frequency transformation of *Bacillus subtilis* protoplasts by plasmid DNA. *Molec. gen. Genet.* **168**, 111–15.

Charnay P., Perricaudet M., Galibert F. & Tiollais P. (1978) Bacteriophage lambda and plasmid vectors allowing fusion of cloned genes in each of the three translational phases. *Nucleic Acids Res.* **5**, 4479–94.

Chilton M-D., Drummond M. H., Merlo D. J., Sciaky D., Montoya A. L., Gordon M. P. & Nester E. W. (1977) Stable incorporation of plasmid DNA into higher plant cells: the molecular basis of crown gall tumorigenesis. *Cell* **11**, 263–71.

Chilton M-D., Tepfer D. A., Petit A., David C., Casse-Delbart F. & Tempe J. (1982) *A. rhizogenes* inserts T-DNA into the genomes of host plant root cells. *Nature, London* **295**, 432–4.

Chu, G. & Sharp P. A. (1981) SV40 DNA transfection of cells in suspension: analysis of the efficiency of transcription and translation of T-antigen. *Gene* **13**, 197–202.

Clarke L. & Carbon J. (1976) A colony bank containing synthetic Col E1 hybrid plasmids representative of the entire *E. coli* genome. *Cell* **9**, 91–9.

Clarke L. & Carbon J. (1980) Isolation of a yeast centromere and construction of functional small circular chromosomes. *Nature, London* **287**, 504–9.

Cohen J. D., Eccleshall T. R., Needleman R. B., Federoff H., Buchferer B. A. & Marmur J. (1980) Functional expression in yeast of the

Escherichia coli plasmid gene coding for chloramphenicol acetyltransferase. *Proc. nat. Acad. Sci. U.S.A.* **77**, 1078–82.

Cohen, S. N., Chang A. C. Y. & Hsu L. (1972) Nonchromosomal antibiotic resistance in bacteria: genetic transformation of *Escherichia coli* by R-factor DNA. *Proc. nat. Acad. Sci. U.S.A.* **69**, 2110–4.

Colbère-Garapin F., Horodniceanu F., Kourilsky P. & Garapin A. C. (1981) A new dominant hybrid selective marker for higher eukaryotic cells. *J. mol. Biol.* **150**, 1–14.

Collins J. & Brüning H. J. (1978) Plasmids usable as gene-cloning vectors in an *in vitro* packaging by coliphage λ: 'cosmids'. *Gene* **4**, 85–107.

Collins J. & Hohn B. (1979) Cosmids: a type of plasmid gene-cloning vector that is packageable *in vitro* in bacteriophage λ heads. *Proc. nat. Acad. Sci. U.S.A.* **75**, 4242–6.

Conner, B. J., Reyes, A. A., Morin C., Itakura K., Teplitz R. & Wallace R. B. (1983) Detection of sickle cell β^s-globin allele by hybridization with synthetic oligonucleotides. *Proc. nat. Acad. Sci. U.S.A.* **80**, 278–82.

Contente S. & Dubnau D. (1979) Characterization of plasmid transformation in *Bacillus subtilis*: kinetic properties and the effect of DNA conformation. *Molec. gen. Genet.* **167**, 251–8.

Cosloy S. D. & Oishi M. (1978) Genetic transformation in *Escherichia coli* K12. *Proc. nat. Acad. Sci. U.S.A.* **70**, 84–7.

Covey S. & Hull R. (1981) Transcription of cauliflower mosaic virus DNA. Detection of transcripts, properties and location of the gene encoding the virus inclusion body. *Virology* **111**, 463–74.

Crabeel M., Huygen R., Cunin R. & Glansdorff N. (1983) The promoter region of the *arg3* gene in *Saccharomyces cerevisiae*. Nucleotide sequence and regulation in an *arg3–lac* Z gene fusion. *EMBO J.* **2**, 205–12.

Crespi R. S. (1982) *Patenting in the Biological Sciences*. John Wiley & Sons, Chichester.

Currier T. C. & Nester E. W. (1976) Evidence for diverse types of large plasmids in tumor-inducing strains of *Agrobacterium*. *J. Bacteriol.* **126**, 157–65.

Dalbadie-McFarland G, Cohen L. W., Riggs A. D., Morin C., Itakura K. & Richards, J. H. (1982) Oligonucleotide-directed mutagenesis as a general and powerful method for studies of protein functions. *Proc. nat. Acad. Sci U.S.A.* **79**, 6409–13.

Dani, G. M. & Zakian U.A. (1983) Mitotic and meiotic stability of linear plasmids in yeast. *Proc. nat. Acad. Sci. U.S.A.* **80**, 3406–10.

Daubert S., Richins R., Shepherd R. J. & Gardner R. C. (1982) Mapping of the coat protein gene of cauliflower mosaic virus by its expression in a prokaryotic system. *Virology* **122**, 444–9.

Daubert S., Shepherd R. J. & Gardner R. C. (1983) Insertional mutagenesis of the cauliflower mosaic virus genome. *Gene* **25**, 201–8.

Davidson E. H. (1976). *Gene Activity in Early Development*, 2nd edn. Academic Press, New York.

Davison J., Brunel F. & Merchez M. (1979) A new host-vector system allowing selection for foreign DNA inserts in bacteriophage λ gt *WES*. *Gene* **8**, 69–80.

de Boer H. A., Comstock L. J. & Vasser M. (1983a) The *tac* promoter: a functional hybrid derived from the *trp* and *lac* promoters. *Proc. nat. Acad. Sci. U.S.A.* **80**, 21–5.

de Boer H. A., Hui A., Comstock L. J., Wong E. & Vasser M. (1983b) Portable Shine-Dalgatno regions: a system for a systematic study of defined alterations of nucleotide sequences within *E.coli* ribosome binding sites. *DNA* **2**, 231–41.

de Framond A. J., Barton K. A. & Chilton M-D. (1983) Mini-Ti: A new vector strategy for plant genetic engineering. *Biotechnology* **1**, 262–9.

De Greve H., Decraemer H., Senrinck J., Van Montagu M. & Schell J. (1981) The functional organization of the octopine *Agrobacterium tumefaciens* plasmid pTi B653. *Plasmid* **6**, 235–48.

De Greve H., Leemans J., Hernalsteens J. P., Thia-Toong L., De Benckeleer M., Willmitzer L., Olten L., Van Montagu M. & Schell J. (1982) Regeneration of normal fertile plants that express octopine synthase from tobacco crown galls after deletion of tumor-controlling functions. *Nature, London* **300**, 752–5.

De Greve H., Phaese P., Seurwick J., Lemmers M., Van Montagu M. & Schell J. (1982) Nucleotide sequence and transcript map of the *Agrobacterium tumefaciens* Ti plasmid-encoded octopine synthase gene. *J. molec. appl. Genet.* **1**, 499–511.

De Maeyer E., Skup D., Prasad K. S. N., De Maeyer-Guignard J., Williams B., Meacock P., Sharpe G., Pioli D., Hennam J., Schuch W. & Atherton K. (1982) Expression of a chemically synthesized human α1 interferon gene. *Proc. nat. Acad. Sci. U.S.A.* **79**, 4256–9.

De Robertis E. M. (1983) Nucleocytoplasmic segregation of proteins and RNAs. *Cell* **32**, 1021–5.

De Robertis E. M., Black P. & Nishikura K. (1981) Intranuclear location of the tRNA splicing enzymes. *Cell* **23**, 89–93.

De Robertis E. M. & Mertz J. E. (1977) Coupled transcription–translation of DNA injected into *Xenopus* oocytes. *Cell* **12**, 175–82.

de Saint Vincent B. R., Delbruck S., Eckhart W., Meinkoth J., Vitto L. & Wahl G. (1981) The cloning and reintroduction into animal cells of a functional CAD gene, a dominant amplifiable genetic marker. *Cell* **27**, 267–77.

de Vos W. M., Venema G., Canosi U. & Trautner T. A. (1981) Plasmid transformation in *Bacillus subtilis*: fate of plasmid DNA. *Molec. gen. Genet.* **181**, 424–33.

Dean D. (1981) A plasmid cloning vector for the direct selection of strains carrying recombinant plasmids. *Gene* **15**, 99–102.

Delseny M. & Hull R. (1983) Isolation and characterization of faithful and altered clones of the genomes of cauliflower mosaic virus isolates Cabb B–J1, CM4–184 and Bari 1. *Plasmid* **9**, 31–41.

Deng G. & Wu R. (1981) An improved procedure for utilizing terminal transferase to add homopolymers to the 3′ termini of DNA. *Nucleic Acids Res.* **9**, 4173–88.

Depicker A., Stachel S., Dhaese P., Zambryski P. & Goodman H. M. (1982) Nopaline synthase: transcript mapping and DNA sequence. *J. molec. appl. Genet.* **1**, 561–74.

Depicker A., Van Montagu M. & Schell J. (1978) Homologous DNA sequences in different Ti-plasmids are essential for oncogenicity. *Nature, London* **275**, 150–3.

Derynck R., Singh A. & Goeddel D. (1983) Expression of the human interferon-γ cDNA in yeast. *Nucleic Acids Res.* **11**, 1819–37.

Ditta G., Stanfield S., Corbin D. & Helinski D. R. (1980) Broad host range DNA cloning system for Gram-negative bacteria. Construction of a gene bank of *Rhizobium meliloti. Proc. nat. Acad. Sci. U.S.A.* **77**, 7347–51.

Dobson M. J., Tuite M. F., Roberts N. A., Kingsman A. J., Kingsman S. M., Perkins R. E., Conroy S. C., Dunbar B. & Fothergill L. A. (1982) Conservation of high efficiency promoter sequences in *Saccharomyces cerevisiae. Nucleic Acids Res.* **10**, 2625–37.

Dobson M. J., Tuite M. F., Mellor J., Roberts N. A., King R. M., Burke D. C., Kingsman A. J. & Kingsman S. M. (1983) Expression in *Saccharomyces cerevisiae* of human interferon-alpha directed by the *TRP* 5′ region. *Nucleic Acids Res.* **11**, 2287–302.

Doel M. T., Eaton M., Cook E. A., Lewis H., Patel T. & Carey N. H. (1980) The expression in *E. coli* of synthetic repeating polymeric genes coding for poly-(L-aspartyl-L-phenylalanine). *Nucleic Acids Res.* **8**, 4575–92.

Donson J. P. & Hull R. (1983) Physical mapping and molecular cloning of caulimovirus DNA. *J. gen. Virol.* **64**, 2281–8.

Doring H. P., Tillmann E. & Starlinger P. (1984) DNA sequence of the maize transposable element Dissociation. *Nature, London* **307**, 127–30.

Drummond M. H. & Chilton M-D. (1978) Tumor-inducing (Ti) plasmids of *Agrobacterium* share extensive regions of DNA homology. *J. Bacteriol.* **136**, 1178–83.

Dugaiczyk A., Boyer H. W. & Goodman H. M. (1975) Ligation of *Eco* RI endonuclease-generated DNA fragments into linear and circular structures. *J. molec. Biol.* **96**, 171–84.

Duncan C. H., Wilson G. A. & Young F. E. (1978) Mechanism of integrating foreign DNA during transformation of *Bacillus subtilis. Proc. nat. Acad. Sci. U.S.A.* **75**, 3664–8.

Durnam D. M. & Palmiter R. D. (1981) Transcriptional regulation of the mouse metallothionein-1 gene by heavy metals. *J. biol. Chem.* **256**, 5712–16.

Dussoix D. & Arber W. (1962) Host specificity of DNA produced by *Escherichia coli*. II. Control over acceptance of DNA from infecting phage λ. *J. molec. Biol.* **5**, 37–49.

Dworkin M. B. & Dawid I. B. (1980) Use of a cloned library for the study of abundant poly(A)$^+$ RNA during *Xenopus laevis* development. *Dev. Biol.* **76**, 449–64.

Edelmann P. & Gallant J. (1977) Mistranslation in *E. coli. Cell* **10**, 131–7.

Edens L., Heslinga L., Klok R., Ledeboer A. M., Maat J., Toonen M. Y., Visser C. & Verrips C. T. (1982) Cloning of cDNA encoding the sweet-tasting plant protein thaumatin and its expression in *Escherichia coli*. *Gene* **18**, 1–12.

Edge M. D., Greene A. R., Heathcliffe G. R., Meacock P. A., Schuch W., Scanlon D. B., Atkinson T. C., Newton C. R. & Markham A. F. (1981) Total Synthesis of a human leukocyt interferon gene. *Nature* **292**, 756–62.

Efstratiadis A., Kafatos F. C., Maxam A. M. & Maniatis T. (1976) Enzymatic *in vitro* synthesis of globin genes. *Cell* **7**, 279–88.

Ehrlich H. A., Cohen S. N. & McDevitt H. O. (1978) A sensitive radioimmunoassay for detecting products translated from cloned DNA fragments. *Cell* **13**, 681–9.

Ehrlich S. D. (1977) Replication and expression of plasmids from *Staphylococcus aureus* in *Bacillus subtilis*. *Proc. nat. Acad. Sci. U.S.A.* **74**, 1680–2.

Ehrlich S. D. (1978) DNA cloning in *Bacillus subtilis*. *Proc. nat. Acad. Sci. U.S.A.* **75**, 1433–6.

Ehrlich S. D., Jupp S., Niaudet B. & Goze A. (1978) *Bacillus subtilis* as a host for DNA cloning. In *Genetic Engineering*, eds Boyer H. W. & Nicosia S., pp. 25–32. Elsevier-North Holland, Amsterdam.

Ehrlich S. D., Niaudet B. & Michel B. (1981) Use of plasmids from *Staphylococcus aureus* for cloning of DNA in *Bacillus subtilis*. *Curr. Topics Microbiol. Immunol.* **96**, 19–29.

Emini E. A., Jameson B. A. & Wimmer E. (1983) Priming for and induction of anti-poliovirus neutralizing antibodies by synthetic peptides. *Nature, London* **304**, 699–703.

Emtage J. S., Angal S., Doel M., Harris T. J. R., Jenkins B., Lilley G. & Lowe P. A. (1983) Synthesis of calf prochymosin (prorennin) in *Escherichia coli*. *Proc. nat. Acad. Sci. U.S.A.* **80**, 3671–5.

Emtage J. S., Tacon W. C. A., Catlin G. H., Jenkins B., Porter A. G. & Carey N. H. (1980) Influenza antigenic determinants are expressed from haemagglutin genes cloned in *Eschericia coli*. *Nature, London* **283**, 171–4.

Engler G., Depicker A., Maenhaut R., Villarroel R., Van Montagu M. & Schell J. (1981) Physical mapping of DNA base sequence homologies between an octopine and a nopaline Ti plasmid of *Agrobacterium tumefaciens*. *J. molec. Biol.* **152**, 183–208.

Ensley B. D., Ratzkin B. J., Osslund T. D., Simon M. J., Wackett L. P. & Gibson D. T. (1983) Expression of napthalene oxidation genes in *Escherichia coli* results in biosynthesis of indigo. *Science* **222**, 167–9.

Falkow S. (1975) *Infectious multiple drug resistance*, London, Pion.

Feinstein S. I., Chernajousky Y., Chen L., Maroteaux L. & Mory Y. (1983) Expression of human interferon genes using the *rec*A promotor of *Escherichia coli*. *Nucleic Acids Res.* **11**, 2927–41.

Federoff N. (1984) Transposable genetic elements in maize. *Am. Sci.* July, **250**, 64–74.

Ferretti L. & Sgaramella V. (1981) Temperature dependence of the joining by T4 DNA ligase of termini produced by type II restriction endonucleases. *Nucleic Acids Res.* **9**, 85–93.

Fitzgerald-Hayes M., Buhler, J. M., Cooper T. G. & Carbon J. (1982)

Isolation and subcloning analysis of functional centromere DNA (cen11) from yeast chromosome XI. *Mol. cell. Biol.* **2**, 82–7.

Flintoff W. F., Davidson S. V. & Siminovitch L. (1976) Isolation and partial characterization of three methotrexate-resistant phenotypes from Chinese hamster ovary cells. *Somat. Cell Genet.* **2**, 245–61.

Forbes D. J., Kirschner M. W. & Newport J. W. (1983) Spontaneous formation of nucleus-like structures around bacteriophage DNA microinjected into *Xenopus* eggs. *Cell* **34**, 13–23.

Fraley R., Rogers S. G., Horsch R. B., Sanders P. R., Flick J. S., Adams S. P., Bittner M. L., Brand L. A., Fink C. L., Fry J. S., Galuppi G. R., Goldberg S. B., Hoffmann N. L. & Woo S. C. (1983) Expression of bacterial genes in plant cells. *Proc. nat. Acad. Sci. U.S.A.* **80**, 4803–7.

Franck A., Guilley H., Jonard G., Richards K. & Hirth L. (1980) Nucleotide sequence of cauliflower mosaic virus DNA. *Cell* **21**, 285–94.

Franklin F. C. H., Bagdasarian M., Bagdasarian M. M. & Timis K. N. (1981) Molecular and functional analysis of the TOL plasmid pWWO from *Pseudomonas putida* and cloning of genes for the entire regulated aromatic ring *meta* cleavage pathway. *Proc. nat. Acad. Sci. U.S.A.* **78**, 7458–62.

Frey J., Bagdasarian M., Feiss D., Franklin C. H. & Deshusses J. (1983) Stable cosmid vectors that enable the introduction of cloned fragments into a wide range of Gram-negative bacteria. *Gene* **24**, 299–308.

Frischauf A-M., Lehrach H., Poustka A. & Murray N. (1983) Lambda replacement vectors carrying polylinker sequences. *J. molec. Biol.* **170**, 827–42.

Galli G., Hofstetter H., Stunnenberg H. G. & Birnstiel M. L. (1983) Biochemical complementation with RNA in the *Xenopus* oocyte: A small RNA is required for the generation of 3′histone mRNA termini. *Cell* **34**, 823–8.

Gallwitz D. & Sures I. (1980) Structure of a split yeast gene. Complete nucleotide sequence of the actin gene in *Saccharomyces cerevisiae. Proc. nat. Acad. Sci., U.S.A.* **77**, 2546–50.

Gardner R. C., Howarth A. J., Hahn P., Brown-Luedi M., Shepherd R. J. & Messing J. (1981) The complete nucleotide sequence of an infectious clone of cauliflower mosaic virus by M13 mp7 shotgun sequencing. *Nucleic Acids Res.* **9**, 2871–87.

Garfinkel D. J. & Nester E. W. (1980) *Agrobacterium tumefaciens* mutants affected in crown gall tumorigenesis and octopine catabolism. *J. Bacteriol.* **144**, 732–43.

Gargiulo G. & Worcel A. (1983) Analysis of the chromatin assembled in germinal vesicles of *Xenopus* oocytes. *J. molec. Biol.* **170**, 699–722.

Gelvin S. B., Thomashaw M. F., McPherson J. C., Gordon M. P. & Nester E. W. (1982) Sizes and map positions of several plasmid DNA encoded transcripts in octopine-type crown gall tumours. *Proc. nat. Acad. Sci. U.S.A.* **79**, 76–80.

Gerbaud C., Fournier P., Blanc H., Aigle M., Heslot H. & Geurineau M. (1979) High frequency of yeast transformation by plasmids carrying part or entire 2 μm yeast plasmid. *Gene* **5**, 233–53.

Gething M-J. & Sambrook J. (1981) Cell-surface expression of influenza haemogglutinin from a cloned DNA copy of the RNA gene. *Nature, London* **293**, 620–5.

Gheysen, D., Iserentant D., Derom C. & Fiers W. (1982) Systematic alteration of the nucleotide sequence preceding the translation initiation codon and the effects on bacterial expression of the cloned SV-40 small-t antigen gene. *Gene* **17**, 55–63.

Gillam S., Astell C. R. & Smith M. (1980) Site-specific mutagenesis using oligodeoxyribonucleotides: isolation of a phenotypically silent ϕX174 mutant, with a specific nucleotide deletion, at very high efficiency. *Gene* **12**, 129–37.

Gluzman Y. (1981) SV40-transformed simian cells support the replication of early SV40 mutants. *Cell* **23**, 175–82.

Goeddel D. V., Heyneker H. L., Hozumi T., Arentzen R., Itakura K., Yansura D. G., Ross M. J., Miozzari G., Crea R. & Seeburg P. H. (1979a) Direct expression in *Escherichia coli* of a DNA sequence coding for human growth hormone. *Nature, London* **281**, 544–8.

Goeddel D. V., Kleid D. G., Bolivar F., Heyneker H. L., Yansura D. G., Crea R., Hirose T., Kraszewski A., Itakura K. & Riggs A. D. (1979b) Expression in *Escherichia coli* of chemically synthesized genes for human insulin. *Proc. nat. Acad. Sci. U.S.A.* **76**, 106–10.

Goff S. P. & Berg P. (1976) Construction of hybrid viruses containing SV40 and λ phage DNA segments and their propagation in cultured monkey cells. *Cell* **9**, 695–705.

Goldberg D. A., Posakony J. W. & Maniatis T. (1983) Correct developmental expression of a cloned alcohol dehydrogenase gene transduced into the *Drosophila* germ line. *Cell* **34**, 59–73.

Goldfarb D. S., Doi R. H. & Rodriguez R. L. (1981) Expression of Tn9-derived chloramphenicol resistance in *Bacillus subtilis*. *Nature, London* **293**, 309–11.

Goldfarb M., Shimizu K., Perucho M. & Wigler M. (1982) Isolation and preliminary characterization of a human transforming gene from T24 bladder carcinoma cells. *Nature, London* **296**, 404–9.

Goodman R. (1981) Gemini viruses. *J. gen. Virol.* **54**, 9–21.

Gordon J. W. & Ruddle F. H. (1981) Integration and stable germ line transmission of genes injected into mouse pronunclei. *Science* **214**, 1244–6.

Gordon M. P. (1980) In *Proteins and Nucleic Acids. The Biochemistry of Plants*, ed. A. Marcus, vol. 6, pp 531–70. Academic Press, New York.

Gordon M. P., Farrand S. K., Sciaky D., Montoya A. L., Chilton M-D., Merlo D. J. & Nester E. W. (1979) In *Molecular Biology of Plants, Symposium*, University of Minnesota, ed. Rubenstein I. Academic Press, London.

Gottesmann M. E. & Yarmolinsky M. D. (1968) The integration and excision of the bacteriophage lambda genome. *Cold Spring Harb. Symp. quant. Biol.*, **33**, 735–47.

Gouy M. & Gautier C. (1982) Codon usage in bacteria. Correlation with gene expressivity. *Nucleic Acids Res.* **10**, 7055–74.

Graham F. L. & van der Eb A. J. (1973) A new technique for the assay of infectivity of human adenovirus 5 DNA. *Virology* **52**, 456–67.

Grinter N. J. (1983) A broad-host-range cloning vector transposable to various replicons. *Gene* **21**, 133–43.

Gronenborn B., Gardner R. C., Schaefer S. & Shepherd R. J. (1981) Propagation of foreign DNA in plants using cauliflower mosaic virus as vector. *Nature, London* **294**, 773–6.

Gronenborn B. & Messing J. (1978) Methylation of single-stranded DNA *in vitro* introduces new restriction endonuclease cleavage sites. *Nature, London,* **272**, 375–7.

Grosjean J. & Fiers W. (1982) Preferential codon usage in prokaryotic genes. The optimal codon-anticodon interaction energy and the selective codon usage in efficiently expressed genes. *Gene* **18**, 199–209.

Grosschedl R. & Birnstiel M. L. (1980a) Identification of regulatory sequences in the prelude sequences of an H2A histone gene by the study of specific deletion mutants *in vivo*. *Proc. nat. Acad. Sci. U.S.A.* **77**, 1432–36.

Grosschedl R. & Birnstiel M. L. (1980b) Spacer DNA sequences upstream of the TATAAATA sequence are essential for promotion of H2A histone gene transcription *in vivo*. *Proc. nat. Acad. Sci. U.S.A.* **77**, 7102–6.

Grosschedl R., Machler M., Rohrer U. & Birnstiel M. L. (1983) A functional component of the sea urchin H2A gene modulator contains an extended sequence homology to a viral enhancer. *Nucleic Acids Res.* **11**, 8123–36.

Grunstein M. & Hogness D. S. (1975) Colony hybridization: a method for the isolation of cloned DNAs that contain a specific gene. *Proc. nat. Acad. Sci. U.S.A.* **72**, 3961–5.

Gruss P., Efstratiadis A., Karathanasis S., Konig M. & Khoury G. (1981a) Synthesis of stable unspliced mRNA from an intronless simian virus 40-rat preproinsulin gene recombinant. *Proc. nat. Acad. Sci. U.S.A.* **78**, 6091–5.

Gruss P., Ellis R. W., Shih T. Y., Konig M., Scolnick E. M. & Khoury G. (1981b) SV40 recombinant molecules express the gene encoding p21 transforming protein of Harvey murine sarcoma virus. *Nature, London* **293**, 486–8.

Gruss P. & Khoury (1981) Expression of simian virus 40-rat preproinsulin recombinants in monkey kidney cells: use in preproinsulin RNA processing signals. *Proc. Nat. Acad. Sci U.S.A.* **78**, 133–7.

Gryczan T. J. & Dubnau D. (1982) Direct selection of recombinant plasmids in *Bacillus subtilis*. *Gene* **20**, 459–69.

Gryczan T. J., Contente S. & Dubnau D. (1980a) Molecular cloning of heterologous chromosomal DNA by recombination between a plasmid vector and a homologous resident plasmid in *Bacillus subtilis*. *Molec. gen. Genet.* **177**, 459–67.

Gryczan T. J., Shivakumar A. G. & Dubnau D. (1980b) Characterization of chimaeric plasmid cloning vehicles in *Bacillus subtilis*. *J. Bacteriol* **141**, 246–53.

Gryzcan T. J. & Dubnau D. (1978) Construction and properties of chimaeric plasmids in *Bacillus subtilis*. *Proc. nat. Acad. Sci. U.S.A.* **75**, 1428–32.

Guarente L. (1984) Yeast promoters: positive and negative elements. *Cell* **36**, 799–800.

Guarente L., Lauer G., Roberts T. M. & Ptashne M. (1980) Improved methods for maximizing expression of a cloned gene: a bacterium that synthesizes rabbit β-globin. *Cell* **20**, 543–53.

Guarente L. & Ptashne M. (1981) Fusion of *Escherichia coli lac Z* to the cytochrome *c* gene of *Saccharomyces cerevisiae. Proc. nat. Acad. Sci. U.S.A.* **78**, 2199–203.

Guilley H., Dudley R. K., Jonard G., Balàzs E. & Richards K. E. (1982) Transcription of cauliflower mosaic virus DNA, detection of promoter sequences and characterization of transcripts. *Cell* **30**, 763–73.

Guilley H., Richards K. E. & Jonard G. (1983) Observation concerning the discontinuous DNAs of cauliflower mosaic virus. *EMBO J.* **2**, 277–82.

Gumport R. I. & Lehman I. R. (1971) Structure of the DNA ligase adenylate intermediate: lysine (ϵ-amino) linked AMP. *Proc. nat. Acad. Sci. U.S.A.* **68**, 2559–63.

Gurdon J. B. (1974) *The Control of Gene Expression in Animal Development*. Clarendon Press, Oxford.

Gurley W. B., Kemp J. D., Albert M. J., Sutton D. W. & Callis J. (1979) Transcription of Ti plasmid-derived sequences in three octopine-type crown gall tumor lines. *Proc. nat. Acad. Sci. U.S.A.* **76**, 2828–32.

Gusella J. F., Keys C., Varsanyi-Breiner A., Kao F-T., Jones C., Puck T. T. & Housman D. (1980) Isolation and localization of DNA segments from specific human chromosomes. *Proc. nat. Acad. Sci. U.S.A.* **77**, 2829–33.

Gutte B., Daumigen M. & Wittschieber E. (1979) Design, synthesis and characterization of a 34-residue polypeptide that interacts with nucleic acids. *Nature, London* **281**, 650–5.

Hadi S. M., Bachi B., Iida S. & Bickle T. A. (1982) DNA restriction-modification enzymes of phage P1 and plasmid p15B. Subunit functions and structural homologies. *J. molec. Biol.* **165**, 19–34.

Hagan C. E. & Warren G. J. (1983) Viability of palindromic DNA is restored by deletions occurring at low but variable frequency in plasmids of *Escherichia coli. Gene* **24**, 317–26.

Hahn P. & Shepherd R. J. (1982) Evidence for a 58-kilodalton polypeptide as a precursor of the coat protein of cauliflower mosaic virus. *Virology* **116**, 480–8.

Hall M. N., Hereford L. & Herskowitz I. (1984) Targeting of *E. coli* β-galactodisase to the nucleus in yeast. *Cell* **36**, 1057–65.

Hamer D. H. (1977) SV40 carrying an *Escherichia coli* suppressor gene. In *Recombinant Molecules: Impact on Science and Society*, eds Beers R. G. & Bassett E. G., pp. 317–35. Raven, New York.

Hamer D. H. & Leder P. (1979a) Expression of the chromosomal mouse β maj-globin gene cloned in SV40. *Nature, London* **281**, 35–40.

Hamer D. H. & Leder P. (1979b) SV40 recombinants carrying a functional RNA splice junction and polyadenylation site from the chromosomal mouse β maj-globin gene. *Cell* **17**, 737–47.

Hamer D. H. & Leder P. (1979c) Splicing and the formation of stable RNA. *Cell* **18**, 1299–302.

Hanahan D. (1983) Studies on transformation of *Escherichia coli* with plasmids. *J. molec. Biol.* **166**, 557–80.

Hanahan D., Lane D., Lipsich L., Wigler M. & Botchan M. (1980) Characteristics of an SV40-plasmid recombinant and its movement into and out of the genome of a murine cell. *Cell* **21**, 127–39.

Hanahan D. & Meselson M. (1980) Plasmid screening at high colony density. *Gene* **10**, 63–7.

Harbers K., Jahner D. & Jaenisch R. (1981) Microinjection of cloned retroviral genomes into mouse zygotes: integration and expression in the animal. *Nature, London* **293**, 540–43.

Hardies S. C. & Wells R. D. (1976) Preparative fractionation of DNA by reversed phase column chromatography. *Proc. nat. Acad. Sci. U.S.A.* **73**, 3117–21.

Hardy K., Stahl S. & Küpper H. (1981) Production in *B. subtilis* of hepatitis B core antigen and of major antigen of foot and mouth disease virus. *Nature, London* **293**, 481–3.

Harland R. M. & Laskey R. A. (1980) Regulated replication of DNA microinjected into eggs of *Xenopus laevis. Cell* **21**, 761–71.

Harland R. M., Weintraub H. & McKnight S. L. (1983) Transcription of DNA injected into *Xenopus* oocytes is influenced by template topology. *Nature, London* **302**, 38–42.

Hashimoto-Gotoh T., Franklin F. C. H., Nordheim A. & Timmis K. N. (1981) Low copy number, temperature-sensitive, mobilization-defective pSC101-derived containment vectors. *Gene* **16**, 227–35.

Hawley D. K. & McClure W. R. (1983) Compilation and analysis of *Escherichia coli* promoter DNA sequences. *Nucleic Acids Res.* **11**, 2237–55.

Hayashi K. (1980) A cloning vehicle suitable for strand separation. *Gene* **11**, 109–15.

Heidecker G. & Messing J. (1983) Sequence analysis of Zein cDNAs by an efficient mRNA cloning method. *Nucleic Acids Res.* **11**, 4891–906.

Helfman D. M., Feramisco J. R., Fiddes J. C., Thomas G. P. & Hughes S. H. (1983) Identification of clones that encode chicken tropomyosin by direct immunological screening of a cDNA expression library. *Proc. nat. Acad. Sci. U.S.A.* **80**, 31–5.

Henikoff S., Tatchell K., Hall B. D. & Nosmyth K. A. (1981) Isolation of a gene from *Drosophila* by complementation in yeast. *Nature, London* **289**, 33–7.

Hennam J. F., Cunningham A. E., Sharpe G. S. & Atherton K. T. (1982) Expression of eukaryotic coding sequences in *Methylophilus methylotrophus. Nature, London* **297**, 80–2.

Hennecke H., Günther I. & Binder F. (1982) A novel cloning vector for the direct selection of recombinant DNA in *E. coli. Gene* **19**, 231–4.

Hentschel C., Probst E. & Birnstiel M. L. (1980) Transcriptional fidelity of histone genes injected into *Xenopus* oocyte nuclei. *Nature, London* **288**, 100–2.

Herrera-Estrella L., Depicker A., Van Montagu M., Schell J. (1983a) Expression of chimaeric genes transferred into plant cells using a Ti-plasmid-derived vector. *Nature, London* **303**, 209–13.

Herrera-Estrella L., DeBlock M., Messens E., Hernalsteens J. P., Van Montagu M. & Schell J. (1983b) Chimeric genes as dominant selectable markers in plant cells. *EMBO J.* **2**, 987–95.

Herrera-Estrella L., Van den Broeck G., Maenhaut R., Van Montagu M., Schell J., Timko M. & Cashmore A. (1984) Light-inducible and chloroplast associated expression of a chimaeric gene introduced into *Nicotiana tabacum* using a Ti-plasmid vector. *Nature, London* **310**, 115–20.

Hershfield V., Boyer H. W., Yanofsky C., Lovett M. A. & Helinski D. R. (1974) Plasmid Col El as a molecular vehicle for cloning and amplification of DNA. *Proc. nat. Acad. Sci. U.S.A.* **71**, 3455–9.

Herskowitz I. (1974) Control of gene expression in bacteriophage lambda. *Ann. Rev. Genet.* **7**, 289–324.

Hicks J. B., Strathern J. N., Klar A. J. S. & Dellaporta S. L. (1982) Cloning by complementation in yeast. The mating type genes. In *Genetic Engineering*, eds Setlow J. K. & Hollaender A., pp. 219–48. Plenum Press, New York.

Higuchi R., Paddock G. V., Wall R. & Salser W. (1976) A general method for cloning eukaryotic structural gene sequences. *Proc. nat. Acad. Sci. U.S.A.* **73**, 3146–50.

Hinnen A., Hicks, J. B. & Fink G. R. (1978) Transformation of yeast. *Proc. nat. Acad. Sci. U.S.A.* **75**, 1929–33.

Hitzeman R. A., Hagie F. E., Levine H. L., Goeddel D. V., Ammerer G. & Hall B. D. (1981) Expression of a human gene for interferon in yeast. *Nature, London* **293**, 717–22.

Hitzeman R. A., Leung D. W., Perry L. J., Kohr W. J., Levine H. L. & Goeddel D. V. (1983) Secretion of human interferons by yeast. *Science* **219**, 620–5.

Hoekstra W. P. M., Bergmans J. E. N. & Zuidweg E. M. (1980) Role of *rec*BC nuclease in *Escherichia coli* transformation. *J. Bacteriol.* **143**, 1031–2.

Hogan B. (1983) Enhancers, chromosome position effects, and transgenic mice. *Nature, London* **306**, 313–14.

Hohn B. (1975) DNA as substrate for packaging into bacteriophage lambda, *in vitro*. *J. molec. Biol.* **98**, 93–106.

Hohn B. & Murray K. (1977) Packaging recombinant DNA molecules into bacteriophage particles *in vitro*. *Proc. nat. Acad. Sci. U.S.A.* **74**, 3259–63.

Hohn T., Hohn B., Lesot A. & Lebeurier G. (1980) Restriction map of native and cloned cauliflower mosaic virus DNA. *Gene* **11**, 21–32.

Hollenberg C. P., Degelmann A., Kustermann-Kuhn B. & Royer H. D. (1976) Characterization of 2-μm DNA of *Saccharomyces cerevisiae* by restriction fragment analysis and integration in an *Escherichia coli* plasmid. *Proc. nat. Acad. Sci. U.S.A.* **73**, 2072–6.

Holsters M., Silva B., Van Vliet F., Genetello C., De Block M., Dhaese P., Depicker A., Inze D., Engler G., Villarroel R., Van Montagu M. &

Schell J. (1980) The functional organization of the nopaline *A. tumefaciens* plasmid pTi C58. *Plasmid* **3**, 212–30.

Hopkins A. S., Murray N. E. & Brammar W. J. (1976) Characterization of λtrp-transducing bacteriophages made *in vitro*. *J. molec. Biol.* **107**, 549–69.

Hopwood D. A., Bibb M. J., Bruton C. J., Chater K. F., Feitelson J. S. & Gil J. A. (1983) Cloning *Streptomyces* genes for antibiotic production. *Trends in Biotechnology* **1**, 42–8.

Hopwood D. A. & Chater K. F. (1982) Cloning in streptomyces: systems and strategies. In *Genetic Engineering*, eds Setlow J. K. & Hollaender A., pp. 119–45.

Horinouchi S. & Weisblum B. (1982a) Nucleotide sequence and functional map of pE194, a plasmid that specifies inducible resistance to macrolide, lincosamide, and streptogramin type B antibiotics. *J. Bacteriol.* **150**, 804–14.

Horinouchi, S. & Weisblum B. (1982b) Nucleotide sequence and functional map of pC194, a plasmid that specifies inducible chloramphenicol resistance. *J. Bacteriol.* **150**, 815–25.

Horsch R. B., Fraley R. T., Rogers S. G., Sanders P. R., Lloyd A. & Hoffmann N. (1984) Inheritance of functional genes in plants. *Science* **223**, 496–8.

Houghton M., Eaton M. A. W., Stewart A. G., Smith J. C., Doel S. M., Catlin G. H., Lewis H. M., Patel T. P., Emtage J. S., Carey N. H. & Porter A. G. (1980) The complete amino acid sequence of human fibroblast interferon as deduced using synthetic oligodeoxyribonucleotide primers of reverse transcriptase. *Nucleic Acids Res.* **8**, 2885–94.

Howarth A. J., Gardner R. C., Messing J. & Shepherd R. J. (1981) Nucleotide sequence of naturally occurring deletion mutants of cauliflower mosaic virus. *Virology* **112**, 678–85.

Howell S. H. & Hull R. (1978) Replication of cauliflower mosaic-virus and transcription of its genome in turnip leaf protoplasts. *Virology* **86**, 468–81.

Howell S. H., Walker L. L. & Dudley R. K. (1980) Cloned cauliflower mosaic virus DNA infects turnips (*Brassica rapa*). *Science* **208**, 1265–7.

Howell S. H., Walker L. L. & Walden R. M. (1981) Rescue of *in vitro* generated mutants of cloned cauliflower mosaic virus genome in infected plants. *Nature, London* **293**, 483–6.

Hsiao C. L. & Carbon J. (1981) Characterization of a yeast replication origin (ars2) and construction of stable minichromosomes containing cloned yeast centromere DNA (CEN 3). *Gene* **15**, 157–66.

Hughes K., Case M. E., Geever R., Vapnek D. & Giles N. H. (1983) Chimeric plasmid that replicates autonomously in both *Escherichia coli* and *Neurospora crassa*. *Proc. nat. Acad. Sci. U.S.A.* **80**, 1053–7.

Hui A., Hayflick J., Dinkelspiel K. & de Boer H. A. (1984) Mutagenesis of the three bases preceding the start codon of the β-galactosidase mRNA and its effect on translation in *Escherichia coli*. *EMBO J.* **3**, 623–9.

Hull R. (1980) Structure of the cauliflower mosaic virus genome. III.

Restriction endonuclease mapping of thirty-three isolates. *Virology* **100**, 76–90.

Hull R. & Covey S. N. (1983a) Replication of cauliflower mosaic virus DNA. *Sci. Progr. Oxf.* **68**, 403–22.

Hull R. & Covey S. N. (1983b) Does cauliflower mosaic virus replicate by reverse transcription? *Trends Biol. Sci.* **8**, 119–21.

Hull R., Covey S. N., Stanley J. & Davies J. W. (1979) The polarity of the cauliflower mosaic virus genome. *Nucleic Acids Res.* **7**, 669–77.

Hull R. & Davies J. W. (1983) Genetic engineering with plant viruses and their potential as vectors. *Adv. Virus Res.* **28**, 1–33.

Hull R. & Donson J. (1982) Physical mapping of the DNAs of carnation etched ring and figwort mosaic viruses. *J. gen. Virol.* **60**, 125–34.

Hull R. & Howell S. H. (1978) Structure of cauliflower mosaic-virus genome. II. Variation in DNA-structure and sequence between isolates. *Virology* **86**, 482–93.

Hull R. & Shepherd R. J. (1977) Structure of cauliflower mosaic-virus genome. *Virology* **79**, 216–30.

Humphries P., Old R., Coggins L. W., McShane T., Watson C. & Paul J. (1978) Recombinant plasmids containing *Xenopus laevis* structural genes derived from complementary DNA. Nucleic Acids Res. **5**, 905–24.

Iida S., Meyer J., Bachi B., Stalhammer-Carlemalm M., Schrickel S., Bickle T. A. & Arber W. (1982) DNA restriction-modification genes of phage P1 and plasmid p15B. Structure and *in vitro* transcription. *J. molec. Biol.* **165**, 1–18.

Ikemura T. (1981a) Correlation between the abundance of *Escherichia coli* transfer RNAs and the occurrence of the respective codons in its protein genes. *J. molec. Biol.* **146**, 1–21.

Ikemura T. (1981b) Correlation between the abundance of *Escherichia coli* transfer RNAs and the occurrence of the respective codons in its protein genes. A proposal for a synonymous codon choice that is optimal for the *E. coli* translational system. *J. molec. Biol.* **151**, 389–409.

Inouye H. & Beckwith J. (1977) Synthesis and processing of an *Escherichia coli* alkaline phosphatase precursor *in vitro*. *Proc. nat. Acad. Sci. U.S.A.* **74**, 1440–4.

Iserentant D. & Fiers W. (1980) Secondary structure of mRNA and efficiency of translation initiation. *Gene* **9**, 1–12.

Ish-Horowicz D. & Burke J. F. (1981) Rapid and efficient cosmid cloning. *Nucleic Acids Res.* **9**, 2989–98.

Israel M. A., Chan H. W., Rowe W. P. & Martin M. A. (1979) Molecular cloning of polyoma virus DNA in an *Escherichia coli* plasmid vector system. *Science* **203**, 883–7.

Itakura K., Hirose T., Crea R., Riggs A. D., Heyneker H. L., Bolivar F. & Boyer H. W. (1977) Expressions in *Escherichia coli* of a chemically synthesized gene for the hormone somatostatin. *Science* **198**, 1056–63.

Ito K., Bassford P. J. & Beckwith J. (1981) Protein localization in *E. coli*: is there a common step in the secretion of periplasmic and outer-membrane proteins? *Cell* **24**, 707–17.

Jackson D. A., Symons R. H. & Berg P. (1972) Biochemical method for inserting new genetic information into DNA of Simian Virus 40:

circular SV40 DNA molecules containing lambda phage genes and the galactose operon of *Escherichia coli. Proc. nat. Acad. Sci. U.S.A.* **69**, 2904–9.

Jacob A. E., Cresswell J. M., Hedges R. W., Coetzee J. N. & Beringer J. E. (1976) Properties of plasmids constructed by *in vitro* insertion of DNA from *Rhizobium leguminosarum* or *Proteus mirabilis* into RP4. *Molec. gen. Genet.* **147**, 315–23.

Jaenisch R. & Mintz B. (1974) Simian virus 40 DNA sequences in DNA of healthy adult mice derived from preimplantation blastocysts injected with viral DNA. *Proc. nat. Acad. Sci. U.S.A.* **71**, 1250–4.

Jalanko A., Palva I. & Soderlund H. (1981) Restriction maps of plasmids pUB110 and pBD9. *Gene* **14**, 325–8.

Jimenez A. & Davies J. (1980) Expression of a transposable antibiotic resistance element in *Saccharomyces. Nature, London* **287**, 869–71.

Johnson I. S. (1983) Human insulin from recombinant DNA technology. *Science* **219**, 632–7.

Johnson P. H. & Grossman L. I. (1977) Electrophoresis of DNA in agarose gels: optimizing separations of conformational isomers of double-stranded and single-stranded DNAs. *Biochemistry* **16**, 4217–25.

Jones I. M., Primrose S. B. & Ehrlich S. D. (1982) Recombination between short direct repeats in a *recA* host. *Molec. gen. Genet.* **188**, 486–9.

Jones I. M., Primrose S. B., Robinson A. & Ellwood D. C. (1980) Maintenance of some Col E1-type plasmids in chemostat culture. *Molec. gen. Genet.* **180**, 579–84.

Jones I. M., Primrose S. B., Robinson A. & Ellwood D. C. (1981) Effect of growth rate and nutrient limitation on the transformability of *Escherichia coli* with plasmid DNA. *J. Bacteriol.* **146**, 841–6.

Jones K. & Murray K. (1975) A procedure for detection of heterologous DNA sequences in lambdoid phage by *in situ* hybridization. *J. molec. Biol.* **51**, 393–409.

Kan Y. W. & Dozy A. M. (1978) Polymorphisms of DNA sequence adjacent to human β-globin structural gene: relation to sickle mutation. *Proc. nat. Acad. Sci. U.S.A.* **75**, 5631–5.

Kangas T. T., Cooney, C. L. & Gomez R. F. (1982) Expression of a proline-enriched protein in *Escherichia coli. Appl. env. Micro.* **43**, 629–35.

Kaper J. B., Lockman H., Baldini M. M. & Levine M. M. (1984) A recombinant live oral cholera vaccine. *Biotechnology* **1**, 345–9.

Karn J., Brenner S., Barnett L. & Cesareni G. (1980) Novel bacteriophage λ cloning vector. *Proc. nat. Acad. Sci. U.S.A.* **77**, 5172–6.

Kastelein R. A., Berkhout B., Overbeek G. P. & van Duin J. (1983) Effect of the sequences upstream from the ribosome-binding site on the yield of protein from the cloned gene for phage MS2 coat protein. *Gene* **23**, 245–54.

Keggins K. M., Lovett P. S. & Duvall E. J. (1978) Molecular cloning of genetically active fragments of *Bacillus* DNA in *Bacillus subtilis* and properties of the vector plasmid pUB110. *Proc. nat. Acad. Sci. U.S.A.* **75**, 1423–7.

Keggins K. M., Lovett P. S., Marrero R. & Hoch S. O. (1979) Insertional inactivation of *trp*C in cloned *Bacillus trp* segments: evidence for a polar effect on *trp*F. *J. Bacteriol.* **139**, 1001–6.

Kelley W. S., Chalmers K. & Murray N. E. (1977) Isolation and characterization of a λ*polA* transducing phage. *Proc. nat. Acad. Sci. U.S.A.* **74**, 5632–6.

Kellogg S. T., Chatterjee D. K. & Chakrabarty A. M. (1981) Plasmid-assisted molecular breeding: new technique for enhanced biodegradation of persistent toxic chemicals. *Science* **214**, 1133–5.

Kelly T. J. & Smith H. O. (1970) A restriction enzyme from *Hemophilus influenzae*. II. Base sequence of the recognition site. *J. molec. Biol.* **51**, 393–409.

Kemp D. J., Coppel R. L., Cowman A. F., Saint R. B., Brown G. V. & Anders R. F. (1983) Expression of *Plasmodium falciparum* blood-stage antigens in *Escherichia coli*: Detection with antibodies from immune humans. *Proc. nat. Acad. Sci. U.S.A.* **80**, 3787–91.

Khorana H. G. (1979) Total synthesis of a gene. *Science* **203**, 614–25.

Khoury G. & Gruss P. (1983) Enhancer elements. *Cell* **33**, 313–14.

Kidd V. J., Wallace R. B., Itakura K. & Woo S. L. C. (1983) α_1-antitrypsin deficiency detection by direct analysis of the mutation in the gene. *Nature, London* **304**, 230–4.

Kingsman A. J., Clarke L., Mortimer R. K. & Carbon J. (1979) Replication in *Saccharomyces cerevisae* of plasmid pBR313 carrying DNA from the yeast *trp* 1 region. *Gene* **7**, 141–52.

Kleid D. G., Yansura D., Small B., Dowbenko D., Moore D. M., Grubman M. J., McKercher P. D., Morgan D. O., Robertson B. H. & Bachrach H. L. (1981) Cloned viral protein vaccine for foot-and-mouth disease: Responses in cattle and swine. *Science* **214**, 1125–9.

Koekman B. P., Ooms G., Klapwijk P. M. & Schilperoort R. A. (1979) Genetic map of an octopine Ti plasmid. *Plasmid* **2**, 347–57.

Kondoh H., Yasuda K. & Okada T. S. (1983) Tissue-specific expression of a cloned chick δ-crystallin gene in mouse cells. *Nature* **301**, 440–2.

Kornberg A. (1980) *DNA replication*. W. H. Freeman & Co., San Francisco.

Koshland D. & Bostein D. (1982) Evidence of post-translation translocation of β-lactamase across the bacterial inner membrane. *Cell* **30**, 893–902.

Kramer R. A., Cameron J. R. & Davis R. W. (1976) Isolation of bacteriophage λ containing yeast ribosomal RNA genes: screening by *in situ* RNA hybridization to plaques. *Cell* **8**, 227–32.

Kramer R. A., De Chiara T. M., Schaber M. D. & Hilliker S. (1984) Regulated expression of a human interferon gene in yeast: control by phosphate concentration or temperature. *Proc. nat. Acad. Sci. U.S.A.* **81**, 367–70.

Kreft J. & Hughes C. (1981) Cloning vectors derived from plasmids and phage of *Bacillus*. *Curr. Topics Microbiol. Immunol.* **96**, 1–17.

Krens F. A., Molendijk L., Wullems G. J. & Schilperoort R. A. (1982) *In vitro* transformation of plant protoplasts with Ti-plasmid DNA. *Nature, London* **296**, 72–4.

Krieg P. A. & Melton D. A. (1984) Formation of the 3′ end of histone mRNA by post-transcriptional processing. *Nature* **308**, 203–6.

Krisch H. M. & Selzer G. B. (1981) Construction and properties of a recombinant plasmid containing gene *32* of bacteriophage T4D. *J. molec. Biol.* **148**, 199–218.

Kroyer J. & Chang S. (1981) The promoter-proximal region of the *Bacillus licheniformis* penicillinase gene. Nucleotide sequence and predicted leader peptide sequence. *Gene* **15**, 343–7.

Kurtz D. T. & Nicodemus C. F. (1981) Cloning of $\alpha_{2\mu}$ globulin cDNA using a high efficiency technique for the cloning of trace messenger RNAs. *Gene* **13**, 145–152.

Lacy E., Roberts S., Evans E. P., Burtenshaw M. D. & Constantini F. D. (1983) A foreign β-globin gene in transgenic mice: Integration at abnormal chromosomal positions and expression in inappropriate tissues. *Cell* **34**, 343–58.

Lamond A. I. & Travers A. A. (1983) Requirement for an upstream element for optimal transcription of a bacterial tRNA gene. *Nature, London* **304**, 248–50.

Land H., Grey M., Hanser H., Lindenmaier W. & Schutz G. (1981) 5′-Terminal sequences of eucaryotic mRNA can be cloned with a high efficiency. *Nucleic Acids Res.* **9**, 2251–66.

Lane C. D., Colman A., Mohun T., Morser J., Champion J., Kourides I., Craig R., Higgins S., James T. C., Appelbaum S. W., Ohlsson R. I., Pauch E., Houghton M., Mathews J. & Miflin B. J. (1980) The *Xenopus* oocyte as a surrogate secretory system. The specificity of protein export. *Eur. J. Biochem.* **111**, 225–35.

Langford C. J. & Gallwitz (1983) Evidence for an intron-contained sequence required for the splicing of yeast RNA polymerase II transcripts. **33**, 519–27.

Langford C. J., Klinz F-J., Donath C. & Gallwitz D. (1984) Point mutations identify the conserved intron-contained TACTAAC box as an essential splicing signal sequence in yeast. *Cell* **36**, 645–53.

Langford C., Nellen W., Niessing J. & Gallwitz D. (1983) Yeast is unable to excise foreign intervening sequences from hybrid gene transcripts. *Proc. nat. Acad. Sci. U.S.A.* **80**, 1496–500.

Larsen J. E. L., Gerdes K., Light J. & Molin S. (1984) Low-copy-number plasmid cloning vectors amplifiable by derepression of an inserted foreign promoter. *Gene* **28**, 45–54.

Laskey R. A., Harland R. M. & Mechali M. (1983) Induction of chromosome replication during maturation of amphibian oocytes. In *Molecular Biology of Egg Maturation*. Ciba Symposium, pp. 25–36. Pitman, London.

Law M-F., Byrne J. & Hawley P. M. (1983) A stable Bovine Papillomavirus hybrid plasmid that expresses a dominant selective trait. *Molec. Cell Biol.* **3**, 2110–15.

Lawn R. M., Fritsch E. F., Parker R. C., Blake G. & Maniatis T. (1978) The isolation and characterization of linked δ and β-globin genes from a cloned library of human DNA. *Cell* **15**, 1157–74.

Lebeurier G., Hirth L., Hohn B. & Hohn T. (1982) *In vivo* recombination of cauliflower mosaic virus DNA. *Proc. nat. Acad. Sci. U.S.A.* **79**, 2932–6.

Lebeurier G., Hirth L., Hohn T. & Hohn B. (1980) Infectivities of native and cloned DNA of cauliflower mosaic virus. *Gene* **12** 139–46.

Lebeurier G., Whitechurch O., Lesot A. & Hirth L. (1978) Physical map of DNA from a new cauliflower mosaic virus strain. *Gene* **4**, 213–26.

Leder P., Tiemeier D. & Enquist L. (1977) EK2 derivatives of bacteriophage lambda useful in the cloning of DNA from higher organisms: the λgt WES system. *Science* **196**, 175–7.

Lederberg S. & Meselson M. (1964) Degradation of non-replicating bacteriophage DNA in non-accepting cells. *J. molec. Biol.* **8**, 623–8.

Lee F., Mulligan R., Berg P. & Ringold G. (1981) Glucocorticoids regulate expression of dihydrofolate reductase cDNA in mouse mammary tumour virus chimaeric plasmids. *Nature, London* **294**, 228–32.

Leemans J., Deblaere R., Willmitzer L., De Greve H., Hernalsteens J. P., Van Montagu M. & Schell J. (1982) Genetic identification of functions of T_L–DNA transcripts in octopine crown galls. *EMBO J.* **1**, 147–52.

Leemans J., Langenakens J., De Greve H., Deblaere R., Van Montagu M. & Schell J. (1982) Broad-host-range cloning vectors derived from the W-plasmid Sa. *Gene* **19**, 361–4.

Leemans J., Shaw C., Deblaere R., De Greve H., Hernalsteens J. B., Maes M., Van Montagu M. & Schell J. (1981) Site-specific mutagenesis of *Agrobacterium* Ti plasmids and transfer of genes to plant cells. *J. molec. appl. Genet.* **1**, 149–64.

Le Hegarat J. C. & Anagnostopoulos C. (1977) Detection and characterization of naturally occurring plasmids in *Bacillus subtilis. Molec. gen. Genet.* **157**, 164–74.

Lemmers M., De Beuckeleer M., Holsters M., Zambryski P., Hernalsteens J. P., Van Montagu M. & Schell J. (1980) Internal organization, boundaries and integration of Ti plasmid DNA in nopaline crown gall tumors. *J. molec. Biol.* **144**, 353–76.

Lis J. T., Simon J. A. & Sutton C. A. (1983) New heat shock puffs and β-galactosidase activity resulting from transformation of *Drosophila* with an *hsp 70-lac Z* Hybrid gene. *Cell* **35**, 403–10.

Lobban P. E. & Kaiser A. D. (1973) Enzymatic end-to-end joining of DNA molecules. *J. molec. Biol.* **78**, 453–71.

Loenen W. A. M. & Brammar W. J. (1980) A bacteriophage lambda vector for cloning large DNA fragments made with several restriction enzymes. *Gene* **10**, 249–59.

Lowy I., Pellicer A., Jackson J. F., Sim G. K., Silverstein S. & Axel R. (1980) Isolation of transforming DNA: Cloning of the hamster aprt gene. *Cell* **22**, 817–23.

Lung M. C. Y. & Pirone T. P. (1974) Acquisition factor required for aphid transmission of purified cauliflower mosaic-virus. *Virology* **60**, 260–4.

Lusky M. & Botchan M. (1981) Inhibitory effect of specific pBR322 DNA sequences upon SV40 replication in simian cells. *Nature, London.* **293**, 79–81.

McClintock B. (1951) Chromosome organization and genic expression. *Cold Spring Harb. Symp. Quant. Biol.* **16**, 13–47.

McKnight G. S., Hammer R. E., Kuenzel E. A. & Brinster R. L. (1983) Expression of the chicken transferrin gene in transgenic mice. *Cell* **34**, 335–41.

McKnight T. D. & Meagher R. B. (1981) Isolation and mapping of small cauliflower mosaic virus DNA fragments active as promoters in *Escherichia coli*. *J. Virol.* **37**, 673–82.

M[c]Laughlin J. R., Murray C. L. & Rabinowitz J. C. (1981) Unique features in the ribosome binding site sequence of the Gram-positive *Staphylococcus aureus* β-lactamase gene. *J. biol. Chem.* **256**, 11283–91.

Malpartida F. & Hopwood D. A. (1984) Molecular cloning of the whole biosynthetic pathway of a *Streptomyces* antibiotic and its expression in a heterologous host. *Nature* **309**, 462–4.

Mandel M. & Higa A. (1970) Calcium-dependent bacteriophage DNA infection. *J. molec. Biol.* **53**, 159–62.

Maniatis T., Hardison R. C., Lacy E., Lauer J., O'Connell C., Quon D., Sim G. K. & Efstratiadis A. (1978) The isolation of structural genes from libraries of eucaryotic DNA. *Cell* **15**, 687–701.

Maniatis T., Sim Gek Kee, Efstratiadis A. & Kafatos F. C. (1976) Amplification and characterization of a β-globin gene synthesized *in vitro*. *Cell* **8**, 163–82.

Martin R. G. (1981) The transformation of cell growth and transmogrification of DNA synthesis by Simian Virus 40. *Adv. Cancer Res.* **34**, 1–68.

Marton L., Wullems G. J., Molendijk L. & Schilperoort R. A. (1979) *In vitro* transformation of cultured cells from *Nicotiana tabacum* by *Agrobacterium tumefaciens*. *Nature, London* **277**, 129–31.

Marx J. L. (1982) Ti plasmids as gene carriers. *Science* **216**, 1305.

Matzke A. J. M. & Chilton M-D. (1981) Site-specific insertion of genes into T-DNA of the *Agrobacterium* tumor-inducing plasmid: an approach to genetic engineering of higher plant cells. *J. molec. appl. Genet.* **1**, 39–49.

Maxam A. M. & Gilbert W. (1977) A new method for sequencing DNA. *Proc. nat. Acad. Sci. U.S.A.* **74**, 560–64.

Maxam A. M. & Gilbert W. (1980) Sequencing end-labelled DNA with base-specific chemical cleavages. *Methods in Enzymology* **65**, 499–560.

Meagher R. B., Shepherd R. J. & Boyer H. W. (1977a) Structure of cauliflower mosaic-virus. I. Restriction endonuclease map of cauliflower mosaic-virus DNA. *Virology* **80**, 367–75.

Meagher R. B., Tait R. C., Betlach M. & Boyer H. W. (1977b) Protein expression in *E. coli* minicells by recombinant plasmids. *Cell* **10**, 521–36.

Mellon P., Parker V., Gluzman Y. & Maniatis T. (1981) Identification of DNA sequences required for transcription of the human α-globin gene in a new SV40 host-vector system. *Cell* **27**, 279–88.

Mercerau-Puijalon O., Royal A., Cami B., Garapin A., Krust A., Gannon F. & Kourilsky P. (1978) Synthesis of an ovalbumin-like protein by *Escherichia coli* K12 harbouring a recombinant plasmid. *Nature, London* **275**, 505–10.

Mertz J. E. & Berg P. (1974) Defective simian virus 40 genomes. Isolation and growth of individual clones. *Virology* **62**, 112–24.

Mertz J. E. & Gurdon J. B. (1977) Purified DNAs are transcribed after microinjection into *Xenopus* oocytes. *Proc. nat. Acad. Sci. U.S.A.* **74**, 1502–6.

Meselson M. & Yuan R. (1968) DNA restriction enzyme from *E. coli. Nature, London* **217**, 1110–14.

Messing J., Crea R. & Seeburg P. H. (1981) A system for shotgun DNA sequencing. *Nucleic Acids Res.* **9**, 309–21.

Messing, J., Gronenborn B., Muller-Hill B. & Hofschneider P. H. (1977) Filamentous coliphage M13 as a cloning vehicle: insertion of a *Hin*d II fragment of the *lac* regulatory region in M13 replicative form *in vitro. Proc. nat. Acad. Sci. U.S.A.* **74**, 3642–6.

Messing J. & Vieira J. (1982) A new pair of M13 vectors for selecting either DNA strand of double-digest restriction fragments. *Gene* **19**, 269–76.

Michel B., Niaudet B. & Ehrlich S. D. (1982) Intramolecular recombination during plasmid transformation of *Bacillus subtilis* competent cells. *EMBO J.* **1**, 1565–71.

Michel B., Palla E., Niaudet B. & Ehrlich S. D. (1980) DNA cloning in *Bacillus subtilis.* III. Efficiency of random-segment cloning and insertional inactivation vectors. *Gene* **12**, 147–54.

Michelson A. M. & Orkin S. H. (1982) Characterization of the homopolymer tailing reaction catalyzed by terminal deoxynucleotidyl transferase. Implications for the cloning of cDNA. *J. biol. Chem.* **256**, 1473–82.

Miller C. A., Tucker W. T., Meacock P. A., Gustafsson P. & Cohen S. N. (1983) Nucleotide sequence of the partition locus of *Escherichia coli* plasmid pSC101. *Gene* **24**, 309–15.

Moir A. & Brammar W. J. (1976) Use of specialized transducing phages in amplification of enzyme production. *Molec. gen. Genet.* **149**, 87–99.

Montaya A. L., Chilton M. D., Gordon M. P., Sciaky D. & Nester E. W. (1977) Octopine and nopaline metabolism in *Agrobacterium tumefaciens* and crown gall tumor cells: role of plasmid genes. *J. Bacteriol.* **129**, 101–7.

Moran C. P., Lang N., Le Grice S. F. J., Lee G., Stephens M., Sonenshein A. L., Pero J. & Losick R. (1982) Nucleotide sequences that signal the initiation of transcription and translation in *Bacillus subtilis. Molec. gen. Genet.* **186**, 339–46.

Moriarty A. M., Hoyer B. H., Shih J. W-K., Gerin J. L. & Hamer D. H. (1981) Expression of the hepatitis B virus surface antigen gene in cell culture by using a Simian virus 40 vector. *Proc. nat. Acad. Sci U.S.A.* **78**, 2606–10.

Morrow J. F., Cohen S. N., Chang A. C. Y., Boyer H. W., Goodman H. M. & Helling R. B. (1974) Replication and transcription of eukaryotic DNA in *Escherichia coli. Proc. nat. Acad. Sci. U.S.A.* **71**, 1743–7.

Mosbach K., Birnbaum S., Hardy K., Davies J. & Bülow L. (1983)

Formation of proinsulin by immobilized *Bacillus subtilis. Nature, London*. **302**, 543–5.

Moser R., Thomas R. M. & Gutte B. (1983) An artificial crystalline DDT-binding polypeptide. FEBS **157**, 247–51.

Moss T. (1982) Transcription of cloned *Xenopus laevis* ribosomal DNA microinjected into *Xenopus* oocytes, and the identification of an RNA polymerase I promoter. *Cell* **30**, 835–42.

Mottes M., Grandi G., Sgaramella V., Canosi U., Morelli G. & Trautner T. A. (1979) Different specific activities of the monomeric and oligomeric forms of plasmid DNA in transformation of *B. subtilis* and *E. coli. Molec. gen. Genet.* **174**, 281–6.

Muesing M., Tamm J., Shepard H. M. & Polisky B. (1981) A single base pair alteration is responsible for the DNA overproduction phenotype of a plasmid copy-number mutant. *Cell* **24**, 235–42.

Mulligan R. C. & Berg P. (1980) Expression of a bacterial gene in mammalian cells. *Science* **209**, 1422–7.

Mulligan R. C. & Berg P. (1981a) Factors governing the expression of a bacterial gene in mammalian cells. *Molec. cell. Biol.* **1**, 449–59.

Mulligan R. C. & Berg P. (1981b) Selection for animal cells that express the *Escherichia coli* gene coding for xanthine-guanine phosphoribosyltransferase. *Proc. nat. Acad. Sci. U.S.A.* **78**, 2072–6.

Mulligan R. C., Howard B. H. & Berg P. (1979) Synthesis of rabbit β-globin in cultured monkey kidney cells following infection with SV40 β-globin recombinant genome. *Nature, London* **277**, 108–114.

Murai N., Sutton D. W., Murray M. G., Slightom J. L., Merlo D. J., Reichert N. A., Sengupta-Gopalan C., Stock C. A., Barker R. F., Kemp J. D. & Hall T. C. (1983) Phaseolin gene from bean is expressed after transfer to Sunflower via tumor-inducing plasmid vectors. *Science* **222**, 476–82.

Murray A. W. & Szostak J. W. (1983a) Pedigree analysis of plasmid segregation in yeast. *Cell* **34**, 961–70.

Murray A. W. & Szostak J. W. (1983b) Construction of artificial chromosomes in yeast. *Nature, London*. **300**, 189–93.

Murray K. & Murray N. E. (1975) Phage lambda receptor chromosomes for DNA fragments made with restriction endonuclease III of *Haemophilus influenzae* and restriction endonuclease I of *Escherichia coli. J. molec. Biol.* **98**, 551–64.

Murray M. J., Shilo B-Z., Chiaho S., Cowing D., Hsu H. W. & Weinberg R. A. (1981) Three different human tumor cell lines contain different oncogenes. *Cell* **25**, 355–61.

Murray N. E. (1983) Phage lambda and molecular cloning. In *The Bacteriophage Lambda*. eds Hendrix R. W., Roberts J. W., Stahl F. W. & Weisberg R. A. Vol 2. Cold Spring Harbor Laboratory. Lambda II (Monograph No 13). Cold Spring Harbor N.Y (U.S.A. 1983.

Murray N. E., Brammar W. J. & Murray K. (1977) Lambdoid phages that simplify recovery of *in vitro* recombinants. *Molec. gen. Genet.* **150**, 53–61.

Murray N. E., Bruce S. A. & Murray K. (1979) Molecular cloning of the

DNA ligase gene from bacteriophage T4. II. Amplification and preparation of the gene product. *J. molec. Biol.* **132**, 493–505.

Murray N. E. & Kelley W. S. (1979) Characterization of λ*pol A* transducing phages. Effective expression of the *E. coli pol A* gene. *Molec. gen. Genet.* **175**, 77–87.

Myers R. M. & Tjian R. (1980) Construction and analysis of simian virus 40 origins defective in tumor antigen binding and DNA replication. *Proc. nat. Acad. Sci. U.S.A.* **77**, 6491–5.

Nagata S., Taira H., Hall A., Johnsrud L., Sreuli M., Ecsodi J., Boll W., Cantell K. & Weissmann C. (1980) Synthesis in *E. coli* of a polypeptide with human leukocyte interferon activity. *Nature, London* **284**, 316–20.

Narang S. A. (1983) DNA Synthesis. *Tetrahedron* **39**, 3–22.

Nasmyth K. A. & Reed S. I. (1980) Isolation of genes by complementation in yeast. Molecular cloning of a cell-cycle gene. *Proc. nat. Acad. Sci. U.S.A.* **77**, 2119–23.

Nathans J. & Hogness D. S. (1983) Isolation, sequence analysis, and intron-exon arrangement of the gene encoding bovine rhodopsin. *Cell* **34**, 807–14.

Nester E. W. & Kosuge T. (1981) Plasmids specifying plant hyperplasias. *Ann. Rev. Microbiol.* **35**, 531–65.

Neugebauer K., Sprengel R. & Schaller H. (1981) Penicillinase from *Bacillus licheniformis.* Nucleotide sequence of the gene and implications for the biosynthesis of a secretory protein in a Gram-positive bacterium. *Nucleic Acids Res.* **9**, 2577–88.

Newman A. J., Linn T. G. & Hayward R. S. (1979) Evidence for cotranscription of the RNA polymerase genes *rpo*BC with a ribosome protein of *Escherichia coli. Molec. gen. Genet.* **169**, 195–204.

Newport J. & Kirschner M. (1982a) A major developmental transition in early *Xenopus* embryos: I. Characterization and timing of cellular changes at the Midblastula stage. *Cell* **30**, 675–86.

Newport J. & Kirschner M. (1982b) A major developmental transition in early *Xenopus* embryos: II. Control of the onset of transcription. *Cell* **30**, 687–96.

Ng R. & Abelson J. (1980) Isolation and sequence of the gene for actin in *Saccharomyces cerevisiae. Proc. nat. Acad. Sci. U.S.A.* **77**, 3912–16.

Niaudet B. & Ehrlich S. D. (1979) *In vitro* genetic labelling of *Bacillus subtilis* cryptic plasmid pHV400. *Plasmid* **2**, 48–58.

Niaudet B., Goze A. & Ehrlich S. D. (1982) Insertional mutagenesis in *Bacillus subtilis*. Mechanism and use in gene cloning. *Gene* **19**, 277–84.

Norrander J., Kempe T. & Messing J. (1983) Construction of improved M13 vectors using oligodeoxynucleotide-directed mutagenesis. *Gene* **27**, 101–6.

Novick P., Field C. & Schekman R. (1980) Identification of 23 complementation groups required for post-translational events in the yeast secretory pathway. *Cell* **21**, 205–15.

Novick R. P., Clowes R. C., Cohen S. N., Curtiss R., Datta N. &

Falkow S. (1976) Uniform nomenclature for bacterial plasmids: a proposal. *Bact. Rev.* **40**, 168–89.

Nugent M. E., Primrose S. B. & Tacon W. C. A. (1983) The stability of recombinant DNA. *Devel. Ind. Microbiol.* **24**, 271–85.

Odell J. & Howell S. H. (1980) The identification, mapping and characterization of mRNA for P66, a cauliflower mosaic virus-coded protein. *Virology* **102**, 349–59.

Odell J. T., Dudley R. K. & Howell S. H. (1981) Structure of the 19S RNA transcript encoded by the cauliflower mosaic virus genome. *Virology* **111**, 377–85.

O'Hare K., Benoist C. & Breathnach R. (1981) Transformation of mouse fibroblasts to methotrexate resistance. *Proc. nat. Acad. Sci. U.S.A.* **78**, 1527–31.

O'Hare K. & Rubin G. M. (1983) Structures of P transposable elements and their sites of insertion and excision in the *Drosophila melanogaster* genome. *Cell* **34**, 25–35.

Okayama H. & Berg P. (1982) High-efficiency cloning of full-length cDNA. *Molec. Cell. Biol.* **2**, 161–70.

Old J. M., Ward R. H. T., Petrov M., Karagozlu F., Modell B. & Weatherall D. J. (1982) First trimester diagnosis for haemoglobinopathies: a report of 3 cases. *Lancet* **2**, 1413–16.

Old R. W., Woodland H. R., Ballantine J. E. M., Aldridge T.C., Newton C. A., Bains W. A. &. Turner P. C. (1982) Organization & expression of cloned histone gene clusters from *Xenopus laevis* and *X. borealis. Nucleic Acids Res.* **10**, 7561–80.

Olivera B. M., Hall Z. W. & Lehman I. R. (1968) Enzymatic joining of polynucleotides. V. A DNA adenylate intermediate in the polynucleotide joining reaction. *Proc. nat. Acad. Sci. U.S.A.* **61**, 237–44.

Olsen R. H., DeBusccher G. & McCombie W. R. (1982) Development of broad-host-range vectors and gene banks. Self-cloning of the *Pseudomonas aeruginosa* PAO chromosome. *J. Bacteriol.* **150**, 60–9.

Olson M. V. (1981) Applications of molecular cloning to *Saccharomyces*. In *Genetic Engineering*, eds Setlow J. K. & Hollaender A. Plenum Press, New York.

Ooms G., Hooykaas P. J. J., Moolenaar G., Schilperoort R. A. (1981) Crown gall plant tumours of abnormal morphology induced by *Agrobacterium tumefaciens* carrying mutated octopine Ti plasmids: analysis of T-DNA functions. *Gene* **14**, 33–50.

Ooms G., Klapwijk P. M., Poulis J. A. & Schilperoort R. A. (1980) Characterization of Tn904 insertion in octopine Ti plasmid mutants of *Agrobacterium tumefaciens. J. Bacteriol.* **144**, 82–91.

Orkin S. H. (1982) Genetic diagnosis of the foetus. *Nature, London.* **296**, 202–3.

Orkin S. H., Little P. F. R., Kazazian H. H. & Boehm C. (1982) Improved detection of the sickle mutation by DNA analysis. *New Eng. J. Med.* **307**, 32–6.

Ostrowski M. C., Richard-Foy H., Wolford R. G., Berard D. S. & Hager

G. L. (1983) Glucocorticoid regulation of transcription at an amplified episomal promoter. *Molec. cell. Biol.* **3**, 2045–57.

Palva I., Pettersson R. F., Nalkkinnen N., Lehtovaara P., Sarvas M., Söderlund H., Takkinen K. & Kääriäinen L. (1981) Nucleotide sequence of the promoter and NH_2-terminal signal peptide region of the α-amylase gene from *Bacillus amyloliquefaciens. Gene* **15**, 43–51.

Palva I., Sarvas M., Lehtovaara P., Sibakov M. & Kääriäinen L. (1982) Secretion of *Escherichia coli* β-lactamase from *Bacillus subtilis* by the aid of α-amylase signal sequence. *Proc. nat. Acad. Sci. U.S.A.* **79**, 5582–6.

Palva I., Lehtovaara P., Kääriäinen L., Sibakov M., Cantell K., Schein C. H., Kashiwagi K. & Weissmann C. (1983) Secretion of interferon by *Bacillus subtilis. Gene* **22**, 229–35.

Palmiter R. D., Bruister R. L., Hammer R. E., Trumbauer M. E., Rosenfeld M. G., Birnberg N. C. & Evans R. M. (1982a) Dramatic growth of mice that develop from eggs microinjected with metallothionein-growth hormone fusion genes. *Nature, London* **300**, 611–15.

Palmiter R. D., Chen H. Y. & Brinster R. L. (1982b) Differential regulation of metallothionein-thymidine kinase fusion genes in transgenic mice and their offspring. *Cell* **29**, 701–10.

Panasenko S. M., Cameron J. R., Davis R. W. & Lehman I. R. (1977) Five hundredfold overproduction of DNA ligase after induction of a hybrid lambda lysogen contructed *in vitro. Science* **196**, 188–9.

Paterson B. M., Roberts B. E. & Kuff E. L. (1977) Structural gene identification and mapping by DNA.mRNA hybrid-arrested cell-free translation. *Proc. nat. Acad. Sci. U.S.A.* **74**, 4370–4.

Peden K. W. C. (1983) Revised sequence of the tetracycline-resistance gene of pBR322. *Gene* **22**, 277–80.

Pennica D., Holmes W. E., Kohr W. J., Harkins R. N., Vehar G. A., Ward C. A., Bennett W. F., Yelverton E., Seeburg P. H., Heyneker H. L. & Goeddel D. V. (1983) Cloning and expression of human tissue-type plasminogen activator cDNA in *E. coli. Nature, London* **301**, 214–21.

Perucho M., Goldfarb M., Shimizu K., Lama C., Fogh J. & Wigler M. (1981) Human-tumor-derived cell lines contain common and different transforming genes. *Cell* **27**, 467–76.

Perucho M., Hanahan D., Lipsich L. & Wigler M. (1980a) Isolation of the chicken thymidine kinase gene by plasmid rescue. *Nature, London* **285**, 207–10.

Perucho M., Hanahan D. & Wigler M. (1980b) Genetic and physical linkage of exogenous sequences in transformed cells. *Cell* **22**, 309–17.

Pfeiffer P. & Hohn T. (1983) Involvement of reverse transcription in the replication of cauliflower mosaic virus. A detailed model and test of some aspects. *Cell* **33**, 781–9.

Plant D. W., Reimers N. J. & Zinder N. D. (1982) *Banbury Report 10: Patenting of Life Forms.* Cold Spring Harbor Laboratory, Cold Spring Harbor.

Powell K. A. & Byrom D. (1983) Culture stability in strains used for single cell protein production. In *Genetics of Industrial Microorganisms*, eds Ikeda Y. & Beppu T., pp. 345–50. Kodansha Ltd., Tokyo, Japan.

Pratt J. M., Boulnois G. J., Darby V., Orr E., Wahle E. & Holland I. B. (1981) Identification of gene products programmed by restriction endonuclease DNA fragments using an *E. coli* in vitro system. *Nucleic Acids Res.* **9**, 4459–74.

Prentki P. & Kirsch H. M. (1982) A modified pBR322 vector with improved properties for the cloning, recovery, and sequencing of blunt-ended DNA fragments. *Gene* **17**, 189–96.

Primrose S. B., Derbyshire P., Jones I. M., Nugent M. E. & Tacon W. C. A. (1983) Hereditary instability of recombinant DNA molecules. In *Bioactive Microbial Products 2: Development and Production* eds Nisbet L. J. and Winstanley D. J., pp. 63–77. Academic Press, London.

Primrose S. B. & Dimmock N. J. (1980) *Introduction to Modern Virology*, 2e. Blackwell Scientific Publications, Oxford.

Primrose S. B. & Ehrlich S. D. (1981) Isolation of plasmid deletion mutants and a study of their instability. *Plasmid* **6**, 193–201.

Probst E., Kressmann A. & Birnstiel M. L. (1979) Expression of sea urchin histone genes in the oocyte of *Xenopus laevis*. *J. molec. Biol.* **135**, 709–32.

Putney S. D., Herlihy W. C. & Schimmel P. (1983) A new troponin T and cDNA clones for 13 different muscle proteins, found by shotgun sequencing. *Nature, London* **302**, 718–21.

Queen C. & Rosenberg, M. (1981) A promoter of pBR322 activated by CAP receptor protein. *Nucleic Acids Res.* **9**, 3365–77.

Radloff R., Bauer W. & Vinograd J. (1967) A dye-buoyant-density method for the detection and isolation of closed circular duplex DNA: the closed circular DNA in HeLa cells. *Proc. nat. Acad. Sci. U.S.A.* **57**, 1514–21.

Rapoport G., Klier A., Billault A., Fargette F. & Dedonder R. (1979) Construction of a colony bank of *E. coli* containing hybrid plasmids representative of the *Bacillus subtilis* 168 genome: expression of functions harbored by the recombinant plasmids in *B. subtilis*. *Molec. gen. Genet.* **176**, 239–45.

Rassoulzadegan M., Binetruy B. & Cuzin F. (1982) High frequency of gene transfer after fusion between bacteria and eukaryotic cells. *Nature, London.* **295**, 257–9.

Ratzkin B. & Carbon J. (1977) Functional expression of cloned yeast DNA in *Escherichia coli*. *Proc. nat. Acad. Sci. U.S.A.* **74**, 487–91.

Razin A. & Riggs A. D. (1980) DNA methylation and gene function. *Science* **210**, 604–10.

Richards K. E., Guilley H. & Jonard G. (1981) Further characterization of the discontinuities in cauliflower mosaic virus DNA. *Febs Letters* **134**, 67–70.

Richins R. D. & Shepherd R. J. (1983) Physical maps of the genomes of dahlia mosaic virus and mirabilis mosaic virus–two members of the caulimovirus group. *Virology* **124**, 208–14.

Rigby P. W. J., Dieckmann M., Rhodes C. & Berg P. (1977) Labelling deoxyribonuclic acid to high specific activity *in vitro* by nick translation with DNA polymerase I. *J. molec. Biol.* **113**, 237–51.

Rine J., Hansen W., Hardeman E. & Davis R. W. (1983) Targeted selection of recombinant clones through gene dosage effects. *Proc. nat. Acad. Sci. U.S.A.*, **80**, 6750–4.

Robbins D. M., Ripley S., Henderson A. & Axel R. (1981) Transforming DNA integrates into the host chromosome. *Cell* **23**, 29–39.

Roberts R. J. (1984) Restriction and modification enzymes and their recognition sequences. *Nucleic Acids Res.* **12**, 167–91.

Roberts T. M. Bikel I., Yocum R. R., Livingston D. M. & Ptashne M. (1979a) Synthesis of simian virus 40 t antigen in *Escherichia coli. Proc. nat. Acad. Sci. U.S.A.* **76**, 5596–600.

Roberts T. M., Kacich R. & Ptashne M. (1979b) A general method for maximizing the expression of a cloned gene. *Proc. nat Acad. Sci. U.S.A.* **76**, 760–4.

Roberts T. M., Swanberg S. L., Poteete A., Riedal G. & Backman K. (1980) A plasmid cloning vehicle allowing a positive selection for inserted fragments. *Gene* **12**, 123–7.

Rochaix J-D., van Dillewijn J. & Rahire M. (1984) Construction and characterization of autonomously replicating plasmids in the green unicellular alga *Chlamydomonas reinhardii. Cell* **36**, 925–31.

Rodriguez R. L., West R. W., Heyneker H. L., Bolivar F. & Boyer H. W. (1979) Characterizing wild-type and mutant promoters of the tetracycline resistance gene in pBR313. *Nucleic Acids Res.* **6**, 3267–87.

Roeder R. G. (1974) Multiple forms of DNA-dependent RNA polymerase in *X. laevis. J. biol. Chem.* **249**, 249–56.

Rood J. I., Sneddon M. K. & Morrison J. F. (1980) Instability in *tyrR* strains of plasmids carrying the tyrosine operon: isolation and characterization of plasmid derivatives with insertions or deletions. *J. Bacteriol.* **144**, 552–9.

Rosamond J., Endlich B. & Linn S. (1979) Electron microscopic studies of the mechanism of action of the restriction endonuclease of *Escherichia coli* B. *J. molec. Biol.* **129**, 619–35.

Rose M., Casadaban M. J. & Botstein D. (1981) Yeast genes fused to β-galactosidase in *Escherichia coli* can be expressed normally in yeast. *Proc. nat. Acad. Sci. U.S.A.* **78**, 2460–4.

Rosteck P. R. Jr & Hershberger C. L. (1983) Selective retention of recombinant plasmids coding for human insulin. *Gene* **25**, 29–38.

Rothstein R. J., Lau L. F., Bahl C. P., Narang S. A. & Wu R. (1979) Synthetic adaptors for cloning DNA. In *Methods in Enzymology, Vol. 68*, ed. Wu R., pp. 98–109. Academic Press, New York.

Roychoudhury R., Jay E. & Wu R. (1976) Terminal labelling and addition of homopolymer tracts to duplex DNA fragments by terminal deoxynucleotidyl transferase. *Nucleic Acids Res.* **3**, 863–77.

Rubin E. M., Wilson G. A. & Young F. E. (1980) Expression of thymidylate synthetase activity in *Bacillus subtilis* upon integration of a cloned gene from *Escherichia coli. Gene* **10**, 227–35.

Rubin G. M., Kidwell M. G. & Bingham P. M. (1982) The molecular basis of P-M hybrid dysgenesis: the nature of induced mutations. *Cell* **29**, 987–94.

Rubin G. M. & Spradling A. C. (1982) Genetic transformation of *Drosophila* with transposable element vectors. *Science* **218**, 348–53.

Rubin J. S., Joyner A. L., Bernstein A. & Whitmore G. F. (1983) Molecular identification of a human DNA repair gene following DNA-mediated gene transfer. *Nature, London.* **306**, 206–8.

Rungger D. & Turler H. (1978) DNAs of simian virus 40 and polyoma direct the synthesis of viral tumor antigens and capsid proteins in *Xenopus* oocytes. **75**, 6073–7.

Rusconi S. & Schaffner W. (1981) Transformation of frog embryos with a rabbit β-globin gene. *Proc. nat. Acad. Sci. U.S.A.* **78**, 5051–5.

Russell D. R. & Bennett G. N. (1981) Characterization of the β-lactamase promoter of pBR322. *Nucleic Acids Res.* **9**, 2517–33.

Russell D. R. & Bennett G. N. (1982) Construction and analysis of in vivo activity of *E. coli* promoter hybrids and promoter mutants that alter the –35 to –10 spacing. *Gene* **20**, 231–43.

Ruvkun G. B. & Ausubel F. M. (1981) A general method for site-directed mutagenesis in prokaryotes. *Nature, London* **289**, 85–88.

Ryoji M. & Worcel A. (1984) Chromatin assembly in *Xenopus* oocytes: in vivo studies. *Cell* **37**, 21–32.

Sancar A., Hack A. M. & Rupp W. D. (1979) Simple method for identification of plasmid-coded proteins. *J. Bacteriol.* **137**, 692–3.

Sanger F., Coulson A. R., Hong G. F., Hill D. F. & Petersen G. B. (1982) Nucleotide sequence of bacteriophage λ DNA. *J. molec. Biol.* **162**, 729–73.

Sanger F., Nicklen S. & Coulson A. R. (1977) DNA sequencing with chain-terminating inhibitors. *Proc. nat. Acad. Sci. U.S.A.* **74**, 5463–7.

Sarver N., Gruss P., Law M-F., Khoury G. & Howley P. M. (1981a) Bovine papilloma virus deoxyribonucleic acid: a novel eucaryotic cloning vector. *Molec. cell. Biol.* **1**, 486–96.

Sarver N., Gruss P., Law M-F., Khoury G. & Howley P. M. (1981b) Rat insulin gene covalently linked to bovine papilloma virus DNA is expressed in transformed mouse cells. In *Developmental Biology Using Purified Genes.* eds Brown D. & Fox C. R., ICN-UCLA Symposia on molecular and cellular biology, vol. 23. Academic Press, New York.

Scalenghe F., Turco E., Edstrom J. E., Pirotta V. & Melli M. (1981) Microdissection and cloning of DNA from a specific region of *Drosophila melanogaster* polytene chromosomes. *Chromosoma (Berl.)* **82**, 205–16.

Scangos G. A., Huttner K. M., Juricek D. K. & Ruddle F. H. (1981) DNA-mediated gene transfer in mammalian cells: molecular analysis of unstable transformants and their progression to stability. *Mol. cell Biol.* **1**, 111–20.

Schaffner W. (1980) Direct transfer of cloned genes from bacteria to mammalian cells. *Proc. nat. Acad. Sci. U.S.A.* **77**, 2163–7.

Scheidereit C., Greisse S., Westphal H. M. & Beato M. (1983) The glucocorticoid receptor binds to defined nucleotide sequences near the promoter of mouse mammary tumour virus. *Nature, London* **304**, 749–52.

Schell J. & Van Montagu M. (1977) The Ti-plasmid of *Agrobacterium tumefaciens*, a natural vector for the introduction of fix genes in plants. In *Genetic Engineering for Nitrogen Fixation*, ed. Hollaender A., pp. 159–79. Plenum Press, New York.

Schell J., Van Montagu M., Holsters M., Hernalsteens J. P., Dhaese P., De Greve H., Leemans J., Joos H., Inze D., Willmitzer L., Otten L., Wostemeyer A., Schroder G. & Schroder J. (1982) Plant cells transformed by modified Ti plasmids: A model system to study plant development. In *Biochemistry of Differentiation and Morphogenesis*, pp. 65–73. Springer-Verlag, Berlin.

Scheller R. H., Dickerson R. E., Boyer H. W., Riggs A. D. & Itakura K. (1977) Chemical synthesis of restriction enzyme recognition sites useful for cloning. *Science* **196**, 177–80.

Schimke R. T., Kaufman R. J., Alt F. W. & Kellems R. F. (1978) Gene amplication and drug resistance in cultured murine cells. *Science* **202**, 1051–5

Schmidhauser T. J., Filutowicz M. & Helinski D. R. (1983) Replication of derivatives of the broad host range plasmid RK2 in two distantly related bacteria. *Plasmid* **9**, 325–30.

Scholnick S. B., Morgan B. A. & Hirsh J. (1983) The cloned Dopa decarboxylase gene is developmentally regulated when reintegrated into the *Drosophila* genome. *Cell* **34**, 37–45.

Schoner R. G., Williams D. M. & Lovett P. S. (1983) Enhanced expression of mouse dihydrofolate reductase in *Bacillus subtilis. Gene* **22**, 47–57.

Schottel J. L., Sninksky J. J. & Cohen S. N. (1984) Effects of alternations in the translation control region on bacterial gene expression: use of *cat* gene constructs transcribed from the *lac* promoter as a model system. *Gene* **28**, 177–93.

Schumann W. (1979) Construction of an *Hpa* I and *Hind* II plasmid vector allowing direct selection of transformants harboring recombinant plasmids. *Molec. gen. Genet.* **174**, 221–4.

Sciaky D., Montoya A. L. & Chilton M. D. (1978) Fingerprints of *Agrobacterium* Ti plasmids. *Plasmid* **1**, 238–54.

Seeburg P. H., Shine J., Martial J. A., Baxter J. D. & Goodman H. M. (1977) Nucleotide sequence and amplification in bacteria of the structural gene for rat growth hormone. *Nature, London* **270**, 486–94.

Seed B. (1983) Purification of genome sequence from bacteriophage libraries by recombination and selection *in vivo. Nucleic Acids Res.* **8**, 2427–45.

Sgaramella V. (1972) Enzymatic oligomerization of bacteriophage P22 DNA and of linear Simian Virus 40 DNA. *Proc. nat. Acad. Sci. U.S.A.* **69**, 3389–93.

Shaw C. H., Leemans J., Shaw C. H., Van Montague M. & Schell J. (1983) A general method for the transfer of cloned genes to plant cells. *Gene* **23**, 315–30.

Shepard H. M., Yelverton E. & Goeddel D. V. (1982) Increased synthesis in *E. coli* of fibroblast and leukocyte interferons through alterations in ribosome binding sites. *DNA* **1**, 125–31.

Shepherd R. J. (1979) DNA plant viruses. *Ann. Rev. Plant Phys.* **30**, 405–23.

Shimotohno K. & Temin H. M. (1981) Formation of infectious progeny virus after insertion of herpes simplex thymidine kinase gene into DNA of an avian retrovirus. *Cell* **26**, 67–77.

Shimotohno K. & Temin H. M. (1982) Loss of intervening sequences in genomic mouse α-globin DNA inserted in an infectious retrovirus vector. *Nature, London.* **299**, 265–8.

Shine J. & Dalgarno L. (1975) Determinant of cistron specificity in bacterial ribosomes. *Nature, London* **254**, 34–8.

Shine J., Fettes I., Lan N. C. Y., Roberts J. L. & Baxter J. D. (1980) Expression of cloned β-endorphin gene sequences by *Escherichia coli. Nature, London* **285**, 456–61.

Shirakawa M., Tsurimoto T. & Matsubara K. (1984) Plasmid vectors designed for high-efficiency expression controlled by the portable *rec*A promoter-operator of *Escherichia coli. Gene* **28**, 127–32.

Shorenstein R. G. & Losick R. (1973) Comparative size and properties of the sigma subunits of ribonucleic acid polymerase from *Bacillus subtilis* and *Escherichia coli. J. biol. Chem.* **248**, 6170–3.

Simon L. D., Randolph B., Irwin N. & Binkowski G. (1983) Stabilization of proteins by a bacteriophage T4 gene cloned in *Escherichia coli. Proc. nat. Acad. Sci. U.S.A.* **80**, 2059–62.

Simpson R., O'Hara P., Lichtenstein C., Montoya A. L., Kwok K., Gordon M. P. & Nester E. W. (1982) The DNA from A6S/Z tumor contains scrambled Ti plasmid sequence near its junction with plant DNA. *Cell* **29**, 1005–14.

Singh H., Bieker J. J. & Dumas L. B. (1982) Genetic transformation of *Saccharomyces cerevisiae* with single-stranded circular DNA vectors. *Gene* **20**, 441–9.

Skalka A. & Shapio L. (1976) *In situ* immunoassays for gene translation products in phage plaques and bacterial colonies. *Gene* **1**, 65–79.

Skogman G., Nilsson J. & Gustafsson P. (1983) The use of a partition locus to increase stability of tryptophan–operon-bearing plasmids in *Escherichia coli. Gene* **23**, 105–15.

Smith D. F., Searle P. F. & Williams J. G. (1979) Characterization of bacterial clones containing DNA sequences derived from *Xenopus laevis. Nucleic Acids Res.* **6**, 487–506.

Smith, E. F. & Townsend C. O. (1907) A plant-tumour of bacterial origin. *Science* **25**, 671–3

Smith G. L., Mackett M. & Moss B. (1983) Infectious vaccinia virus recombinants that express hepatitis B virus surface antigen. *Nature, London* **302**, 490–5.

Smith G. L., Murphy B. R. & Moss B. (1983) Construction and characterization of an infectious vaccinia virus recombinant that expresses the influenza hemagglutinin gene and induces resistance to influenza virus infection in hamsters. *Proc. nat. Acad. Sci. U.S.A.* **80**, 7155–9.

Smith H. O & Nathans D. (1973) A suggested nomenclature for bacterial host modification and restriction systems and their enzymes. *J. molec Biol.* **81**, 419–23.

Smith H. O. & Wilcox K. W. (1970) A restriction enzyme from

Hemophilus influenzae. I. Purification and general properties. *J. molec. Biol.* **51**, 379–91.

Smith J., Cook E., Fotheringham I., Pheby S., Derbyshire R., Eaton M. A. W., Doel M., Lilley D. M. J., Patel T., Lewis H. & Bell L. D. (1982) Chemical synthesis and cloning of gene for human β-urogastrone. *Nucleic Acids Res.* **10**, 4467–82.

Southern E. M. (1975) Detection of specific sequences among DNA fragments separated by gel electrophoresis. *J. molec. Biol.*, **98**, 503–17.

Southern E. M. (1979a) Measurement of DNA length by gel electrophoresis. *Analytical Biochem*, **100**, 319–23.

Southern E. M. (1979b) Gel electrophoresis of restriction fragments. *Methods Enzymol.* **68**, 152–76.

Sorge J. & Hughes S. (1982) The splicing of intervening sequences introduced into an infectious retrovirus vector. *J. molec. applied genet.* **1**, 547–59.

Spoerel N., Herlich P. & Bickle T. A. (1979) A novel bacteriophage defence mechanism: the anti-restriction protein. *Nature, London* **278**, 30–4.

Spradling A. C. & Rubin G. M. (1982) Transposition of cloned P. elements into *Drosophila* germ line chromosomes. *Science* **218**, 341–7.

Spradling A. C. & Rubin G. M. (1983) The effect of chromosomal position on the expression of the *Drosophila* xanthine dehydrogenase gene. *Cell* **34**, 47–57.

Sprague K. V., Faulds D. H. & Smith G. R. (1978) A single base-pair change creates a *chi* recombinational hotspot in bacteriophage λ. *Proc. nat. Acad. Sci. U.S.A.* **75**. 6182–6.

Stahl U., Tudzynski P., Kuck U. & Esser K. (1982) Replication and expression of a bacterial-mitochondrial hybrid plasmid in the fungus *Podospora anserina. Proc. nat. Acad. Sci. U.S.A.* **79**, 3641–5.

Stallcup M. R., Sharrock W. J. & Rabinowitz, J. C. (1974) Ribosome and messenger specificity in protein synthesis by bacteria. *Biochem. biophys. Res. Commun.* **58**, 92–8.

Stanley J. & Gay M. R. (1983) Nucleotide sequence of cassava latent virus DNA. *Nature, London.* **301**, 260–2.

Steitz J. A. (1979) Genetic signals and nucleotide sequences in messenger RNA. In *Biological Regulation and Development. 1. Gene Expression*, ed. Goldberger R. F., pp. 349–99. Plenum Press, New York.

Stepien P. P., Brousseau R., Wu R., Narang S. & Thomas D. Y. (1983) Synthesis of a human insulin gene VI. Expression of the synthetic pro-insulin gene in yeast. *Gene* **24**, 289–97.

Stetler G. L. & Thorner J. (1984) Molecular cloning of hormone-responsive genes from the yeast *Sacchoromyces cerevisiae. Proc. nat. Acad. Sci. U.S.A.* **81**, 1144–8.

Stewart A. G., Richards H., Roberts S., Warwick J., Edwards K., Bell L., Smith J. & Derbyshire R. (1983) Cloning and expression of a porcine prorelaxin gene in E. coli. *Nucleic Acids Res.* **11**, 6597–609.

Stinchcomb D. T., Mann C. & Davis R. W. (1982) Centromeric DNA from *Saccharomyces cerevisiae. J. molec. Biol.* **158**, 157–79.

Stinchcomb D. T., Struhl K. & Davis R. W. (1979) Isolation and characterization of a yeast chromosomal replicator. *Nature, London* **282**, 39–43.

Stohl L. L. & Lambowitz A. M. (1983) Construction of a shuttle vector for the filamentous fungus *Neurospora crassa. Proc. nat. Acad. Sci. U.S.A.* **80**, 1058–62.

Stoker N. G., Fairweather N. F. & Spratt B. G. (1982) Versatile low-copy-number plasmid vectors for cloning in *Escherichia coli. Gene* **18**, 335–41.

Storb U., O'Brien R. L., McMullen M. D., Gollahon K. A. & Brimster R. L. (1984) High expression of cloned immunoglobulin kappa gene in transgenic mice is restricted to B-lymphocytes. *Nature, London* **310**, 238–48.

Stormo G. D., Schneider T. D. & Gold L. M. (1982) Characterization of translational initiation sites in *E. coli. Nucleic Acids Res.* **10**, 2971–96.

Storms R. K., McNeil J. B., Khandekar P. S., An G., Parker J. & Friesen J. D. (1979) Chimaeric plasmids for cloning of deoxyribonucleic acid sequences in *Saccharomyces cerevisiae. J. Bacteriol.* **140**, 73–82.

Strueli M., Hall A., Boll W., Stewart W. E. II, Nagata S. & Weissmann C. (1982) Target cell specificity of two species of human interferon-α produced in *Escherichia coli* and of hybrid molecules derived from them. *Proc. nat. Acad. Sci. U.S.A.* **78**, 2848–52.

Struhl K. (1983) The new yeast genetics. *Nature, London* **305**, 391–7.

Struhl K., Cameron J. R. & Davis R. W. (1976) Functional genetic expression of eukaryotic DNA in *Escherichia coli. Proc. nat. Acad. Sci. U.S.A.* **73**, 1471–5.

Struhl K. & Davis R. D. (1980) A physical, genetic and transcriptional map of the cloned his3 gene region of *Saccharomyces cerevisiae. J. molec. Biol.* **136**, 309–32.

Struhl K., Stinchcomb D. T., Scherer S. & Davis R. W. (1979) High-frequency transformation of yeast: autonomous replication of hybrid DNA molecules. *Proc. nat. Acad. Sci. U.S.A.* **76**, 1035–9.

Stüber D. & Bujard H. (1981) Organization of transcriptional signals in plasmids pBR322 and pACYC184. *Proc. nat. Acad. Sci. U.S.A.* **78**, 167–71.

Stueber D. & Bujard H. (1982) Transcription from efficient promoters can interfere with plasmid replication and diminish expression of plasmid specified genes. *EMBO J.* **1**, 1399–404.

Stunnenberg H. G. & Birnstiel M. L. (1982) Bioassay for components regulating eukaryotic gene expression: A chromosomal factor involved in the generation of histone mRNA 3′-termini. *Proc. nat. Acad. Sci. U.S.A.* **79**, 6201–4.

Summers D. K. & Sherratt D. J. (1984) Multimerization of high copy number plasmids causes instability: Col E1 encodes a determinant essential for plasmid monomerization and stability. *Cell* **36**, 1097–103.

Sutcliffe J. G. (1978) pBR322 restriction map derived from the DNA sequence: accurate DNA size markers up to 4361 nucleotide pairs long. *Nucleic Acids Res.* **5**, 2721–28.

Sutcliffe J. G. (1979) Complete nucleotide sequence of the *Escherichia coli* plasmid pBR322. *Cold Spring Harb. Symp. quant. Biol.* **43** (1), 77–90.

Sveda M. M. & Lai C-J. (1981) Functional expression in primate cells of cloned DNA coding for the hemagluttinin surface glycoprotein of influenza virus. *Proc. nat. Acad. Sci U.S.A.* **78**, 5488–92.

Swyryd E. A., Seaver S. & Stark G. R. (1974) N-(phosphonacetyl)-L-aspartate, a potent transition state analog inhibitor of aspartate transcarbamylase, blocks proliferation of mammalian cells in culture. *J. biol. Chem.* **249**, 6945–50.

Szeto W., Hamer D. H., Carlson P. S. & Thomas C. A. (1977) Cloning of cauliflower mosaic virus (CaMV)DNA in *Escherichia coli. Science* **196**, 210–12.

Szostak J. W. & Blackburn E. H. (1982) Cloning yeast telomeres on linear plasmid vectors. *Cell* **29**, 245–55.

Szybalska E. H. & Szybalski W. (1962) Genetics of human cell lines, IV. DNA-mediated heritable transformation of a biochemical trait. *Proc. nat. Acad. Sci. U.S.A.* **48**, 2026–34.

Szybalski E. H. & Szybalski W. (1979) A comprehensive molecular map of bacteriophage lambda. *Gene* **7**, 217–70.

Tabin C. J., Hoffman J. W., Groff S. P. & Weinberg R. A. (1982) Adaption of a retrovirus as a eucaryotic vector transmitting the herpes simplex virus thymidine kinase gene. *Molec. cell. Biol.* **2**, 426–36.

Tacon W. C. A., Bonass W. A., Jenkins B. & Emtage J. S. (1983) Plasmid expression vectors containing tryptophan promoter transcriptional regulons lacking the attenuator. *Gene* **23**, 255–65.

Tacon W., Cary N. & Emtage S. (1980) The construction and characterization of plasmid vectors suitable for the expression of all DNA phases under the control of the *E. coli* tryptophan promoter. *Molec. gen. Genet.* **177**, 427–38.

Tait R. C., Close T. J., Lundquist R. C., Hagiya M., Rodriguez R. L. & Kado C. I. (1983) Construction and characterization of a versatile broad host range DNA cloning system for Gram-negative bacteria. *Biotechnology* **1**, 269–75.

Tait R. C., Lundquist R. C. & Kado C. I. (1982) Genetic map of the crown gall suppressive *Inc*W plasmid pSa. *Molec. gen. Genet.* **186**, 10–15.

Talmadge K. & Gilbert W. (1980) Construction of plasmid vectors with unique *Pst*I cloning sites in a signal sequence coding region. *Gene* **21**, 235–41.

Talmadge K. & Gilbert W. (1982) Cellular location affects protein stability in *Escherichia coli. Proc. nat. Acad. Sci. U.S.A.* **79**, 1830–3.

Talmadge K., Stahl S. & Gilbert W. (1980) Eukaryotic signal sequence transports insulin antigen in *Escherichia coli. Proc. nat. Acad. Sci. U.S.A.* **77**, 3369–73.

Taniguchi T., Guarente L., Roberts T. M., Kimelman D., Douhan J. & Ptashne M. (1980) Expression of the human fibroblast interferon gene in *Escherichia coli. Proc. nat. Acad. Sci. U.S.A.* **77**, 5230–3.

Taniguchi T., Matsui H., Fujita T., Takaoka C., Kashima N., Yoshimoto

R. & Hamuro J. (1983) Structure and expression of a cloned cDNA for human interleukin-2. *Nature, London* **302**, 305–10.

Teem J. L. & Rosbash M. (1983) Expression of a β-galactosidase gene containing the ribosomal protein 51 intron is sensitive to the *rna* 2 mutation of yeast. *Proc. nat. Acad. Sci. U.S.A.* **80**, 4403–7.

Thomas C. M., Stalker D. M., Guiney D. G. & Helinski D. R. (1979) Essential regions for the replication and conjugal transfer of the broad host range plasmid RK2. In *Plasmids of Medical, Environmental and Commercial Importance*, ed. Timmis K. N. & Puhler A., pp. 375–85. Elsevier/North Holland Biomedical Press, Amsterdam.

Thomas C. M., Stalker D. M. & Helinski D. R. (1981) Replication and incompatibility properties of segments of the origin region of replication of the broad host range plasmid RK2. *Molec. gen. Genet.* **81**, 1–7.

Thomas M., Cameron J. R. & Davis R. W. (1974) Viable molecular hybrids of bacteriophage lambda and eukaryotic DNA. *Proc. nat. Acad. Sci. U.S.A.* **71**, 4579–83.

Thomas P. S. (1980) Hybridization of denatured RNA and small DNA fragments transferred to nitrocellulose. *Proc. nat. Acad. Sci. U.S.A.* **77**, 5201–5.

Thomashow M. F., Nutter R., Montoya A. L., Gordon M. P. & Nester E. W. (1980) Integration and organization of Ti plasmid sequences in crown gall tumours. *Cell* **19**, 729–39.

Tikchonenko T. I., Karamov E. V., Zavizion B. A. & Naroditsky B. S. (1978) *Eco*RI* activity: enzyme modification or activation of accompanying endonuclease? *Gene* **4**, 195–212.

Tilghman S. M., Tiemeier D. C., Polsky F., Edgell M. H., Seidman J. G., Leder A., Enquist L. W., Norman B. & Leder P. (1977) Cloning specific segments of the mammalian genome: bacteriophage λ containing mouse globin and surrounding gene sequences. *Proc. nat. Acad. Sci. U.S.A.* **74**, 4406–10.

Tomizawa J-I. & Itoh T. (1981) Plasmid ColE1 incompatibility determined by interaction of RNAI with primer transcript. *Proc. nat. Acad. Sci. U.S.A.* **78**, 6096–100.

Tomizawa J-I. & Itoh T. (1982) The importance of RNA secondary structure in ColE1 primer formation. *Cell* **31**, 575–83.

Tommassen J., van Tol H. & Lugtenberg B. (1983) The ultimate localization of an outer membrane protein of *Escherichia coli* K-12 is not determined by the signal sequence. *EMBO J.* **2**, 1275–9.

Tooze J. (1980) *Molecular Biology of Tumor Viruses. Part 2. DNA Tumor Viruses*, 2nd edn. Cold Spring Harbor Laboratory, Cold Spring Harbor, New York.

Traboni C., Cortese R., Cilibert G. & Cesarini G. (1983) A general method to select M13 clones carrying base pair substitution mutants constructed in vitro. *Nucleic Acids Res.* **11**, 4229–39.

Travers A. A. (1984) Conserved features of coordinately regulated *E. coli* promoters. *Nucleic Acids Res.* **12**, 2605–18.

Tuite M. F., Dobson M. J., Roberts N. A., King R. M., Burke D. C.,

Kingsman S. M. & Kingsman A. J. (1982) Regulated high efficiency expression of human interferon-alpha in *Saccharomyces cerevisiae. EMBO J.* **1**, 603–8.

Turgeon R., Wood H. N. & Braun A. C. (1976) Studies on the recovery of crown gall tumor cells. *Proc. nat. Acad. Sci. U.S.A.* **73**, 3562–4.

Twigg A. J. & Sherratt D. (1980) Trans-complementable copy-number mutants of plasmid ColE1. *Nature, London* **283**, 216–18.

Uhlin B. E., Molin S., Gustafsson P. & Nordstrom K. (1979) Plasmids with temperature-dependent copy number for amplification of cloned genes and their products. *Gene* **6**, 91–106.

Uhlin B. E., Schweickart V. & Clark A. J. (1983) New runaway–replication–plasmid cloning vectors and suppression of runaway replication by novobiocin. *Gene* **22**, 255–65.

Valuenzuela P., Medina A., Rutter W. J., Ammerer G. & Hall B. D. (1982) Synthesis and assembly of hepatitis B virus surface antigen particles in yeast. *Nature, London.* **298**, 347–50.

van Randen J. & Venema G. (1984) Direct plasmid transfer from replica-plated *E. coli* colonies to competent *B. subtilis* cells. Identification of an *E. coli* clone carrying the *his*H and *tyr*A genes of *B. subtilis. Molec. gen. Genet.* **195**, 57–61.

Van Larbeke N., Engler G., Holsters M., van den Elsacker S., Zaenen I., Schilperoort R. A. & Schell J. (1974) Large plasmid in *Agrobacterium tumefaciens* essential for crown gall-inducing ability. *Nature* **252**, 169–70.

van Wezenbeek P. M. G. F., Hulsebos T. J. M. & Schoenmakers J. G. G. (1980) Nucleotide sequence of the filamentous bacteriophage M13 DNA genome: comparison with phage fd. *Gene* **1**, 129–48.

Vapnek D., Hautala J. A., Jacobson J. W., Giles N. H. & Kushner S. R. (1977) Expression in *Escherichia coli* K12 of the structural gene for catabolic dehydroquinase of *Neurospora crassa. Proc. nat. Acad. Sci. U.S.A.* **74**, 3508–12.

Velten J., Fukada K. & Abelson J. (1976) *In vitro* construction of bacteriophage λ and plasmid DNA molecules containing DNA fragments from bacteriophage T4. *Gene* **1**, 93–106.

Vieira J. & Messing J. (1982) The pUC plasmids, an M13mp7-derived system for insertion mutagenesis and sequencing with synthetic universal primers. *Gene* **19**, 259–68.

Villafranca J. E., Howell E. E., Voet D. H., Strobel M. S., Ogden R. C., Abelson J. N. & Kraut J. (1983) Directed mutagenesis of dihydrofolate reductase. *Science* **222**, 782–8.

Villa-Komaroff L., Efstratiadas A., Broome S., Lomedico P., Tizard R., Naber S. P., Chick W. L. & Gilbert W. (1978) A bacterial clone synthesizing pro-insulin. *Proc. nat. Acad. Sci. U.S.A.* **75**, 3727–31.

Volvovitch M., Drugeon G. & Yot P. (1978) Studies on the single-stranded discontinuities of the cauliflower mosaic virus genome. *Nucleic Acids Res.* **5**, 2913–25.

Wagner T. E., Hoppe P. C., Jollick J. D., Scholl D. R., Hodinka R. L. & Gault J. B. (1981) Microinjection of a rabbit β-globin gene into zygotes and its subsequent expression in adult mice and their offspring. *Proc. nat. Acad. Sci. U.S.A.* **78**, 6376–80.

Wahl G. M., de Saint Vincent B. R. & DeRose M. L. (1984) Effect of chromosomal position on amplification of transfected genes in animal cells. *Nature, London*, **307**, 516–20.

Walden R. M. & Howell S. H. (1982) Intergenomic recombination events among pairs of defective cauliflower mosaic virus genomes in plants. *J. mol. appl. Genet.* **1**, 447–56.

Walker M. D., Edlund T., Boulet A. M. & Rutter W. J. (1983) Cell-specific expression controlled by the 5′-flanking region of insulin and chymotrypsin genes. *Nature, London* **306**, 557–61.

Wallace R. B., Johnson P. F., Tanaka S., Schold M., Itakura K. & Abelson J. (1980) Directed deletion of a yeast transfer RNA intervening sequence. *Science* **209**, 1396–400.

Wallace R. B., Schold M., Johnson M. J., Dembek P. & Itakura K. (1981) Oligonucleotide directed mutagenesis of the human β-globin gene: a general method for producing specific point mutations in cloned DNA. *Nucleic Acids Res.* **9**, 3647–56.

Watson J. D. (1972) Origin of concatameric T7 DNA. *Nature New Biol.* **239**, 197–201.

Watson B., Currier T. C., Gordon M. P., Chilton M. D. & Nester E. W. (1975) Plasmid requirement for virulence of *Agrobacterium tumefaciens. J. Bacteriol.* **123**, 255–64.

Weatherall D. J. & Clegg J. B. (1982) Thalassaemia revisited. *Cell* **29**, 7–9.

Weatherall D. J. & Old J. M. (1983) Antenatal diagnosis of the haemoglobin disorders by analysis of foetal DNA. *Molec. biol. Med.* **1**, 151–5.

Weck P. K., Apperson S., Stebbing N., Gray P. W., Leung D., Shepard H. M. & Goeddel D. V. (1982) Antiviral activities of hybrids of two major human leukocyte interferons. *Nucleic Acids Res.* **9**, 6153–66.

Wei C-M., Gibson P., Spear P. G. & Scolnick E. M. (1981) Construction and isolation of a transmissible retrovirus containing the *src* gene of Harvey Sarcoma Virus and the thymidine kinase gene of herpes simplex virus type 1. *J. Virol.* **39**, 935–44.

Wensink P. C., Finnegan D. J., Donelson J. E. & Hogness D. S. (1974) A system for mapping DNA sequences in the chromosomes of *Drosophila melanogaster Cell* **3**, 315–25.

Weston A., Humphreys G. O., Brown M. G. M. & Saunders J. R. (1979); Simultaneous transformation of *Escherichia coli* by pairs of compatible and incompatible plasmid DNA molecules. *Molec. gen. Genet.* **172**, 113–18.

Wetzel R., Kleid D., G., Crea R., Heyneber H. L., Yansura D. G., Hirose, T., Kraszewski A., Riggs A. D., Itakura K. & Goeddel D. V. (1981) Expression in *Escherichia coli* of a chemically synthesized gene for a "mini-C" analog of human proinsulin. *Gene* **16**, 63–71.

White F. F. & Nester E. W. (1980) Relationship of plasmids responsible for hairy root and crown gall tumorigenicity. *J. Bacteriol.* **144**, 710–20.

Wickens M. P. & Gurdon J. B. (1983) Post-transcriptional processing of Simian Virus 40 late transcripts in injected frog oocytes. *J. molec. Biol.* **163**, 1–26.

Wickens M. P., Woo S., O'Malley B. W. & Gurdon J. B. (1980)

Expression of a chicken chromosomal ovalbumin gene injected into frog oocyte nuclei. *Nature, London.* **285**, 628–34.

Widera G., Gautier F., Lindenmaier W. & Collins J. (1978) The expression of tetracycline resistance after insertion of foreign DNA fragments between the *Eco* RI and *Hind* III sites of the plasmid cloning vector pBR322. *Molec. gen. Genet.* **163**, 301–5.

Wigler M., Perucho M., Kurtz, D., Dana S., Pellicer A., Axel R. & Silverstein S. (1980) Transformation of mammalian cells with an amplifiable dominant-acting gene. *Proc. nat. Acad. Sci. U.S.A.* **77**, 3567–70.

Wigler M., Silverstein S., Lee L. S., Pellicer A., Cheng Y. C. & Axel R. (1977) Transfer of purified herpes virus thymidine kinase gene to cultured mouse cells. *Cell* **11**, 223–32.

Wigler M., Sweet R., Sim G. K., Wold B., Pellicer A., Lacy E., Maniatis T., Silverstein S. & Axel R. (1979) Transformation of mammalian cells with genes from procaryotes and eucaryotes. *Cell* **16**, 777–85.

Wilkinson A. J., Fersht A. R., Blow D. M., Carter P. & Winter G. (1984) A large increase in enzyme-substrate affinity by protein engineering. *Nature, London.* **307**, 187–8.

Williams B. G. & Blattner F. R. (1980) Bacteriophage lambda vectors for DNA cloning. In *Genetic Engineering* Vol. 2, ed. Setlow J. K. & Hollaender A., pp. 201–81. Plenum Press, New York.

Williams D. C., Van Frank R. M., Muth W. L. & Burnett J. P. (1982) Cytoplasmic inclusion bodies in *Escherichia coli* producing biosynthetic human insulin proteins. *Science* **215** 687–8.

Williams D. M., Duvall E. J. & Lovett P. S. (1981a) Cloning restriction fragments that promote expression of a gene in *Bacillus subtilis. J. Bacteriol* **146** 1162–5.

Williams D. M., Schoner R. G., Duvall E. J., Preis L. H. & Lovett P. S. (1981b) Expression of *Escherichia coli trp* genes and the mouse dihydrofolate reductase gene cloned in *Bacillus subtilis. Gene* **16**, 199–206.

Williams J. G. & Lloyd M. M. (1979) Changes in the abundance of polyadenylated RNA during slime mould development measured using cloned molecular hybridization probes. *J. molec. Biol.* **129**, 19–35.

Williamson R., Eskdale J., Coleman D. V., Niazi M., Loeffler F. E. & Modell B. (1981) Direct gene analysis of chorionic villi: a possible technique for first trimester diagnosis of haemoglobinopathies. *Lancet* **2**, 1127.

Willmitzer L., Simons G. & Schell J. (1982) The Ti DNA in octopine crown gall tumours codes for seven well-defined polyadenylated transcripts. *EMBO J.* **1**, 139–46.

Wilson G. G. & Murray N. E. (1979) Molecular cloning of the DNA ligase gene from bacteriophage T4. I. Characterization of the recombinants. *J. molec. Biol.* **132**, 471–91.

Windass J. D., Worsey M. J., Pioli E. M., Pioli D., Barth P. T., Atherton K. T., Dart E. C., Byrom D., Powell K. & Senior P. J. (1980) Improved conversion of methanol to single-cell protein by *Methylophilus methylotrophus. Nature* **287**, 396–401.

Winter G., Fersht A. R., Wilkinson A. J., Zoller M. & Smith M. (1982) Redesigning enzyme structure by site-directed mutagenesis: tyrosyl tRNA synthetase and ATP binding. *Nature, London* **299**, 756–8.

Woolston C. J., Covey S. N., Penswick J. R. & Davies J. W. (1983) Aphid transmission and a polypeptide are specified by a defined region of the cauliflower mosaic virus genome. *Gene* **23**, 15–23.

Wu R., Bahl C. P. & Narang S. A. (1978) Chemical synthesis of oligonucleotides. *Progr. Nucleic Acid Res. Molec. Biol.* **21**, 101–38.

Wullems G. J., Molendijk L., Ooms G. & Schilperoort R. (1981a) Differential expression of crown gall tumor markers in transformants obtained after *in vitro Agrobacterium tumefaciens* induced transformation of cell wall regenerating protoplasts derived from *Nicotiana tabacum. Proc. nat. Acad. Sci. U.S.A.* **78**, 4344–8.

Wullems G. J., Molendijk L., Ooms G. & Schilperoort R. A. (1981b) Retention of tumor markers in F1 progeny plants formed from *in vitro* induced octopine and nopaline tumor tissues. *Cell* **24**, 719–28.

Yadav N. S., Vanderleyden J., Bennet D., Barnes W. M. & Chilton M-D. (1982) Short direct repeats flank the T-DNA on a nopaline Ti plasmid. *Proc. nat. Acad. Sci. U.S.A.* **79**, 6322–26.

Yagi Y., McLellan T., Frez W. & Clewell D. (1978) Characterization of a small plasmid determining resistance to erythromycin, lincomycin and vernamycin B_{α} in a strain of *Streptococcus sanguis* isolated from dental plaque. *Antimicrobial agents and chemotherapy* **13**, 884–7.

Yang F., Montoya A., Merlo D., Drummond H., Chilton M-D., Nester E. & Gordon M. (1980) Foreign DNA sequences in crown gall teratomas and their fate during the loss of tumorous traits. *Molec. gen. Genet.* **177**, 707–14.

Yansura D. G. & Henner D. J. (1984) Use of the *Escherichia coli lac* repressor and operator to control gene expression in *Bacillus subtilis. Proc. nat. Acad. Sci. U.S.A.* **81**, 439–43.

Yasuda S. & Takagi T. (1983) Overproduction of *Escherichia coli* replication proteins by the use of runaway-replication plasmids. *J. Bacteriol.* **154**, 1153–61.

Yelton M. M., Hamer J. E. & Timberlake W. E. (1984) Transformation of *Aspergillus nidulans* by using a *trp*C plasmid. *Proc. nat. Acad. Sci. U.S.A.* **81**, 1470–4.

Young B. D., Birnie G. D. & Paul J. (1976) Complexity and specificity of polysomal poly $(A)^+$ RNA in mouse tissues. *Biochemistry* **15**, 2823–8.

Young R. A. & Davis R. W. (1983) Efficient isolation of genes by using antibody probes. *Proc. nat. Acad. Sci. U.S.A.* **80**, 1194–8.

Zaenen I., Van Larebeke N., Teuchy H., Van Montagu M. & Schell J. (1974) Super-coiled circular DNA in crown-gall inducing *Agrobacterium* strains. *J. molec. Biol.* **86**, 109–127.

Zambryski P., Depicker A., Kruger H. & Goodman H. (1982) Tumor induction by *Agrobacterium tumefaciens*: analysis of the boundaries of T-DNA. *J. molec. appl. Genet.* **1**, 361–70.

Zambryski P., Holsters M., Kruger K., Depicker A., Schell J., Van Montagu M. & Goodman H. (1980) Tumor DNA structure in plant cells transformed by *A. tumefaciens. Science* **209**, 1385–91.

Zimmerman S. B. & Pheiffer B. (1983) Macromolecular crowding allows blunt end ligation by DNA ligases from rat liver or *Escherichia coli. Proc. nat. Acad. Sci U.S.A.* **80**, 5852–6.

Zoller M. J. & Smith M. (1983) Oligonucleotide-directed mutagenesis of DNA fragments cloned into M13 vectors. *Meth. Enzymol.* **100**, 468–500.

Index

Page numbers in *italics* refer to pages on which illustrations/tables appear.